Epigenetic Aspects of Chronic Diseases

Helmtrud I. Roach[†] • Felix Bronner
Richard O.C. Oreffo
(Editors)

Epigenetic Aspects of Chronic Diseases

Springer

Editors
Helmtrud I. Roach[†], BA, MSc, PhD
Bone and Joint Research Group
Institute of Developmental Sciences
School of Medicine
University of Southampton
Hampshire
UK

Felix Bronner, PhD, Dr (h.c.)
Department of Reconstructive Sciences
The University of Connecticut
Health Center
Farmington, CT
USA

Richard O.C. Oreffo, BSc, DPhil
Bone and Joint Research Group
Institute of Developmental Sciences
School of Medicine
University of Southampton
Hampshire
UK

ISBN 978-1-4471-5910-0 ISBN 978-1-84882-644-1 (eBook)

DOI 10.1007/978-1-84882-644-1

A catalogue record for this book is available from the British Library

Printed on acid-free paper

Springer Science+Business Media (www.springer.com)

This book is dedicated to Trudy Roach, who initiated and developed this treatise. Shortly after formulating the outline, Trudy was diagnosed with bowel cancer. For over a year Trudy responded to treatment and continued her research, teaching, and editorial activities. In her last few months, even though treatment was no longer effective, Trudy remained optimistic, a source of inspiration, and gained pleasure from her varied scholarly activities and time with her family.

In the name of Trudy, we also dedicate this book to the many victims of the diseases discussed here, with the hope that in the not too distant future effective treatment, based in part on epigenetic insights, will improve the quality of life of patients with these diseases.

Felix Bronner and Richard Oreffo

Preface

Epigenetics refers to processes that alter gene expression without changes in DNA sequence. In development, for example, genes are turned on and off, causing changes in the phenotype, from fetus to adult. Epigenetic mechanisms also mediate environmentally induced changes that may result in disease states, as when people on a low-sugar diet are exposed to high sugar intakes in a new environment and develop diabetes. Indeed, as will be discussed in detail in this book, many chronic diseases are the result of epigenetically induced structural changes in the DNA, resulting in DNA hypo- and hypermethylation, which cause genes not normally expressed to be expressed, whereas other genes are no longer expressed.

Rebecca Smith and Jonathan Mill, in Chap. 1, provide an overview of the relationship between epigenetics and chronic disease. The chapter describes and discusses epigenetics and the epigenome, DNA methylation and changes in histone structure, and interactions between the two processes. Attention is called to the increasing evidence that external influences interact directly with the epigenome, leading to changes in epigenetic processes and gene expression of the individual. Human pathologies, e.g., cancer or imprinting disorders, involve epigenetic changes. Also, discordances between monozygotic twins have been ascribed to nongenetic factors, with epigenetics linking the environment to changes in the phenotype. Prenatal and early-life environmental factors appear to play a role in the etiology of chronic disease. An example cited by the authors is the finding that serious hunger during the Dutch winter of 1944 led to a higher incidence of disease in the children of pregnant women when they became adults. The chapter proceeds to discuss epigenetic mechanisms and how these may affect chronic disease treatment, and raises the question of transgenerational epigenetic inheritance, an issue of intense current research.

In Chap. 2, Ester Lara, Vincenzo Calvanese, Agustin F. Fernandez, and Mario Fraga describe in detail techniques to study DNA methylation and histone modification, both at the global and gene-specific level. DNA methylation can be studied by a variety of methods. This may involve a locus-specific study of the methylation status of a specific gene, or a genome-wide study, involving many genes, or analysis of the methylation status of a cell or tissue, i.e., a global study. One of the oldest techniques to study methylation is reverse-phase high-performance liquid chromatography. Other methods described in the chapter are immunochemical methods, the methyl group acceptance assay, the chloroacetaldehyde assay, the bisulfate assay, among others. The authors then describe several locus-specific assays, including the use of melting analysis. In addition to the many specific methylation methods, the authors proceed to histone analysis, both global and locus-specific. Throughout the chapter, the various methods are comprehensively described, analyzed, evaluated, tabulated, and reference-supported.

Chapter 3, by Marie-Pierre Lambert and Zdenko Herceg, discusses the mechanisms of epigenetic silencing. Recognizing that dysregulation of epigenetic mechanisms contributes to human disease, the authors describe the structure of chromatin, essential for the maintenance of geometric stability, and point out that permanent silencing of the transposable elements and non-coding sequences of DNA is mainly due to epigenetic mechanisms, notably DNA methylation, which regulates chromatin. The chapter details the steps and mechanisms involved in DNA methylation and indicates that regulation of gene expression requires a dynamic equilibrium between promoter regions, chromatin structure, and access for transcriptional regulatory factors. This involves histones, the principal proteins of chromatin, with the *N*-terminal region, the histone "tail," constituting the major site for epigenetic regulation. The authors then examine the role of microRNAs in the regulation of epigenetic silencing and the equilibrium between active and repressive marks in the chromatin structure, an equilibrium modulated by the interplay of epigenetic mechanisms. Epigenetic plasticity and genomic imprinting are discussed, as is X chromosome inactivation. The chapter concludes by calling attention to the fundamental role of gene silencing, with unscheduled silencing responsible for disease.

In vertebrates, cytosine methylation constitutes an epigenetic DNA modification involved in genome stability, gene repression, and gene imprinting, with incorrect DNA methylation patterns associated with various pathological situations. In Chap. 4, Thierry Grange and Edio E. Lourenço analyze the mechanisms of gene activation, with emphasis on the dynamics of DNA methylation and demethylation. The chapter describes what the authors call the methylation landscape, i.e., large-scale and genome-scale methylation maps, the regulation of promoter activity by DNA methylation, transcription-dependent gene-body methylation, and the link to chromatin. The authors then analyze epigenetic reprogramming of DNA demethylation, both global and local, and the mechanisms of DNA demethylation, pointing out that active demethylation mechanisms are akin to DNA repair mechanisms and may involve recruitment of DNA repair machinery. The controversial proposal that in vertebrates demethylation by a base excision repair pathway is initiated by a DNA glycosylase is discussed at length, with comparisons made to the process in plants. Demethylation by the nucleotide excision repair pathways is then taken up and analyzed, as are links to other pathways, with the chapter concluding that demethylation may function in the recruitment of methylation-sensitive transcription factors, rather than in regulating chromatin switches.

Paul Cloos, in Chap. 5, describes in molecular detail the essential role of histone demethylases, enzymes that catalyze the removal of methylation from histones. It is dysregulation or inappropriate positioning of these enzymes that appears to contrib ute or cause disease, particularly cancer. The chapter discusses the role of transcription factors pRB and p53 as tumor repressors and in inducing senescence, another repressor of carcinogenesis. Several other suppressor proteins are discussed in detail. For example, the question is asked whether JHDMIB histone demethylase is a tumor suppressor or an oncoprotein. Possible roles of demethylases in a variety of cancers are discussed – prostate, breast, hematopoietic, brain, and renal cancers – as well as the relationship of these enzymes to neural disorders (mental retardation, schizophrenia, autism, epilepsy, and neuropathy). Other conditions or disorders that may involve the demethylases are male infertility, obesity, congenital heart disease, and alopecia. The author concludes by anticipating that future research will uncover more insights into the pathogenetic role of these important enzymes.

Epigenetic mechanisms play a central regulatory role in the immune system, as discussed by Travis Hughes and Amr H. Sawalha in Chap. 6. The widely varying characteristics of T cell populations are largely the result of epigenetic modifications at key regulatory loci. The chapter discusses the regulatory, T helper as well as follicular T cells, and then proceeds to an evaluation of epigenetic dysregulation in systemic lupus erythematosus. The effect of promoter demethylation on gene overexpression in lupus is discussed, as is demethylation of the inactive X chromosome in lupus, the effect of DNA methylation inhibitors, signaling in the ERK pathway, histone modification, and chromatin remodeling. The chapter concludes with an analysis of the role of methyl-binding domain proteins, a brief discussion of the role of microRNAs, and reference to rheumatoid arthritis. As in other chapters, helpful figures and a thorough list of references complete the text.

Rheumatoid arthritis and the role of epigenetics are discussed in Chap. 7 by Alec M. Grabiec, Paul P. Tak, and Kris A. Reedquist. The disease is the outcome of a combination of genetic susceptibility factors, autoantibody production due to aberrant regulation of the immune system, and environmental factors, such as smoking or inappropriate nutrition. Change in the activation of a number of intracellular signaling pathways resulting from aberrant epigenetic modifications has, as discussed by the authors, led to better understanding of the pathobiology of the disease and to intensive search for new therapeutic targets. The chapter discusses global and promoter-specific modulation of DNA methylation, aberrant microRNA expression, the perturbation of histone acetyl transferase and histone deacetylase activity in rheumatoid arthritis, and includes findings on the targeting of that activity in animal models of the disease. The chapter concludes by emphasizing the relevance of epigenetic mechanisms for the pathogenic alterations in gene expression that lead to chronic inflammation and its persistence in rheumatoid arthritis, even though the mechanisms involved are not yet fully understood.

Osteoarthritis, a chronic disease that affects some two-thirds of the elderly population, is, as discussed by Helmtrud I. Roach[†] in Chap. 8, a prominent example of how changes in the epigenetic status, specifically DNA methylation, affect disease evolution and progress. The chapter analyzes degradation and matrix changes of the articular cartilage, cellular and molecular changes, and evaluates the role of genetics in osteoarthritis. Evidence from monozygotic twin studies suggests a role for epigenetic changes induced by environmental factors. The chapter then discusses hypo- and demethylation at specific CpG sites in the DNA, in vitro studies that mimic changes in gene expression that occur in osteoarthritis, the role of 1L1-b expression, and describes secondary arthritis that results if developmental dysplasia of the hip in the young is not corrected. The energy hormone leptin is aberrantly expressed in osteoarthritis, varying inversely with DNA methylation status. Factors that affect DNA methylation are described and analyzed, with the chapter concluding that methylation of genomic DNA is a significant mechanism for regulating tissue-specific gene expression, but that the molecular steps that lead to changes in the methylation status remain unknown.

Type 2 diabetes mellitus, the incidence of which is rapidly increasing worldwide because of urbanization, physical inactivity, and increasing obesity in both adults and youngsters, is the outcome of a complicated interaction between genome and epigenome. Charlotte Ling, Tina Rőnn, and Marloes D. Nitert, in Chap. 9, describe the role of epigenetics in the evolution of type 2 diabetes, how undernutrition and low birth weight enhance the disease risk in later life, and how chromatin structure and DNA methylation affect beta cell lines. The chapter discusses oxidative phosphorylation in relation to the disease, the decline in beta cell proliferation, and the change in DNA methylation with increasing age. The authors then devote much attention to the importance and role of nutrition and obesity, both in human and animal model studies. The interaction between fat metabolism and type 2 diabetes is described, as is the role of leptin. Low physical activity is a risk factor for the disease, and the authors discuss the still limited number of studies that deal with the role of epigenetics in exercise. The chapter concludes with a discussion of diabetic complications and their relationship to epigenetic changes and emphasizes the importance epigenetic mechanisms play in the pathogenesis of type 2 diabetes.

Andrew L. Durham and Ian M. Adcock, in Chap. 10, discuss epigenetic regulation of asthma and allergic diseases. The chapter begins with a discussion of the four classes of hypersensitive responses: IgE mediated, IgG or IgM mediated, immune complex mediated, and cell-mediated hypersensitivity. The authors then turn their attention to asthma, to how allergy develops, and the importance of maternal imprinting, and analyze epigenetic mechanisms, imprinting, and epigenetic regulation of the immune response, with histone acetylation playing an important role. They discuss how environmental factors alter the epigenetic profile and how diet affects epigenetics and allergy. The chapter then details how exposure to environmental tobacco smoke is associated with impaired respiratory function and increases the risk of asthma. A similar correlation applies to air pollution. Exposure to particulate pollutants and to diesel exhaust has been shown to increase proinflammatory cytokines in utero, in animal models, and in human airway epithelial cells. Asthma is treated with glucocorticoids which modulate the epigenetic environment; the molecular mechanisms of the response are described in detail. Molecules that modulate histone acetyltransferase and histone deacetylase have been developed and used to restore glucocorticoid responsiveness. The authors conclude that epigenetics has immense potential to understand and treat preventable environmental disease.

Attempts at identifying genetic mechanisms for major mental diseases have not been successful. Indeed, analyses of the brain transcriptome by microarray assays have shown that hundreds of genes are differentially expressed in the affected brain regions of patients with these diseases, but not in most of their other tissues. Having made these points in their introduction to Chap. 11, Hamid Mostafavi-Abdomaleky, Stephen J. Glatt, and Ming T. Tsuang call attention to the fact that most psychiatric disorders are episodic and may have long-lasting periods of remission. Therefore genetic mutations alone cannot be responsible for the disease phenotypes, inasmuch as spontaneous remission and fluctuation do not occur in purely genetic diseases. Dysregulation of epigenetic machinery due to environmental factors can cause periods of remission. The authors then describe aberrant DNA methylation in psychiatric disease, as in Rett's syndrome, schizophrenia, in association with suicide and childhood abuse, in post-traumatic stress disorder, and as a result of smoking and in alcoholism. Aberrant histone modification also occurs in major mental diseases and is described next. A third molecular mechanism is dysregulation of microRNA, associated with schizophrenia and schizo-affective disorder. A number of highly expressed microRNAs in the superior temporal gyrus and the dorsolateral prefrontal cortex are known to be involved in the pathogenesis of schizophrenia. After discussing these mechanisms, the authors deal with epigenetic aberrations in relation to paternal effects, gender differences, and brain laterality. The chapter concludes by highlighting the importance and interrelationship of signal pathways and the need to study the epigenetic effects therein, in the hope that discoveries will lead to novel strategies to deal with these devastating diseases.

The relationship between epigenetics and mental disease is further developed in Chap. 12, where Axel Schumacher, Syed Bihaqi, and Nasser H. Zawia discuss late-onset Alzheimer's disease. The authors show that the molecular transition from memory encoding and initial consolidation to progressive long-term storage, retrieval, and reconsolidation involves complex layers of local and system-wide epigenetic modifications. In Alzheimer's disease gene expression in the brain is altered, with multiple functional and molecular pathways affected. Late-onset Alzheimer's disease is characterized by many non-Mendelian anomalies that suggest an epigenetic component. After listing and discussing these, the authors describe methylation homeostasis in this condition, for example, that some genes that play a central role in amyloid processing display significant epigenetic variability. Other components of the methylation pathway are also abnormal. Attention is called to the low folate level in the spinal fluid of Alzheimer patients and its effects. In the next section the authors discuss the effect of epigenetic drift, probably caused by high epigenetic turnover, and then summarize the evidence for an epigenetic fingerprint in Alzheimer's disease. The effects of DNA methylation and oxidation on the disease are reviewed in detail, with the conclusion raising the question whether epigenetic changes precede late-onset Alzheimer's disease, conferring disease risk, or whether epigenetic drift is the result. The authors favor the possibility that predisposition to the disease is related to DNA methylation profiles and influenced by epigenetic drift.

Epigenetic effects are particularly significant in development and the life course. The importance of epigenetic modification of fetal development and the resulting change in susceptibility in later life to noncommunicable diseases, such as metabolic syndrome, cardiovascular disease, osteoporosis, obesity, is discussed in detail in Chap. 13 by Keith M. Godfrey, Karen A. Lillycrop, Mark A. Hanson and Graham C. Burdge. The authors point out that links between prenatal growth and later disease risk reflect variations in the quality of the intrauterine environment, with prenatal nutrition as a major factor. The concept of predictive adaptive response is defined; it constitutes an integrated regulator in early life to meet the predicted later environment. The resulting phenotype implies a relatively constant postnatal environment. Otherwise the individual is "mismatched", resulting in a phenotype no longer appropriate for the actual environment the individual inhabits. The authors discuss the mismatch and then proceed to an evaluation of epigenetics during development and aging. Genomic imprinting is discussed, with most of the 53 human genes known to be imprinted located in CpG-rich domains where methylation of the CpG dinucleotides represses the maternal or paternal allele. Changes in the epigenetic regulation of genes due to nutrition and the effects of altered, nutrition-induced transcription are, it is pointed out, related to later diseases, both in humans and animal models. Interventions to prevent or reverse induced phenotypes are evaluated, transgenerational effects are discussed, and the chapter concludes with an analysis of the relevance of epigenetic processes to the risk of adult disease.

Active inflammatory genes are suppressed by histone deacetylases. Corticosteroids recruit deacetylases to switch off inflammatory genes. Histone deacetylases may therefore be a target for developing anti-inflammatory treatments, especially needed in diseases with active corticosteroid resistance, as in chronic obstructive pulmonary disease. This is the topic developed by Peter J. Barnes, in Chap. 14. After describing the disease with its progressive airflow limitation, due to remodeling and narrowing of small airways and the destruction of the lung parenchyma, the author analyzes the inflammation typical of the disease and then proceeds to a description of histone acetylation and deacetylation, the role of the histone deacetylase enzymes, the reduced activity of which in the alveolar macrophages of cigarette smokers is correlated with increased inflammatory gene expression. How glucocorticoids suppress inflammation is described in cellular and molecular detail, as is corticosteroid resistance in chronic obstructive pulmonary disease and the mechanisms of histone deacetylase reduction in the disease. Interference with specific signal pathways is an attractive therapeutic option, according to the author, and theophylline is one molecule that in low concentrations can restore histone deacetylase activity. New therapeutic targets are antioxidants; new theophylline derivatives; curcurin, a polyphenol found in curry powder; macrolides; and as yet elusive histone deacetylase activators.

Myeloid malignancies, including acute myeloid leukemia and the myelodysplastic syndrome, are discussed in Chap. 15, by Lauren C. Suarez and Steven D. Gore. Similar to other malignancies, myelodysplastic syndromes exhibit chromosomal abnormalities and mutations. In addition, epigenetic modifications such as DNA methylation play a prominent role both in the acute disease and in the myelodysplastic syndrome, its precursor lesion. The results of pilot studies on the effects of azacitidine, a DNA methyltransferase inhibitor, are discussed in terms of dosage and treatment cycles. Parallel studies with decitabine, an azacitidine congener, are analyzed and compared with azacitidine studies. The authors then raise the question whether these two agents act through epigenetic mechanisms and from their evaluation of clinical studies think that changes in epigenetic profiles may be linked to treatment response and thus constitute a response indicator. The role of histone deacetylase inhibitors is then taken up, with discussion of short chain fatty acids, benzamides, and romidepsin. Other histone deacetylase inhibitors such as hydroxamic acids (vorinostat, panobinostat, belinostat) are evaluated. The authors then raise the question about combining epigenetic drugs and discuss studies that have dealt with this important therapeutic approach. The chapter concludes with an analysis of the future of epigenetic therapies in the treatment of the myelodysplastic syndrome.

Study of the epigenome has over recent years become an important field of research and insights are finding increasing application in medical science and practice. We are grateful to the authors for their willingness to share their knowledge and experience with a wider professional audience, thereby reinforcing the link between developing knowledge and its practical application, at the same time emphasizing the as yet wide gap between promise and therapy. We also thank Springer, our publisher, for their help in assuring the intellectual and aesthetic quality of this treatise.

Southampton, UK — Helmtrud I. Roach†
Farmington, CT, USA — Felix Bronner
Southampton, UK — Richard O.C. Oreffo (Januray 2011)

Contents

Contributors

Ian Adcock, PhD
Department of Airways Disease, National Heart and Lung Institute, Imperial College London, London, UK

Peter J. Barnes, DM, DSc, FRCP, FMedSci, FRS
National Heart and Lung Institute, Imperial College London, London, UK

Syed W. Bihaqi, PhD
Department of Biomedical and Pharmaceutical Sciences, College of Pharmacy, University of Rhode Island, Kingston, RI, USA

Felix Bronner, PhD
Department of Reconstructive Sciences, The University of Connecticut Health Center, Farmington, CT, USA

Graham C. Burdge
Institute of Human Nutrition, School of Medicine, University of Southampton, Southampton, UK

Vincenzo Calvanese, BS
Department of Immunology and Oncology, National Center for Biotechnology, CNB – CSic, Madrid, Spain

Paul Cloos, PhD
Biotech Research & Innovation Centre, Copenhagen, Denmark

Andrew L. Durham, PhD
Department of Airways Disease, National Heart and Lung Institute, Imperial College London, London, UK

Agustin F. Fernandez, PhD
Department of Cancer Epigenetics, Instituto Universitario de Oncología del Principado de Asturias (IUOPA), Asturias, Spain

Mario F. Fraga, PhD
Department of Immunology and Oncology, National Center for Biotechnology, (CNB – CSic), Madrid, Spain

Stephen J. Glatt, PhD
Director, Psychiatric Genetic Epidemiology and Neurobiology Laboratory (PsychGENe Lab) Assistant Professor, Department of Psychiatry and Behavioral Sciences and Department of Neuroscience and Physiology Associate Director,

Medical Genetics Research Center (MGRC) SUNY Upstate Medical University
750 East Adams Street Syracuse, NY 13210, USA

Keith M. Godfrey, PhD, FRCP, CCST, MRCP, BM (Hons)
MRC Epidemiology Resource Centre, Southampton General Hospital, University of Southampton, Southampton, UK

Steven D. Gore, MD
Department of Oncology/Hematologic Malignancies, Johns Hopkins University, Baltimore, MD, USA

Aleksander M. Grabiec, MSc
Division of Clinical Immunology and Rheumatology, Academic Medical Center, University of Amsterdam, Amsterdam, The Netherlands

Thierry Grange, PhD
Department of Genomes and Epigenomes, Institut Jacques Monod, Paris, France

Mark A. Hanson, MA, DPhil, Cert Ed, FRCOG
DOHaD Division, School of Medicine, University of Southampton, Institute of Developmental Sciences, Southampton General Hospital, Southampton, UK

Zdenko Herceg, PhD
Epigenetics Group, International Agency for Research on Cancer, Lyon, France

Travis Hughes, BS
Department of Arthritis and Immunology, Oklahoma Medical Research Foundation, Oklahoma City, OK, USA

Marie-Pierre Lambert, PhD
Epigenetics Group, International Agency for Research on Cancer, Lyon, France

Ester Lara, BS, PhD
Department of Department of Immunology and Oncology, National Center for Biotechnology, CNB – CSic, Madrid, Spain

Karen A. Lillycrop, BSc (Hons), PhD
School of Biological Sciences, University of Southampton, Southampton, UK

Charlotte Ling, MSc, PhD
Department of Clinical Sciences, Lund University, Malmö, Sweden

Edio Eligio Lourenço, MSc, PhD
Department of Genomes and Epigenomes, Institut Jacques Monod, Paris, France

Jonathan Mill, BA, PhD
Institute of Psychiatry, King's College London, London, UK

Hamid Mostafavi-Abdolmaleky, MD
Harvard Institute of Psychiatric Epidemiology and Genetics, Harvard Medical School, Boston, MA2- Mental Health Research Center and Tehran Psychiatrist Institute, Department of Psychiatry, Tehran University of Medical Sciences, Tehran, Iran

Marloes Dekker Nitert, MSc, PhD
Department of Clinical Services, Lund University, Malmö, Sweden

Richard O.C. Oreffo, BSc, DPhil
Bone and Joint Research Group, Institute of Developmental Sciences, School of Medicine, University of Southampton, Southampton, UK

Kris A. Reedquist, PhD
Division of Clinical Immunology and Rheumatology, Academic Medical Center, University of Amsterdam, Amsterdam, The Netherlands

Helmtrud I. Roach†, BA, MSc, PhD
Bone and Joint Research Group, Institute of Developmental Sciences, School of Medicine, University of Southampton, Southampton, UK

Tina Rönn, MSc
Department of Clinical Sciences, Lund University, Malmö, Sweden

Amr H. Sawalha, MD
Department of Medicine, Division of Rheumatology, and
Department of Pathology, University of Oklahoma Health Sciences Center, and
Department of Veterans Affairs Medical Center, and
Arthritis and Immunology Program, Oklahoma Medical Research Foundation, Oklahoma City, OK, USA

Axel Schumacher, PhD
The Krembil Family Epigenetics Laboratory, Centre for Addiction and Mental Health, Toronto, ON, Canada

Rebecca Smith, BSc
Institute of Psychiatry, King's College London, London, UK

Lauren C. Suarez
Sidney Kimmel Comprehensive Cancer Center, Johns Hopkins University, Baltimore, MD, USA

Paul P. Tak, MD, PhD
Division of Clinical Immunology and Rheumatology, Academic Medical Center, University of Amsterdam, Amsterdam, The Netherlands

Ming T. Tsuang, MD, PhD, DSc
Center for Behavioral Genomics, Department of Psychiatry, Institute of Genomic Medicine, University of California, San Diego, CA, USA and
Veterans Affairs San Diego Healthcare System, San Diego, CA, USA and
Harvard Institute of Psychiatric Epidemiology and Genetics, Harvard Departments of Epidemiology and Psychiatry, Boston, MA, USA

Nasser H. Zawia, PhD
Department of Biomedical and Pharmaceutical Sciences, University of Rhode Island, Kingston, RI, USA

Epigenetics and Chronic Diseases: An Overview

1

Rebecca Smith and Jonathan Mill

1.1 Introduction

According to the World Health Organization, chronic diseases account for an estimated 35 million deaths per year, representing ~60% of worldwide mortality.[163] These disorders, including heart disease, obesity, arthritis, cancer, diabetes, psychiatric illness and dementia, confer a major economic, social, and healthcare burden. In the developed world, for example, the treatment of chronic disease accounts for the major proportion of public healthcare spending. As demographic factors shift and the population ages, the prevalence of chronic disease is likely to increase significantly, especially in the developing world. For instance, the prevalence of adult obesity is on a dramatic upward trajectory, increasing from 12% in 1989 to 27% in 2008 in the USA (http://www.cdc.gov/brfss/). Likewise, as the population ages, the number of cases of Alzheimer's Disorder is projected to increase from an estimated 24 million in 2001 to >80 million by 2040, with rates in countries such as India and China increasing by more than 300% over this period.[41] The possibility of understanding the biology underpinning human chronic illness is therefore one of the most exciting perspectives of contemporary biomedical research, and the focus of considerable research effort across the world.

Traditional approaches to understanding the molecular etiology of chronic disorders have put particular emphasis on the role of DNA sequence variation in mediating susceptibility. While many chronic diseases are familial and highly heritable, with higher levels of concordance observed between genetically identical monozygotic (MZ) twin pairs than between dizygotic (DZ) twin pairs, it has been difficult to pinpoint specific genetic risks for most of these disorders. Recent years have seen tremendous progress in our ability to interrogate genetic variation, and since the completion of the first draft human genome sequence in 2001, there have been major advances in genotyping and sequencing technology. It is now economically feasible to perform genome-wide association (GWA) studies using high-resolution microarrays tagging all common variants across the genome, and the development of next-generation sequencing technology means that we are at the dawn of an era where it will be possible to sequence entire genomes on an industrial scale. The advent of GWA studies has allowed a systematic, hypothesis-free exploration of the genes associated with chronic diseases, but for many of these disorders, associations are characterized by small effect sizes, considerable heterogeneity, and await convincing replication. We are thus still a long way from realizing the postgenomic promises of novel diagnostic and therapeutic strategies for many chronic illnesses.

It is clear that most chronic disorders do not conform to a simple Mendelian transmission pattern. Instead, they are multifactorial and polygenic, with genetic factors rarely the sole causes of increased susceptibility. Quantitative genetic analyses have shown that environmental influences are likely to be important across the spectrum of chronic diseases, with accumulating epidemiological evidence now linking exposure to various environmental insults, occurring at specific stages of development, with an increased risk of illness. The "developmental origins of disease hypothesis" postulates that the cause of many common chronic diseases is not only genetic, but results from early-life events, especially occurring during

J. Mill (✉)
Institute of Psychiatry, King's College London, London, UK
e-mail: jonathan.mill@iop.kcl.ac.uk

H.I. Roach et al. (eds.), *Epigenetic Aspects of Chronic Diseases*,
DOI: 10.1007/978-1-84882-644-1_1, © Springer-Verlag London Limited 2011

gestation, which have profound long-term developmental consequences.[50] The mechanisms by which such environmental factors mediate susceptibility in the course of development, however, are not well understood, although it has been widely speculated that epigenetic processes are likely to be important. In this chapter, we introduce the role that epigenetic mechanisms play in dynamically mediating gene expression and summarize current evidence supporting the contribution of epigenetic processes to the etiology of chronic disease. The aim is not to focus on specific chronic diseases – these are the focus of other chapters in this book – but to give a conceptual introduction to the role that epigenetic processes may play in chronic disease and highlight the promise epigenetic approaches hold for our future understanding of these conditions.

1.2 Epigenetics and the Epigenome

With the exception of a few rare somatic mutation events, the sequence of nucleotides comprising an individuals' genome is identical across all cells in the body and remains unchanged from the moment of conception onward. But DNA is structurally much more complex than a simple string of nucleotides and, at a functional level, the genome is anything but static. While every cell in our bodies contains the same DNA sequence, each has its own unique phenotype characterized by a specific pattern of gene expression that is in constant flux. It is not only the gene-encoding DNA sequence that is important in determining the phenotype of a cell, but the degree to which specific genes are functionally active at any particular time in development. Sequencing the genome was therefore only the first step in the quest to understand how genes are expressed and regulated. Sitting above the DNA sequence is a second layer of information (the "epigenome") that regulates several genomic functions, including when and where genes are turned on or off.

The British biologist Conrad Waddington first coined the term "epigenetics" (literally meaning "above genetics") in the mid-twentieth century, introducing the concept of an epigenetic landscape to describe the ways in which cell fates are established during development, enabling the tissues and organs of complex organisms to develop from an initially "undifferentiated mass" of cells. A contemporary definition regards epigenetics as the reversible regulation of gene expression, occurring independently of DNA sequence, mediated principally through changes in DNA methylation and chromatin structure.[62] Structurally, we know that DNA is much more than a sequence of DNA bases; in its mitotic state, for example, each DNA molecule is packaged into a chromosome that is 10,000 times shorter than its extended length. DNA is coiled tightly around histone proteins to form nucleosomes, which in turn are condensed to form chromatin, the complex combination of DNA, RNA, and protein that makes up chromosomes. Chemical modifications to both DNA and histone proteins are important in regulating how accessible the genome is to the cells' transcriptional machinery (see Fig. 1.1). Condensed chromatin (heterochromatin), in which the DNA and histone proteins are tightly packed, acts to block the access of transcription factors and other instigators of gene expression to DNA and is thus associated with repressed transcription. Conversely, an open chromatin conformation (euchromatin) allows the transcriptional machinery of the cells to access DNA and drive transcription. Epigenetic processes are thus essential for normal cellular development and differentiation and allow the long-term regulation of gene function through nonmutagenic mechanisms.[58] For a glossary of epigenetic mechanisms, and a basic description of the genomic functions they perform, see Table 1.1.

1.2.1 DNA Methylation

DNA methylation involves the transfer of a methyl group to position 5 of the cytosine pyrimidine (C) ring of a cytosine–guanine dinucleotide (CpG) catalyzed by a group of enzymes called DNA methyltransferases (DNMTs).[76] DNA methylation is one of the most well-characterized and stable epigenetic mechanisms modulating the transcriptional plasticity of mammalian genomes,[64] making it the focus of most epigenetic studies of chronic disease performed to date. The methylation of CpG sites acts to disrupt the binding of transcription factors and to attract methyl-binding proteins that initiate chromatin compaction and gene silencing.[76] Because methylated

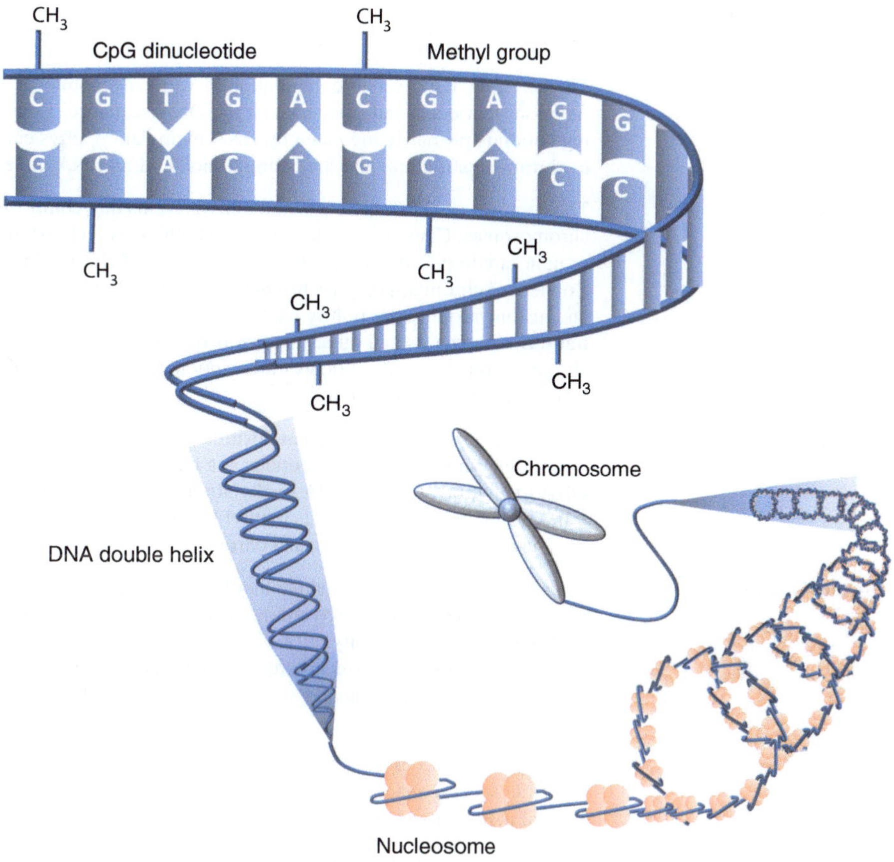

Fig. 1.1 Epigenetic modifications and chromatin structure. DNA is wrapped around a histone core, forming a nucleosome. DNA methylation at CpG dinucleotides and covalent modifications to histone tails alter the physical structure of the genome, attracting methyl-binding proteins and mediating the access of transcription factors that drive gene expression. DNA methylation at CpG dinucleotides also disrupts transcription directly by blocking binding sites of certain transcription factors

cytosines are liable to spontaneous deamination, CpG dinucleotides are less common in the genome than would be predicted by chance, primarily occurring in clusters called "CpG islands" which are often found around gene promoters and are typically unmethylated.[13] These CG-rich regions cover only 0.57% of the genome but contain 5.5% of all CpG dinucleotides.[124] DNA methylation at these CpG islands has been the predominant focus of epigenetic research and is associated with chromatin remodeling and reduced gene expression at proximal genes.[62] When highly methylated, CpG islands have a closed chromatin structure that restricts transcription. Low levels of DNA methylation are associated with an open chromatin structure, allowing the binding of transcription factors and RNA polymerase complexes that drive expression. It is increasingly apparent, however, that DNA methylation at CpGs occurs outside CpG islands (~70% of total CpGs); it also has an important role in regulating genomic function, particularly during development.[134]

1.2.2 Histone Modifications

DNA is wrapped around an octamer of histones (two molecules of H2A, H2B, H3, and H4) in 146-bp segments which make up a nucleosome.[82] Histone proteins are thus essential for packaging DNA inside the nucleus, and play an important role in regulating gene expression. The N-terminal tails of the histone particles extend from the nucleosome and are subject to post-synthesis modifications.[140] A number of these covalent histone modifications, occurring at specific residues, have been described, including acetylation, methylation, phosphorylation, SUMOylation, and ubiquitylation; together these are believed to constitute a complex "histone code" that modulates gene expression via alterations in chromatin structure.[11] These alterations affect the access of the cell's transcriptional machinery to the DNA. Condensed chromatin (heterochromatin), in which the DNA and histone proteins are tightly packed, acts to block the access of transcription factors and other instigators of gene expression to DNA.

Table 1.1 Glossary of epigenetic terms

Term	Definition	Key reference
Epigenetics	The heritable, but reversible, regulation of various genomic functions that occur independently of the DNA sequence. Epigenetic regulation is primarily mediated by DNA methylation, physical changes to chromatin structure, and the action of non-coding RNA molecules.	58
Chromatin	The complex of DNA, histones, and other proteins that comprise chromosomes. Chemical modifications to both DNA and histone proteins are important in regulating the structure of chromatin. Condensed chromatin (heterochromatin), in which the DNA and histone proteins are tightly packed, acts to block the access of transcription factors and other instigators of gene expression. Open chromatin (euchromatin) allows transcriptional factors to access DNA and drive transcription.	11
DNA methylation	The addition of a methyl group at position 5 of the cytosine pyrimidine ring in CpG dinucleotides in a reaction catalyzed by DNA methyltransferases. DNA methylation disrupts the binding of transcription factors and attracts methyl-binding proteins that are associated with gene silencing and chromatin compaction.	62
Histone modifications	Covalent posttranslational histone modifications that occur at specific residues include acetylation, methylation, phosphorylation, SUMOylation, and ubiquitylation. These modulate gene expression via alterations in chromatin structure. Like DNA methylation, histone modifications are dynamic and actively regulated by a host of catalytic enzymes.	11
Genomic imprinting	An epigenetic process that alters the expression of genes in a parent of-origin specific manner. Genomic imprinting is fundamental to normal mammalian development and growth.	27
X-inactivation	X-chromosome inactivation silences genes on one X-chromosome in females to ensure dosage compensation with males via a process involving hypermethylation of CpG islands. X-inactivation in any given cell is typically random and, once established, is maintained so that the inactivated allele is transcriptionally silenced for the lifetime of that cell.	5
Epigenetic inheritance	Epigenetic signals are transmitted mitotically through cell lineages, but are generally assumed to be reset during gametogenesis and thus not transmitted meiotically. Evidence is mounting, however, that the epigenetic marks of at least some mammalian genes are not fully erased during meiosis and may constitute a possible mechanism for transgenerational epigenetic inheritance.	117

Heterochromatin is thus associated with repressed transcription. Conversely, an open chromatin conformation (euchromatin) allows the cell's transcriptional machinery to access DNA and drive transcription[53] (See Fig. 1.1). The addition or removal of an acetyl group (CH_3CO), for example, is associated with replication and nucleosome assembly, higher-order chromatin packing, and interactions between nonhistone proteins and nucleosomes.[52] While acetylation of histone 3 at lysines 9 and 14 is commonly associated with gene activation, other types of modification have different effects on gene expression.[3] Another example is H3K4 methylation, which is generally associated with gene activation, whereas methylation of H3K9 and 27 is associated with decreased gene expression.[3] Like DNA methylation, histone modifications are highly dynamic and actively regulated by a host of catalytic enzymes, such as histone acetyltransferases (HATs) and histone deacetylases (HDACs), which add and remove acetyl groups, respectively.[126]

1.2.3 Interactions Between DNA Methylation and Histone Modifications

While often investigated independently, epigenetic modifications to DNA and histones are not mutually exclusive and interact in a number of ways. In addition, it is apparent that the classification of epigenetic mechanisms in terms of either gene activation or suppression is too simplistic.[11] The methyl-binding protein MeCP2, for example, binds specifically to methylated cytosines, attracting HDACs which hypoacetylate histones.[122] Another study has shown that unmethylated histone H3 residues at the lysine 4 position recruit DNMTs, which results in de novo DNA methylation.[100] In addition to catalyzing the DNA methylation reaction, DNMTs also interact with HDACs and histone methyltransferases (HMTs) to alter chromatin structure and thus gene expression.[62]

1.3 The Dynamic Epigenome

Like the DNA sequence, the epigenetic profile of somatic cells is transmitted from maternal to daughter chromatids during mitosis. Unlike the DNA sequence, which is stable and strongly conserved, epigenetic processes are tissue-specific, developmentally regulated, and can be highly dynamic. A growing body of evidence is now available, for example, indicating that epigenetic variation is strongly associated with age.[14,22,41,43,136] This is an important observation given that the prevalence of many chronic diseases increases with advancing age.

As will be discussed below, and in other chapters in this book, increasing evidence suggests that influences external to the organism can directly interact with the epigenome, altering epigenetic processes and thus gene expression. In addition, random stochastic and developmental epigenetic changes are also important. Experiments tracking the inheritance of epigenetic marks through generations of genetically identical cells in tissue-culture, for example, have indicated that there is considerable infidelity in the maintenance of methylation patterns in mammalian cells, and that de novo methylation events are fairly common during mitosis.[119,154] As epigenetic processes are integral in determining when and where specific genes are expressed, epigenetic metastability, environmentally or stochastically induced, may have profound phenotypic effects on gene expression in the cell. The dynamic nature of the epigenome calls into question many of our basic assumptions about the origins of phenotypic variance, and offers new insights about the non-Mendelian patterns of inheritance often observed for a wide range of complex traits and diseases.[94,107,108]

1.4 Epigenetics and Chronic Disease

It is well established that the regulation of gene activity is critically important for normal functioning of the genome. It follows that even genes that carry no mutations or disease predisposing polymorphisms can be rendered harmful if they are not expressed at the appropriate level in the correct type of cell at the right time of the cell cycle. Because epigenetic processes are integral for cellular development and function, aberrant DNA methylation signatures and histone modifications are thought to be involved in a diverse range of complex human pathologies,[56] including cancer[65] and several rare imprinting disorders.[38] Numerous epidemiological, clinical, and molecular peculiarities associated with many *chronic* illnesses are also strongly suggestive of an epigenetic contribution to etiology (see Table 1.2). These include the incomplete concordance between MZ twins, a fluctuating disease course with periods of remission and relapse, sexual dimorphism, peaks of susceptibility to disease that coincides with major hormonal rearrangements and parent-of-origin effects.[108]

1.4.1 Discordance Between MZ Twins

Numerous examples of phenotypic differences between genetically identical inbred animals,[47] genetically cloned animals,[23,149] and MZ twins[162] have been reported, supporting the notion that nongenetic factors can strongly influence phenotype. While MZ twins

Table 1.2 Examples of disorders that may involve an epigenetic etiology

Condition	Symptoms/Characteristics	Known/Suspected risk factors	Evidence for an epigenetic etiology	Selected references
Alzheimer's disease (AD)	• Senile plaques and neurofibrillary tangles • Neuronal loss leading to dementia and death	• Exposure to lead • Advancing age	• May display high homocysteine and low B12 and folate in blood. This may be due to dysregulation of the S-adenosylmethionine cycle that is required for DNA methylation • Evidence for AD-associated epigenetic changes in several genes • Evidence for accelerated age-specific epigenetic drift in affected patients	130, 137, 155, 170
Asthma	• Chronic inflammation of the lungs in which the bronchi are reversibly narrowed after exposure to an allergen	• Exposure to endotoxins • Viruses • Environmental tobacco smoke • In utero environmental stressors	• Many studies published on G X E interactions with risk factors • Many epigenetic changes seen after exposure to risk factors	97, 99
Autism spectrum disorder (ASD)	• Impairment in social interaction and language • Communication deficits • Repetitive behavior	• Advanced paternal age • Low birth weight or premature birth	• Mutations in the methyl-binding protein (MECP2) seen in some ASDs • Increased MECP2 promoter methylation seen in frontal cortex of male patients	26, 77, 98
"Bipolar disorder stressful life events" into the list of 'Known/ Suspected risk factors'	• Experiences of depression and at least one episode of mania	• Advanced paternal age	• Parental-origin effects, e.g., linkage with Chr18 found with paternal transmission. • Numerous loci with altered DNA methylation identified in case-control study using brain tissue • Evidence for skewed X-inactivation in affected individuals	24, 44, 48, 89, 96, 125
Cardiovascular disease (CVD)	• Diseases include hypertension, congenital heart disease and coronary artery disease	• High ingestion of cholesterol, sugar and processed foods	• Indications of global hypermethylation in CVD cases	51, 142
Chronic obstructive pulmonary disease	• Progressive airway obstruction	• Chronic cigarette smoking	• Insufficient HDAC2 in airway cells	9, 10

Table 1.2 (continued)

Condition	Symptoms/Characteristics	Known/Suspected risk factors	Evidence for an epigenetic etiology	Selected references
Diabetes mellitus	• Autoimmune disease • Inability of the body to regulate blood sugar • Immune system destroys insulin secreting β cells in the pancreatic islets (Type 1 diabetes) • insulin resistance combined with relatively reduced insulin secretion (Type 2 diabetes)	• Viral infections such as mumps • Lifestyle factors and the prenatal environment may be critical for Type 2 diabetes	• Promoter of the Ins1/2 gene in pancreatic β cells displays H3 hyperacetylation and hypermethylation at H3K4 • Evidence for fetal programming mediated by epigenetic changes	8,19, 49, 59, 83
Emphysema	• Reduction in alveolar walls and lung capillary blood vessels • Inflammation in bronchioles resulting in breathing difficulties	• Tobacco smoke exposure	• Elevated global levels of H3 acetylation and H4 seen in smokers and emphysema sufferers	72, 147
Endometriosis	• Presence and growth of functional endometrial-like tissues outside the uterine cavity • Responsible for dysmenorrhoea and pelvic pain • Reduction in fertility	• Hormonal • Autoimmune • Exposure to environmental toxins	• Promoters of PR-B and HOXA10 in sufferers is hypermethylated • DNMT1, DNMT3A, and DNMT3B, over-expressed in sufferers	54, 164, 165
Leukemia	• Abnormal lymphoid cells or myeloid cells resulting in either lymphoid or myeloid leukemia • Diminished immune system • Painful joints • Easily bruised and bleeding	• Exposure to radiation • Smoking • Certain blood disorders • Human T-cell leukemia virus type I	• Global DNA hypomethylation in cancer cells • Hypermethylation of promoter region of CDKN2B which encodes tumor suppressors • Cells from bone marrow of sufferers exhibit hypermethylation in a variety of associated genes	21, 92
Lupus erythematosus	• Autoimmune disease • Immune system attacks the body's connective tissue	• Viruses • UV light • Certain medications	• Treating cells with DNA Methyltransferase inhibitors 5-Azacytidine or Procainamide causes lupus like symptoms in mouse models	61, 113
Major depressive disorder (MDD)	• Episodes of depressed mood • Disturbed Sleep • Reduced appetite • Reduced concentration and energy • Suicidal thoughts	• Stressful life events	• Critical periods of exposure to life-stress • High female prevalence of MDD suggests potential skewed X-inactivation or hormone-specific modifications	79, 94

(continued)

Table 1.2 (continued)

Condition	Symptoms/Characteristics	Known/Suspected risk factors	Evidence for an epigenetic etiology	Selected references
Malignant hyperthermia (MH)	• On exposure to certain drugs used for general anesthesia, increase in skeletal muscle oxidative metabolism	• RYR1 mutation	• Mono-allelic silencing of RYR1 gene is of importance in MH inheritance • RYR1 mutations account for most MH cases	123, 171
Osteoarthritis	• Degradation of articular cartilage	• Age, mechanical trauma with temporary inflammation	• Increased and continuous expression of catabolic genes due to loss of DNA methylation at specific CpG sites in the promoters of catabolic genes • DNA hypermethylation of some CpG sites of anabolic genes, e.g., COL9A1, silences the gene	60, 120, 121
Ovarian cancer	• Cancerous cells in ovaries or germ cells	• Advancing age • Hormone therapy • Endometriosis • Smoking	• Hypomethylation of DNA adjacent to the centromeres of Chr1 in cancer cells • Hypomethylation more prevalent in tumors of an advanced stage • Hypermethylation seen in many genes including BRCA1, tumor suppressor, pro-apoptotic and cell adhesion genes	6, 159
Prostate cancer	• Malignant cancerous growth from cells in the prostate gland	• Advancing age • Exposure to toxins and chemical agents • Viruses and infections of prostate	• Expression of EZH2 and BMI-1 increases as cancer progresses • Hypermethylation in many genes of cancer cells	28, 132
Rett syndrome	• Neurodevelopmental disorder (ASD) • Cognitive and communicative impairment • Microcephaly • Abnormal growth	• Mutations in the MECP2	• MECP2, a protein essential for nerve cell function, contains a methylated DNA binding domain	4, 105
Rheumatoid arthritis (RA)	• Autoimmune disease • Progressive and irreversible destruction by osteoclasts of the trabecular bone and cartilage at joints	• Chronic Inflammation	• Global methylation less in RA sufferers • DR3 shows less expression due to hypermethylation of promoter	59, 71, 84, 85, 118, 147, 153

show higher concordance for most chronic diseases than DZ twins, the overall level of concordance for all such disorders is well below 100% (see Fig. 1.2). For example, concordance of MZ twins reaches only ~25% in multiple sclerosis and hypertension, and ~50% in schizophrenia and coronary heart disease. Such discordance between MZ twins suggests that non-genetic factors influence susceptibility to chronic disease.[7,55]

It has been demonstrated that fairly profound epigenetic differences across the genome arise during the lifetime of MZ twins, highlighting the dynamic nature of epigenetic processes.[43] Kaminsky and colleagues investigated genome-wide epigenetic differences in

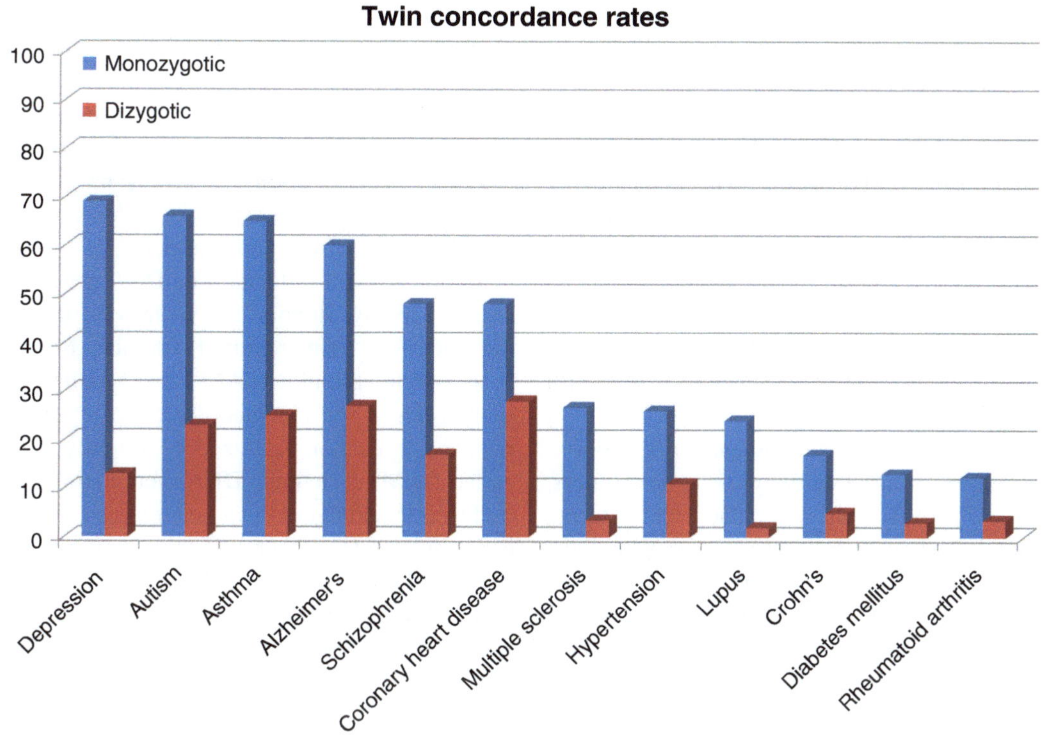

Fig. 1.2 Monozygotic (MZ) and dizygotic (DZ) twin concordance rates for a range of common complex disorders. The higher MZ concordance rates reflect that these disorders are heritable to some extent, but not fully, as all MZ values are considerably below 100%[2,29,33,34,37,70,110,150]

MZ and DZ twin pairs using microarrays that can simultaneously assess DNA methylation across thousands of promoter regions.[68] They found a large degree of MZ co-twin DNA methylation variation in all three tissue types investigated. Moreover, there was significantly higher epigenetic difference between DZ co-twins compared to MZ co-twins. This implicates that molecular mechanisms of heritability may not be limited to DNA sequence differences, but to some extent are underlined by epigenetic differences (see below). Interestingly, specific MZ twin DNA methylation differences have been reported between MZ twins discordant for a number of common chronic disorders, including systemic lupus erythematosus,[63] dementia,[87] and several psychiatric conditions including bipolar disorder.[124]

Traditionally, most of this discordance between MZ twins has been attributed to "non-shared environmental" influences, but the dynamic nature of the epigenome and its significant role in regulating gene expression have led to speculation that epigenetic variation is likely to mediate nongenetic phenotypic differences. Many of these epigenetic changes are likely to be environmentally mediated (see below), but because of random epigenetic drift, stochastic variation may also contribute to MZ twin discordance.[161]

1.4.2 Epigenetics: Linking the Environment to Phenotype in Chronic Disease

There is mounting evidence to suggest that epigenetic processes are induced in response to environmental insults. DNA methylation, for example, varies as a function of nutritional, chemical, physical, and psychosocial factors. Because epigenetic changes can be stably inherited through mitosis in somatic cell lineages, they provide a mechanism by which the environment can lead to long-term alterations in the cellular phenotype. Fraga and colleagues examined DNA methylation and histone acetylation in 80 pairs of MZ twins, ranging from 3 to 74 years of age, using a combination of global and locus-specific methods.[43] They found that one-third of MZ twins had a significantly dissimilar epigenetic profile, with older twins and those with a history of nonshared environments being the most disparate. This suggests that environmental factors may shape the epigenome over the life-course. While this study highlighted mounting epigenetic discordance with age, several studies also have reported significant DNA methylation differences that were already deleted in MZ twin pairs when they were still very young.[93]

Given the role that prenatal and early-life environmental factors play in the etiology of chronic illness, the epigenome is likely to be particularly labile during key developmental periods.[156] This is especially the case during embryogenesis, when the rate of DNA synthesis is high and the epigenetic marks needed for normal tissue differentiation and development are being established.[30] It is commonly thought that exposure to an adverse in utero environment, particularly during certain sensitive developmental periods, dramatically increases the risk of disease later in life. Such a link is well established for coronary heart disease, obesity, and type 2 diabetes, but also exists for neurological and mental illness. The concept of the developmental origin of health and disease is based on the proposition that a poor in utero environment manifests itself in changes to metabolic processes. Such changes are initially adaptive, but may lead to an increased risk of chronic disease later in life. Epidemiological studies show, for example, that reduced birth weight, which is strongly correlated with fetal under-nutrition, is associated with a range of cardiovascular and metabolic diseases.[8,49] One limitation to this hypothesis is that the biological mechanisms underlying such fetal programming are not yet well understood, although mounting evidence suggests that epigenetic changes in response to an adverse in utero environment may affect normal patterns of development and increase susceptibility to disease after birth (see Fig. 1.3).

Epigenetic mechanisms may represent a major link between genes and the environment, with environmentally induced epigenetic changes accounting for disease discordance between MZ twins. Environmental influences mediated through epigenetic effects are speculated to cause a diverse range of chronic diseases, including type 1 diabetes,[83] chronic kidney disease,[36] respiratory disease,[132] chronic fatigue syndrome,[81] alcohol dependence,[67] major psychosis,[95] and Parkinson's disease.[150] A growing number of disease-associated environmental factors mediate the epigenome during key developmental periods. Some examples will be discussed briefly below, but for a comprehensive discussion see reviews by Dolinoy and Jirtle,[30] and Jirtle and Skinner.[64]

Major external influences on DNA methylation are external methyl donors and co-factors, usually from the diet, that are required for the formation of S-adenosylmethionine (SAM). SAM, in turn, acts as a methyl-donor for the methylation of cytosine DNA residues, a reaction catalyzed by a family of enzymes called DNMTs. De novo DNA methylation is initiated by Dnmt3a and Dnmt3b and maintained in mitotic lineages by Dnmt1.[12] Dietary factors required for the formation of SAM include folate, methionine, choline, vitamin B12, vitamin B6, and vitamin B2. Given the role of DNA methylation in coordinating the correct pattern of gene expression during embryogenesis and development, exposure to a diet lacking such components at specific developmental time-points could have profound phenotypic effects. Extreme folate deficiency, for example, causes depletion of SAM; this results in genome-wide DNA hypomethylation, the activation of oncogenes and cancer.[133] Interestingly, a dietary deficiency of vitamin B12 and folate is also linked to impaired central nervous system development and to several psychiatric conditions.[116] Futhermore of interest is the epidemiological evidence from two large independent population samples in the Netherlands and China that have suggested that in utero nutritional deficiency, due to maternal exposure to severe famine during pregnancy, is associated with an increase in schizophrenia in adult life.[141,144,145] Exposure to famine during pregnancy has also been associated with loss of imprinting at IGF2,[57] an epigenetic change associated with various growth and developmentally-related phenotypes, including brain size.[109]

Other environmental factors that epigenetically alter gene expression in animals include toxins and chemicals associated with chronic illness. For example, the offspring of pregnant rats exposed to the endocrine disruptor vinclozolin, an agrochemical used as a fungicide in crops, have altered DNA methylation profiles that correlate with adverse phenotypic changes.[20] It is not just the chemical environment that can cause long-lasting epigenetic changes. Even the

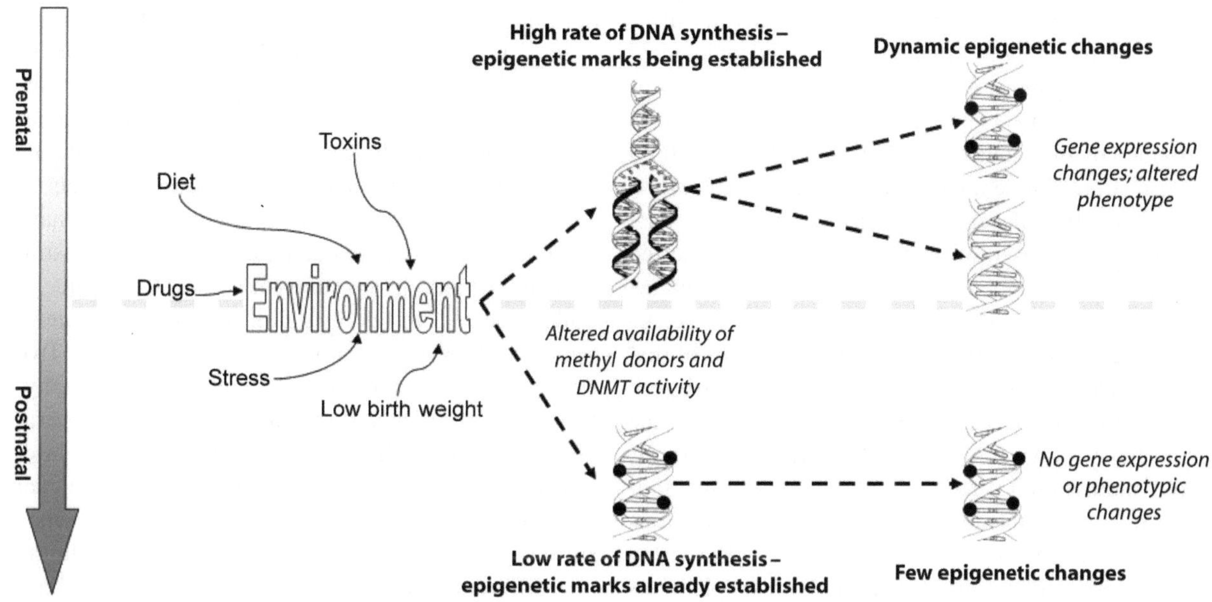

Fig. 1.3 Environmental insults during key developmental periods such as embryogenesis, when DNA synthesis is high and when the epigenetic marks needed for normal tissue development are being established, may alter gene expression and phenotype via changes to epigenetic processes such as DNA methylation[94]

psychosocial environment during key developmental periods early in life can epigenetically mediate gene expression.[146] Early psychosocial adversity has been linked to an increased risk of developing neuropsychiatric conditions. Research by Weaver and colleagues,[157] for example, has shown that immediate postnatal maternal care in rats, as measured by increased pup licking, grooming, and arch-backed nursing, leads to epigenetic modification across the promoter region of the glucocorticoid receptor gene (NR3C1), specifically at CpG sites located in a NGF1 transcription factor-binding motif. This site directly affects gene expression and stress-related phenotypes in offspring. Interestingly, like the other examples discussed above, these epigenetic changes occur only during a specific critical period – in this case, immediately after birth. Subsequent transcriptomic studies by the same group identified over 900 genes in the hippocampus that are stably regulated by maternal care.[158] This indicates a profound effect of the early social environment on gene expression throughout. A study in humans has investigated epigenetic differences in a homologous *NR3C1* promoter region, comparing DNA methylation in postmortem hippocampus samples obtained from suicide victims with a history of childhood abuse with samples from suicide victims with no childhood abuse or from controls. In line with the animal findings, abused suicide victims had increased CpG methylation of the NR3C1 promoter with concomitant changes in mRNA.[90]

1.4.3 Epigenetic Mechanisms and Gene–Environment Interactions

Environmental modulation of the epigenome provides a mechanism for the gene-environment interactions currently being identified across the spectrum of chronic diseases for phenotypes as diverse as the metabolic syndrome[1] and depression.[17] The pathogenic effect of a polymorphism associated with disrupted gene function is likely to depend upon the extent of actual expression of that particular variant. It is thus plausible that genetic risk could be enhanced or minimized under conditions when environmental factors acting via processes like DNA methylation directly influence gene expression. Of particular interest are so-called metastable epialleles, i.e., loci that can be epigenetically modified to produce a range of phenotypes from genetically identical cells.[115] Many of these loci have been shown to be environmentally sensitive and to be particularly affected by the environment of the developing fetus. An example of how such a mechanism could explain gene–environment interactions is provided by the agouti viable yellow allele (A^{vy}) inbred mouse strain. These mice have a range of coat colors, each phenotype depending upon the epigenetic state of a large transposable element that is inserted upstream of the agouti gene. The transposon contains a cryptic promoter which expresses a phenotype characterized by yellow fur and metabolic features of disorders such as diabetes or obesity. When the transposon is methylated, the phenotype is not expressed; the mice have brown fur, and are healthy. Interestingly, DNA methylation across this region can be manipulated by altering the diet of pregnant mothers and thus changing the phenotype of the offspring.[25,32] Enriching the maternal diet with methyl-donor supplements increases offspring DNA methylation, inducing changes in gene expression that lead to brown fur and metabolic health. Gene–environment interactions may also result when genetic polymorphisms alter the ability of a specific region of the genome to be epigenetically altered in response to an environmental pathogen. The interplay between the genome, the environment, and epigenetic processes is complicated by the fact that some DNA alleles and haplotypes are themselves associated with a specific epigenetic profile.[131] For example, allele-specific epigenetic modifications have been associated with "risk" polymorphisms in psychiatric candidate genes, including the serotonin receptor gene (*5HTR2A*)[111] and the gene that encodes the brain-derived neurotrophic factor (BDNF).[96] Given the known influence of environmental factors on epigenetic regulation, the cis-regulation of DNA methylation by genetic variation may reflect the existence of a common pathway that acts on both genetic and environmental effects and represents a potential mechanism for G X E interaction.

1.4.3.1 Sex Effects

A common characteristic of many chronic disorders is a high level of sexual dimorphism – i.e., differential disease susceptibility between males and females (see Fig. 1.4). While sex hormones are the usual explanation for gender effects in complex diseases, no underlying mechanisms have been proposed as to how sex hormones predispose to or protect individuals from a disease involving specific molecular mechanisms of hormone action.[67] A potential mechanism of sex hormone impact would be to alter the epigenetic signatures of particular chromosomal regions that modulate access of transcription factors. The direct effects of sex hormone administration on epigenetic states have been demonstrated. An example is the effect of estradiol on the methylation status of various CpG dinucleotides located in an estradiol-mediated regulatory region of the avian vitellogenin II gene.[91,127,159,160] Two cytosines in the gene promoter are actively demethylated in a strand-specific manner in response to estradiol treatment.[127] The demethylation persists after transcription has ended[121,160,161] and leads to a quicker induction of vitellogenin II mRNA synthesis when subsequently stimulated by estradiol.[132] In addition, sex differences in DNA methylation patterns have been detected in numerous studies.[15,73,80,87,128,135,136,168] These findings suggest that sex hormones may mediate a long-lasting epigenetic effect on gene transcription.

Similar to their effect on DNA methylation, members of the nuclear hormone receptor (NHR) superfamily modify chromatin. In this connection, the steroid receptor (SR) subset of NHR is of particular interest, as it contains the androgen and estrogen receptors.[74] These sex hormone receptors can activate or repress transcriptional activity, depending on whether a ligand is present or absent.[45,166] Steroid hormone–mediated transcriptional activation or repression is the result of the SR having recruited the co-activator and co-repressor complexes. These protein complexes associate with the epigenetic modifiers, such as HDACs, HATs, and HMTs.[45,46,139] It is these "co-regulatory complexes" that bring about the epigenetic modification needed to remodel chromatin because they allow or restrict access of transcription factors and RNA polymerase II, thereby mediating the epigenetic effects of the sex hormones.

1.4.3.2 Parental-Origin Effects

Another non-Mendelian feature, commonly observed in chronic disease, is a parent-of-origin effect, with disease susceptibility mediated by parental factors in a sex-specific manner. Parent-of-origin effects are sometimes seen in patterns of familial disease transmission, with the risk to offspring depending on whether it is the

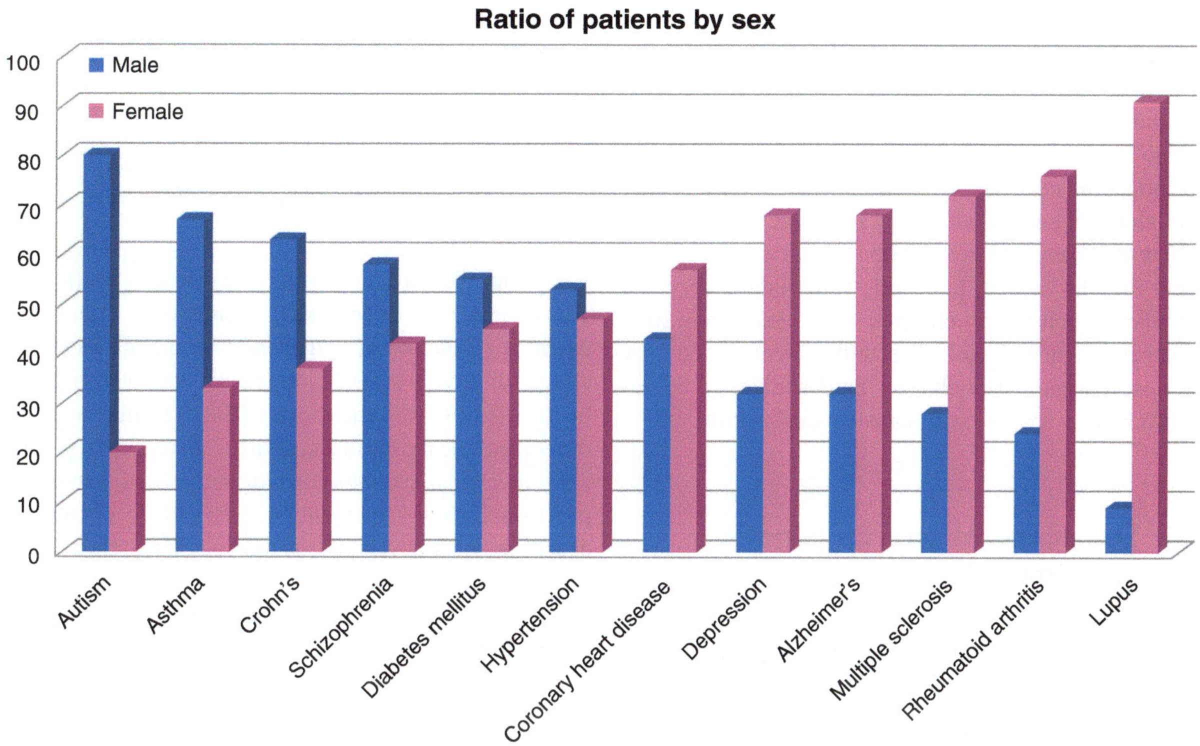

Fig. 1.4 Many complex disease phenotypes are characterized by large sex ratio differences[18,67,69,101]

mother or father who is affected. At the molecular level, risk alleles sometimes confer increased susceptibility only if transmitted from the affected parent. Asthma, bipolar disorder, and epilepsy are more often transmitted from the mother, but type 1 diabetes is more likely to be transmitted from an affected father.[106] Genetic studies of complex chronic disease phenotypes have shown that susceptibility variants often depend on the parent from whom they are inherited.[78] The most likely mechanism of parent of origin effects is genomic imprinting,[39,102,143] the differential regulation of gene activity depending upon parental origin, mediated by allele-specific epigenetic modifications. In "classical" cases, imprinted genes exhibit "on/off" regulation, i.e., only maternally- *or* paternally-inherited alleles are expressed, but not both. There is now evidence to suggest that imprinting is more widespread and less clear-cut.[112] Interestingly, imprinted genes are often involved in regulating growth and development and therefore are particularly vulnerable to adverse prenatal insults. It appears that interindividual differences in IGF2 methylation for example, are environmentally mediated, particularly by prenatal exposure to famine.[57] This is of considerable relevance, given the known link between prenatal factors (including nutrition) and the risk of developing chronic illness as an adult.

1.4.3.3 Paternal Age Effects

Convincing evidence from epidemiologic studies implicates that advanced paternal age is linked to increased risk for several chronic conditions, including schizophrenia and autism.[16,77,86,138,169] This may be due to several mechanisms.[104] Initially explanations focused solely on mutagenesis, attributing it to the accumulation in successive generations of spermatozoa of de novo mutations, increases in DNA repeats, and chromosomal breaks that occur with advanced paternal age. However, mutations in sperm cannot fully explain the association for many diseases.[152] An alternative explanation is that epigenetic dysfunction alters normal patterns and genomic imprinting.[42,104] Flanagan et al.[41] have described significant intra- and inter-individual epigenetic variability in the male germline and identified genes with age-related DNA methylation changes.[41]

In addition to de novo mutations/epimutations that occur in the course of a large number of germline cell divisions, accumulated exposure to various environmental toxins throughout life may also bring about germline alterations in older men. Toxins have been shown to induce in mice germline mutations, DNA damage, and global hypermethylation.[167] It is therefore plausible that these changes increase with age and are more prevalent in older fathers. In a study using inbred animals, significant behavioral changes were associated with advancing paternal age. The authors concluded that de novo chromosomal changes and/or inherited epigenetic changes are the most plausible explanation.[139]

1.5 Transgenerational Epigenetic Inheritance?

Epigenetic profiles are thought to become fully erased and reset during gametogenesis. This prevents the meiotic transmission of epigenetic information between generations. After having been documented in plants, however, evidence is mounting that the epigenetic marks of at least some mammalian genes are not fully erased during meiosis and can thus be potentially transmitted from generation to generation.[75,114,117,128] Epigenetic transgenerational inheritance blurs the demarcation between epigenetic- and DNA sequence-based inheritance and, though still a controversial notion, challenges the assumption that the "heritable" component to complex chronic disorders is entirely genetic. In fact, given the growing evidence that environmental factors influence epigenetic modifications, this so-called "soft inheritance" supports the concept that environmental exposures in one generation could alter phenotype phenotype in subsequent generations.[117] A growing body of epidemiological evidence in humans supports the notion that environmental exposures in ancestral generations have phenotypic effects in later generations, many related to chronic disease.[103] However, there is as yet no direct empirical evidence for transgenerational epigenetic inheritance in humans. Meiotic transmission of epigenetic marks is not necessary for transgenerational transmission of environmentally induced phenotypes. For example, rat pups cared for by high-licking and -grooming mothers themselves go on to provide high levels of care to their offspring, an effect apparently mediated by epigenetic changes upstream of the glucocorticoid receptor gene.[155] In this way, epigenetic effects can be transmitted across generations through a purely experiential effect on gene expression.

1.6 Implications for the Treatment of Chronic Disease

A primary aim of all etiological research is to identify novel targets for therapeutic intervention. Given the dynamic regulation of epigenetic phenomena and their effects, discovery of epigenomic dysfunction is bound to be important for disease therapy. The dynamic nature of the epigenome means that epigenetic disruption is potentially reversible and thus constitutes a realistic target for pharmacological intervention. So-called epigenetic drugs are being developed for a range of disorders, most notably cancer.[35] Numerous agents have been discovered that can alter DNA methylation and histone modifications, and several of these are already being tested in clinical trials. Inhibitors of DNA methylation can reactivate the expression of genes that are influenced by epigenetic silencing. Decitabine (5-aza-2-deoxycytidine), for example, hypomethylates DNA by forming irreversible covalent bonds with DNA methyltransferases at cytosine sites targeted for methylation. It has been used successfully in the treatment of acute myelogenous leukemia, myelodysplasia, and chronic myelogenous leukemia.[61] Similarly, numerous HDAC inhibitors are being developed to reduce the function of HDACs, either specifically or generally. One potential obstacle to the widespread use of these agents is that drugs which target the epigenome can have potentially pathogenic effects on expression of genes that are not the intervention target.

1.7 Looking Forward

It is now apparent that most chronic diseases result from the interaction of genetic, environmental, and epigenetic factors (see Fig. 1.5). Technological advances in epigenomic profiling mean it is now feasible to investigate ways in which environmental factors act upon the genome, bringing about epigenetic changes in gene expression and in the risk for chronic illness. As understanding of epigenetic processes involved in chronic disease advances, more opportunities for preventative and therapeutic intervention will arise. It is important to establish when such disease-associated epigenetic

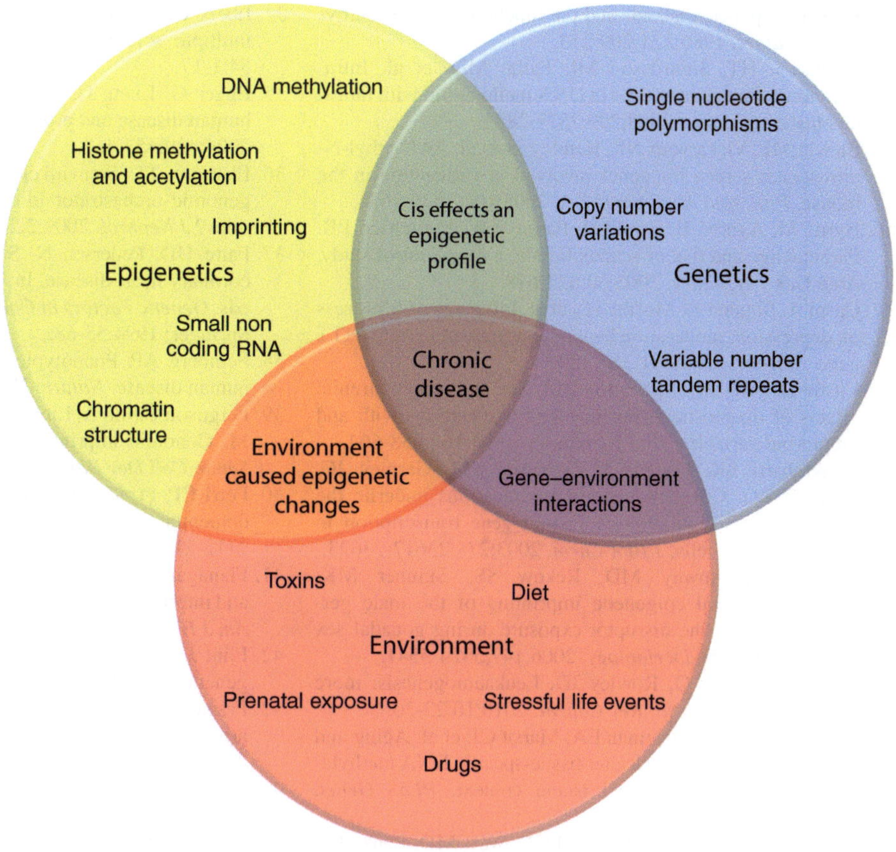

Fig. 1.5 Chronic disease results from the interaction of genetic, environmental, and epigenetic factors

changes occur. They may arise prior to the illness and contribute to the disease phenotype. They could also be a secondary effect of the disorder or medication. Such uncertainties can be resolved by combining longitudinal epigenetic studies with more traditional methods of research, such as GWA studies. Epigenetic studies are currently limited by insufficient knowledge about the "normal" epigenetic patterns that characterize tissues and cell types. It is hoped that current initiatives such as the NIH Epigenomic Roadmap, which aims to catalog patterns of epigenetic variation across different cell and tissue types, will be useful in this regard.

References

1. Adamo KB, Tesson F. Gene–environment interaction and the metabolic syndrome. *Novartis Found Symp*. 2008;293: 103-119; discussion 119-127.
2. Aho K, Koskenvuo M, Tuominen J, Kaprio J. Occurrence of rheumatoid arthritis in a nationwide series of twins. *J Rheumatol*. 1986;13:899-902.
3. Akbarian S, Huang HS. Epigenetic regulation in human brain-focus on histone lysine methylation. *Biol Psychiatry*. 2009;65:198-203.
4. Amir RE, Van den Veyver IB, Wan M, Tran CQ, Francke U, Zoghbi HY. Rett syndrome is caused by mutations in X linked MECP2, encoding methyl CpG binding protein 2. *Nat Genet*. 1999;23:185-188.
5. Avner P, Heard E. X-chromosome inactivation: counting, choice and initiation. *Nat Rev Genet*. 2001;2:59-67.
6. Balch C, Fang F, Matei DE, Huang TH, Nephew KP. Minireview: epigenetic changes in ovarian cancer. *Endocrinology*. 2009;150:4003-4011.
7. Ballestar E, Ballestar E. Epigenetics lessons from twins: prospects for autoimmune disease. *Clin Rev Allergy Immunol*. 2010;39(1):30-41.
8. Barker DJ. In utero programming of chronic disease. *Clin Sci (Lond)*. 1998;95:115-128.
9. Barnes PJ, Adcock IM, Ito K. Histone acetylation and deacetylation: importance in inflammatory lung diseases. *Eur Respir J*. 2005;25:552-563.
10. Barnes PJ, Ito K, Adcock IM. Corticosteroid resistance in chronic obstructive pulmonary disease: inactivation of histone deacetylase. *Lancet*. 2004;363:731-733.
11. Berger SL. The complex language of chromatin regulation during transcription. *Nature*. 2007;447:407-412.
12. Bestor TH. The DNA methyltransferases of mammals. *Hum Mol Genet*. 2000;9:2395-2402.

13. Bird AP. CpG-rich islands and the function of DNA methylation. *Nature*. 1986;321:209-213.
14. Bjornsson HT, Sigurdsson MI, Fallin MD, et al. Intra-individual change over time in DNA methylation with familial clustering. *Jama*. 2008;299:2877-2883.
15. Blewitt ME, Vickaryous NK, Hemley SJ, et al. An N-ethyl-N-nitrosourea screen for genes involved in variegation in the mouse. *Proc Natl Acad Sci U S A*. 2005;102:7629-7634.
16. Byrne M, Agerbo E, Ewald H, Eaton WW, Mortensen PB. Parental age and risk of schizophrenia: a case-control study. *Arch Gen Psychiatry*. 2003;60:673-678.
17. Caspi A, Sugden K, Moffitt TE, et al. Influence of life stress on depression: moderation by a polymorphism in the 5-HTT gene. *Science*. 2003;301:386-389.
18. Castle DJ, Wessely S, Murray RM. Sex and schizophrenia: effects of diagnostic stringency, and associations with and premorbid variables. *Br J Psychiatry*. 1993;162:658-664.
19. Chakrabarti SK, Francis J, Ziesmann SM, Garmey JC, Mirmira RG. Covalent histone modifications underlie the developmental regulation of insulin gene transcription in pancreatic beta cells. *J Biol Chem*. 2003;278:23617-23623.
20. Chang HS, Anway MD, Rekow SS, Skinner MK. Transgenerational epigenetic imprinting of the male germline by endocrine disruptor exposure during gonadal sex determination. *Endocrinology*. 2006;147:5524-5541.
21. Chen J, Odenike O, Rowley JD. Leukaemogenesis: more than mutant genes. *Nat Rev Cancer*. 2010;10:23-36.
22. Christensen BC, Houseman EA, Marsit CJ, et al. Aging and environmental exposures alter tissue-specific DNA methylation dependent upon CpG island context. *PLoS Genet*. 2009;5:e1000602.
23. Cibelli JB, Campbell KH, Seidel GE, West MD, Lanza RP. The health profile of cloned animals. *Nat Biotechnol*. 2002;20:13-14.
24. Connor CM, Akbarian S. DNA methylation changes in schizophrenia and bipolar disorder. *Epigenetics*. 2008; 3:55-58.
25. Cooney CA, Dave AA, Wolff GL. Maternal methyl supplements in mice affect epigenetic variation and DNA methylation of offspring. *J Nutr*. 2002;132:2393S-2400S.
26. Currenti SA. Understanding and determining the etiology of autism. *Cell Mol Neurobiol*. 2010;30:161-171.
27. Davies W, Isles AR, Wilkinson LS. Imprinted gene expression in the brain. *Neurosci Biobehav Rev*. 2005;29:421-430.
28. De Marzo AM, Platz EA, Sutcliffe S, et al. Inflammation in prostate carcinogenesis. *Nat Rev Cancer*. 2007;7:256-269.
29. Deapen D, Escalante A, Weinrib L, et al. A revised estimate of twin concordance in systemic lupus erythematosus. *Arthritis Rheum*. 1992;35:311-318.
30. Dolinoy DC, Jirtle RL. Environmental epigenomics in human health and disease. *Environ Mol Mutagen*. 2008;49:4-8.
31. Dolinoy DC, Weidman JR, Jirtle RL. Epigenetic gene regulation: linking early developmental environment to adult disease. *Reprod Toxicol*. 2007;23:297-307.
32. Dolinoy DC, Weidman JR, Waterland RA, Jirtle RL. Maternal genistein alters coat color and protects Avy mouse offspring from obesity by modifying the fetal epigenome. *Environ Health Perspect*. 2006;114:567-572.
33. Duffy DL, Martin NG, Battistutta D, Hopper JL, Mathews JD. Genetics of asthma and hay fever in Australian twins. *Am Rev Respir Dis*. 1990;142:1351-1358.
34. Ebers GC, Sadovnick AD. The role of genetic factors in multiple sclerosis susceptibility. *J Neuroimmunol*. 1994; 54:1-17.
35. Egger G, Liang G, Aparicio A, Jones PA. Epigenetics in human disease and prospects for epigenetic therapy. *Nature*. 2004;429:457-463.
36. Ekstrom TJ, Stenvinkel P. The epigenetic conductor: a genomic orchestrator in chronic kidney disease complications? *J Nephrol*. 2009;22:442-449.
37. Faire UD, Pedersen N. Studies of twins and adoptees in coronary heart disease. In: Goldbourt U, Faire UD, Berg K, eds. *Genetic Factors in Coronary Heart Disease*. New York: Springer; 1994:55-68.
38. Feinberg AP. Phenotypic plasticity and the epigenetics of human disease. *Nature*. 2007;447:433-440.
39. Ferguson-Smith A, Lin SP, Tsai CE, Youngson N, Tevendale M. Genomic imprinting – insights from studies in mice. *Semin Cell Dev Biol*. 2003;14:43-49.
40. Ferri CP, Prince M, Brayne C, et al. Global prevalence of dementia: a Delphi consensus study. *Lancet*. 2005;366: 2112-2117.
41. Flanagan JM, Popendikyte V, Pozdniakovaite N, et al. Intra- and interindividual epigenetic variation in human germ cells. *Am J Hum Genet*. 2006;79:67-84.
42. Flint J. Implications of genomic imprinting for psychiatric genetics. *Psychol Med*. 1992;22:5-10.
43. Fraga MF, Ballestar E, Paz MF, et al. Epigenetic differences arise during the lifetime of monozygotic twins. *Proc Natl Acad Sci U S A*. 2005;102:10604-10609.
44. Frans EM, Sandin S, Reichenberg A, Lichtenstein P, Langstrom N, Hultman CM. Advancing paternal age and bipolar disorder. *Arch Gen Psychiatry*. 2008;65:1034-1040.
45. Fu M, Rao M, Wang C, et al. Acetylation of androgen receptor enhances coactivator binding and promotes prostate cancer cell growth. *Mol Cell Biol*. 2003;23:8563-8575.
46. Fu M, Wang C, Zhang X, Pestell RG. Acetylation of nuclear receptors in cellular growth and apoptosis. *Biochem Pharmacol*. 2004;68:1199-1208.
47. Gartner K. A third component causing random variability beside environment and genotype. A reason for the limited success of a 30 year long effort to standardize laboratory animals? *Lab Anim*. 1990;24:71-77.
48. Gershon ES, Badner JA, Detera-Wadleigh SD, Ferraro TN, Berrettini WH. Maternal inheritance and chromosome 18 allele sharing in unilineal bipolar illness pedigrees. *Am J Med Genet*. 1996;67:202 207.
49. Gluckman PD, Hanson MA. Developmental origins of disease paradigm: a mechanistic and evolutionary perspective. *Pediatr Res*. 2004;56:311-317.
50. Gluckman PD, Hanson MA, Beedle AS. Early life events and their consequences for later disease: a life history and evolutionary perspective. *Am J Hum Biol*. 2007;19:1-19.
51. Gluckman PD, Hanson MA, Buklijas T, Low FM, Beedle AS. Epigenetic mechanisms that underpin metabolic and cardiovascular diseases. *Nat Rev Endocrinol*. 2009;5: 401-408.
52. Grant PA, Berger SL. Histone acetyltransferase complexes. *Semin Cell Dev Biol*. 1999;10:169-177.
53. Grewal SI, Jia S. Heterochromatin revisited. *Nat Rev Genet*. 2007;8:35-46.

54. Guo SW. Epigenetics of endometriosis. *Mol Hum Reprod.* 2009;15:587-607.
55. Haque FN, Gottesman II, Wong AH. Not really identical: epigenetic differences in monozygotic twins and implications for twin studies in psychiatry. *Am J Med Genet C Semin Med Genet.* 2009;151C:136-141.
56. Hatchwell E, Greally JM. The potential role of epigenomic dysregulation in complex human disease. *Trends Genet.* 2007;23:588-595.
57. Heijmans BT, Tobi EW, Stein AD, et al. Persistent epigenetic differences associated with prenatal exposure to famine in humans. *Proc Natl Acad Sci U S A.* 2008;105:17046-17049.
58. Henikoff S, Matzke MA. Exploring and explaining epigenetic effects. *Trends Genet.* 1997;13:293-295.
59. Hewagama A, Richardson B. The genetics and epigenetics of autoimmune diseases. *J Autoimmun.* 2009;33:3-11.
60. Imagawa K, de Andes MC, Hashimoto K, Itoi E, Oreffo R, Roach H. Reduced Expression of Collagen Type IX in Human Osteoarthritic Chondrocytrs is Associated with Epigenetic Silencing by DNA Hypermethylation. *Osteoarthritis and Cartilage.* 2009;18(Suppl. 2):S36-S36 .
61. Issa JP, Garcia-Manero G, Giles FJ, et al. Phase 1 study of low-dose prolonged exposure schedules of the hypomethylating agent 5-aza-2′-deoxycytidine (decitabine) in hematopoietic malignancies. *Blood.* 2004;103:1635-1640.
62. Jaenisch R, Bird A. Epigenetic regulation of gene expression: how the genome integrates intrinsic and environmental signals. *Nat Genet.* 2003;33(Suppl):245-254.
63. Javierre BM, Fernandez AF, Richter J, et al. Changes in the pattern of DNA methylation associate with twin discordance in systemic lupus erythematosus. *Genome Res.* 2010;20:170-179.
64. Jirtle RL, Skinner MK. Environmental epigenomics and disease susceptibility. *Nat Rev Genet.* 2007;8:253-262.
65. Jones PA, Baylin SB. The epigenomics of cancer. *Cell.* 2007;128:683-692.
66. Kalsi G, Prescott CA, Kendler KS, Riley BP. Unraveling the molecular mechanisms of alcohol dependence. *Trends Genet.* 2009;25:49-55.
67. Kaminsky Z, Wang SC, Petronis A. Complex disease, gender and epigenetics. *Ann Med.* 2006;38:530-544.
68. Kaminsky ZA, Tang T, Wang SC, et al. DNA methylation profiles in monozygotic and dizygotic twins. *Nat Genet.* 2009;41:240-245.
69. Kanner L. Autistic disturbances of affective contact. *Acta Paedopsychiatr.* 1968;35:100-136.
70. Kaprio J, Tuomilehto J, Koskenvuo M, et al. Concordance for type 1 (insulin-dependent) and type 2 (non-insulin-dependent) diabetes mellitus in a population-based cohort of twins in Finland. *Diabetologia.* 1992;35:1060-1067.
71. Karouzakis E, Gay RE, Gay S, Neidhart M. Epigenetic control in rheumatoid arthritis synovial fibroblasts. *Nat Rev Rheumatol.* 2009;5:266-272.
72. Karrasch S, Holz O, Jorres RA. Aging and induced senescence as factors in the pathogenesis of lung emphysema. *Respir Med.* 2008;102:1215-1230.
73. Kawakami K, Ruszkiewicz A, Bennett G, et al. DNA hypermethylation in the normal colonic mucosa of patients with colorectal cancer. *Br J Cancer.* 2006;94:593-598.
74. Kinyamu HK, Archer TK. Modifying chromatin to permit steroid hormone receptor-dependent transcription. *Biochim Biophys Acta.* 2004;1677:30-45.
75. Klar AJ. Propagating epigenetic states through meiosis: where Mendel's gene is more than a DNA moiety. *Trends Genet.* 1998;14:299-301.
76. Klose RJ, Bird AP. Genomic DNA methylation: the mark and its mediators. *Trends Biochem Sci.* 2006;31:89-97.
77. Kolevzon A, Gross R, Reichenberg A. Prenatal and perinatal risk factors for autism: a review and integration of findings. *Arch Pediatr Adolesc Med.* 2007;161:326-333.
78. Kong A, Steinthorsdottir V, Masson G, et al. Parental origin of sequence variants associated with complex diseases. *Nature.* 2009;462:868-874.
79. Krishnan V, Nestler EJ. The molecular neurobiology of depression. *Nature.* 2008;455:894-902.
80. Lai JC, Cheng YW, Chiou HL, Wu MF, Chen CY, Lee H. Gender difference in estrogen receptor alpha promoter hypermethylation and its prognostic value in non-small cell lung cancer. *Int J Cancer.* 2005;117:974-980.
81. Lin E, Hsu SY. A Bayesian approach to gene-gene and gene-environment interactions in chronic fatigue syndrome. *Pharmacogenomics.* 2009;10:35-42.
82. Luger K, Mader AW, Richmond RK, Sargent DF, Richmond TJ. Crystal structure of the nucleosome core particle at 2.8 A resolution. *Nature.* 1997;389:251-260.
83. Macfarlane AJ, Strom A, Scott FW. Epigenetics: deciphering how environmental factors may modify autoimmune type 1 diabetes. *Mamm Genome.* 2009;20(9–10):624-632.
84. Maciejewska-Rodrigues H, Karouzakis E, Strietholt S, et al. Epigenetics and rheumatoid arthritis: the role of SENP1 in the regulation of MMP-1 expression. *J Autoimmun.* 2010;35(1):15-22.
85. Maciejewska HR, Jungel A, Gay RE, Gay S. Innate immunity, epigenetics and autoimmunity in rheumatoid arthritis. *Mol Immunol.* 2009;47:12-18.
86. Malaspina D, Harlap S, Fennig S, et al. Advancing paternal age and the risk of schizophrenia. *Arch Gen Psychiatry.* 2001;58:361-367.
87. Mani ST, Thakur MK. In the cerebral cortex of female and male mice, amyloid precursor protein (APP) promoter methylation is higher in females and differentially regulated by sex steroids. *Brain Res.* 2006;1067:43-47.
88. Mastroeni D, McKee A, Grover A, Rogers J, Coleman PD. Epigenetic differences in cortical neurons from a pair of monozygotic twins discordant for Alzheimer's disease. *PloS One.* 2009;4:e6617.
89. McGowan PO, Kato T. Epigenetics in mood disorders. *Environ Health Prev Med.* 2008;13:16-24.
90. McGowan PO, Sasaki A, D'Alessio AC, et al. Epigenetic regulation of the glucocorticoid receptor in human brain associates with childhood abuse. *Nat Neurosci.* 2009;12:342-348.
91. Meijlink FC, Philipsen JN, Gruber M, Ab G. Methylation of the chicken vitellogenin gene: influence of estradiol administration. *Nucleic Acids Res.* 1983;11:1361-1373.
92. Melki JR, Vincent PC, Clark SJ. Concurrent DNA hypermethylation of multiple genes in acute myeloid leukemia. *Cancer Res.* 1999;59:3730-3740.
93. Mill J, Dempster E, Caspi A, Williams B, Moffitt T, Craig I. Evidence for monozygotic twin (MZ) discordance in methylation level at two CpG sites in the promoter region of the catechol-O-methyltransferase (COMT) gene. *Am J Med Genet B Neuropsychiatr Genet.* 2006;141B:421-425.

94. Mill J, Petronis A. Molecular studies of major depressive disorder: the epigenetic perspective. *Mol Psychiatry.* 2007;12(9):799-814.
95. Mill J, Petronis A. The relevance of epigenetics to major psychosis. In: Ferguson-Smith A, Greally J, Martienssen R, eds. *Epigenomics*. New York: Springer; 2009.
96. Mill J, Tang T, Kaminsky Z, et al. Epigenomic profiling reveals DNA methylation changes associated with major psychosis. *Am J Hum Genet.* 2008;82(3):696-711.
97. Miller RL, Ho SM. Environmental epigenetics and asthma: current concepts and call for studies. *Am J Respir Crit Care Med.* 2008;177:567-573.
98. Nagarajan RP, Patzel KA, Martin M, et al. MECP2 promoter methylation and X chromosome inactivation in autism. *Autism Res.* 2008;1:169-178.
99. Ober C, Thompson EE. Rethinking genetic models of asthma: the role of environmental modifiers. *Curr Opin Immunol.* 2005;17:670-678.
100. Ooi SL, Henikoff S. Germline histone dynamics and epigenetics. *Curr Opin Cell Biol.* 2007;19:257-265.
101. Orton SM, Herrera BM, Yee IM, et al. Sex ratio of multiple sclerosis in Canada: a longitudinal study. *Lancet Neurol.* 2006;5:932-936.
102. Paulsen M, Ferguson-Smith AC. DNA methylation in genomic imprinting, development, and disease. *J Pathol.* 2001;195:97-110.
103. Pembrey ME, Bygren LO, Kaati G, et al. Sex-specific, male-line transgenerational responses in humans. *Eur J Hum Genet.* 2006;14:159-166.
104. Perrin MC, Brown AS, Malaspina D. Aberrant epigenetic regulation could explain the relationship of paternal age to schizophrenia. *Schizophr Bull.* 2007;33:1270-1273.
105. Petel-Galil Y, Benteer B, Galil YP, et al. Comprehensive diagnosis of Rett's syndrome relying on genetic, epigenetic and expression evidence of deficiency of the methyl-CpG-binding protein 2 gene: study of a cohort of Israeli patients. *J Med Genet.* 2006;43:e56.
106. Petronis A. Human morbid genetics revisited: relevance of epigenetics. *Trends Genet.* 2001;17:142-146.
107. Petronis A. Epigenetics and bipolar disorder: new opportunities and challenges. *Am J Med Genet C Semin Med Genet.* 2003;123:65-75.
108. Petronis A. The origin of schizophrenia: genetic thesis, epigenetic antithesis, and resolving synthesis. *Biol Psychiatry.* 2004;55:965-970.
109. Pidsley R, Dempster EL, Mill J. Brain weight in males is correlated with DNA methylation at IGF2. *Mol Psychiatry.* 2010;15(9):880-881.
110. Plomin R, Owen MJ, McGuffin P. The genetic basis of complex human behaviors. *Science.* 1994;264:1733-1739.
111. Polesskaya OO, Aston C, Sokolov BP. Allele C-specific methylation of the 5-HT2A receptor gene: evidence for correlation with its expression and expression of DNA methylase DNMT1. *J Neurosci Res.* 2006;83:362-373.
112. Preis JI, Downes M, Oates NA, Rasko JE, Whitelaw E. Sensitive flow cytometric analysis reveals a novel type of parent-of-origin effect in the mouse genome. *Curr Biol.* 2003;13:955-959.
113. Quddus J, Johnson KJ, Gavalchin J, et al. Treating activated CD4+ T cells with either of two distinct DNA methyltransferase inhibitors, 5-azacytidine or procainamide, is sufficient to cause a lupus-like disease in syngeneic mice. *J Clin Invest.* 1993;92:38-53.
114. Rakyan V, Whitelaw E. Transgenerational epigenetic inheritance. *Curr Biol.* 2003;13:R6.
115. Rakyan VK, Blewitt ME, Druker R, Preis JI, Whitelaw E. Metastable epialleles in mammals. *Trends Genet.* 2002;18: 348-351.
116. Reynolds E. Vitamin B12, folic acid, and the nervous system. *Lancet Neurol.* 2006;5:949-960.
117. Richards EJ. Inherited epigenetic variation – revisiting soft inheritance. *Nat Rev Genet.* 2006;7:395-401.
118. Richardson B, Scheinbart L, Strahler J, Gross L, Hanash S, Johnson M. Evidence for impaired T cell DNA methylation in systemic lupus erythematosus and rheumatoid arthritis. *Arthritis Rheum.* 1990;33:1665-1673.
119. Riggs AD, Xiong Z, Wang L, LeBon JM. Methylation dynamics, epigenetic fidelity and X chromosome structure. *Novartis Found Symp.* 1998;214:214-225; discussion 225-232.
120. Roach HI, Aigner T. DNA methylation in osteoarthritic chondrocytes: a new molecular target. *Osteoarthr Cartil.* 2007;15:128-137.
121. Roach HI, Yamada N, Cheung KS, et al. Association between the abnormal expression of matrix-degrading enzymes by human osteoarthritic chondrocytes and demethylation of specific CpG sites in the promoter regions. *Arthritis Rheum.* 2005;52:3110-3124.
122. Robertson KD, Wolffe AP. DNA methylation in health and disease. *Nat Rev Genet.* 2000;1(1):11-19.
123. Robinson RL, Carpenter D, Halsall PJ, et al. Epigenetic allele silencing and variable penetrance of malignant hyperthermia susceptibility. *Br J Anaesth.* 2009;103:220-225.
124. Rollins RA, Haghighi F, Edwards JR, et al. Large-scale structure of genomic methylation patterns. *Genome Res.* 2006;16:157-163.
125. Rosa A, Picchioni MM, Kalidindi S, et al. Differential methylation of the X-chromosome is a possible source of discordance for bipolar disorder female monozygotic twins. *Am J Med Genet B Neuropsychiatr Genet.* 2008;147B: 459-462.
126. Saha RN, Pahan K. HATs and HDACs in neurodegeneration: a tale of disconcerted acetylation homeostasis. *Cell Death Differ.* 2006;13:539-550.
127. Saluz HP, Jiricny J, Jost JP. Genomic sequencing reveals a positive correlation between the kinetics of strand-specific DNA demethylation of the overlapping estradiol/glucocorticoid-receptor binding sites and the rate of avian vitellogenin mRNA synthesis. *Proc Natl Acad Sci U S A.* 1986;83:7167-7171.
128. Sandovici I, Kassovska-Bratinova S, Loredo-Osti JC, et al. Interindividual variability and parent of origin DNA methylation differences at specific human Alu elements. *Hum Mol Genet.* 2005;14:2135-2143.
129. Sarter B, Long TI, Tsong WH, Koh WP, Yu MC, Laird PW. Sex differential in methylation patterns of selected genes in Singapore. *Chinese Hum Genet.* 2005;117:402-403.
130. Scarpa S, Cavallaro RA, D'Anselmi F, Fuso A. Gene silencing through methylation: an epigenetic intervention on Alzheimer disease. *J Alzheimers Dis.* 2006;9:407-414.
131. Schalkwyk LC, Meaburn EL, Smith R, et al. Allelic skewing of DNA methylation is widespread across the genome. *Am J Hum Genet.* 2010;86:196-212.

132. Schulz WA, Hoffmann MJ. Epigenetic mechanisms in the biology of prostate cancer. *Semin Cancer Biol*. 2009;19: 172-180.
133. Schwartz DA. Gene-environment interactions and airway disease in children. *Pediatrics*. 2009;123(Suppl 3):S151-S159.
134. Shen Y, Chow J, Wang Z, Fan G. Abnormal CpG island methylation occurs during in vitro differentiation of human embryonic stem cells. *Hum Mol Genet*. 2006;15:2623-2635.
135. Shimabukuro M, Jinno Y, Fuke C, Okazaki Y. Haloperidol treatment induces tissue- and sex-specific changes in DNA methylation: a control study using rats. *Behav Brain Funct*. 2006;2:37.
136. Shimabukuro M, Sasaki T, Imamura A, et al. Global hypomethylation of peripheral leukocyte DNA in male patients with schizophrenia: a potential link between epigenetics and schizophrenia. *J Psychiatr Res*. 2007;41(12): 1042-1046.
137. Siegmund KD, Connor CM, Campan M, et al. DNA methylation in the human cerebral cortex is dynamically regulated throughout the life span and involves differentiated neurons. *PloS One*. 2007;2:e895.
138. Sipos A, Rasmussen F, Harrison G, et al. Paternal age and schizophrenia: a population based cohort study. *BMJ*. 2004;329:1070.
139. Smith RG, Kember RL, Mill J, et al. Advancing paternal age is associated with deficits in social and exploratory behaviors in the offspring: a mouse model. *PloS One*. 2009;4:e8456.
140. Spencer VA, Davie JR. Role of covalent modifications of histones in regulating gene expression. *Gene*. 1999;240:1-12.
141. St Clair D, Xu M, Wang P, et al. Rates of adult schizophrenia following prenatal exposure to the Chinese famine of 1959-1961. *JAMA*. 2005;294:557-562.
142. Stenvinkel P, Karimi M, Johansson S, et al. Impact of inflammation on epigenetic DNA methylation – a novel risk factor for cardiovascular disease? *J Intern Med*. 2007;261:488-499.
143. Surani MA, Sasaki H, Ferguson-Smith AC, et al. The inheritance of germline-specific epigenetic modifications during development. *Philos Trans R Soc Lond*. 1993;339:165-172.
144. Susser E, Neugebauer R, Hoek HW, et al. Schizophrenia after prenatal famine. Further evidence. *Arch Gen Psychiatry*. 1996;53:25-31.
145. Susser ES, Lin SP. Schizophrenia after prenatal exposure to the Dutch Hunger Winter of 1944–1945. *Arch Gen Psychiatry*. 1992;49:983-988.
146. Szyf M, Weaver I, Meaney M. Maternal care, the epigenome and phenotypic differences in behavior. *Reprod Toxicol*. 2007;24(1):9-19.
147. Szulakowski P, Crowther AJ, Jimenez LA, et al. The effect of smoking on the transcriptional regulation of lung inflammation in patients with chronic obstructive pulmonary disease. *Am J Respir Crit Care Med*. 2006;174:41-50.
148. Takami N, Osawa K, Miura Y, et al. Hypermethylated promoter region of DR3, the death receptor 3 gene, in rheumatoid arthritis synovial cells. *Arthritis Rheum*. 2006;54: 779-787.
149. Tamashiro KL, Wakayama T, Yamazaki Y, et al. Phenotype of cloned mice: development, behavior, and physiology. *Exp Biol Med (Maywood)*. 2003;228:1193-1200.
150. Tansey MG, McCoy MK, Frank-Cannon TC. Neuroinflammatory mechanisms in Parkinson's disease: potential environmental triggers, pathways, and targets for early therapeutic intervention. *Exp Neurol*. 2007;208:1-25.
151. Thompson NP, Driscoll R, Pounder RE, Wakefield AJ. Genetics versus environment in inflammatory bowel disease: results of a British twin study. *BMJ*. 1996;312:95-96.
152. Tiemann-Boege I, Navidi W, Grewal R, et al. The observed human sperm mutation frequency cannot explain the achondroplasia paternal age effect. *Proc Natl Acad Sci U S A*. 2002;99:14952-14957.
153. Trenkmann M, Brock M, Ospelt C, Gay S. Epigenetics in rheumatoid arthritis. *Clin Rev Allergy Immunol*. 2010;39(1):10-19.
154. Ushijima T, Watanabe N, Okochi E, Kaneda A, Sugimura T, Miyamoto K. Fidelity of the methylation pattern and its variation in the genome. *Genome Res*. 2003;13:868-874.
155. Wang SC, Oelze B, Schumacher A. Age-specific epigenetic drift in late-onset Alzheimer's disease. *PloS One*. 2008;3:e2698.
156. Waterland RA, Michels KB. Epigenetic epidemiology of the developmental origins hypothesis. *Annu Rev Nutr*. 2007;27:363-388.
157. Weaver IC, Cervoni N, Champagne FA, et al. Epigenetic programming by maternal behavior. *Nat Neurosci*. 2004;7:847-854.
158. Weaver IC, Meaney MJ, Szyf M. Maternal care effects on the hippocampal transcriptome and anxiety-mediated behaviors in the offspring that are reversible in adulthood. *Proc Natl Acad Sci U S A*. 2006;103:3480-3485.
159. Widschwendter M, Jiang G, Woods C, et al. DNA hypomethylation and ovarian cancer biology. *Cancer Res*. 2004;64:4472-4480.
160. Wilks A, Seldran M, Jost JP. An estrogen-dependent demethylation at the 5′ end of the chicken vitellogenin gene is independent of DNA synthesis. *Nucleic Acids Res*. 1984;12:1163-1177.
161. Wilks AF, Cozens PJ, Mattaj IW, Jost JP. Estrogen induces a demethylation at the 5′ end region of the chicken vitellogenin gene. *Proc Natl Acad Sci U S A*. 1982;79:4252-4255.
162. Wong AH, Gottesman II, Petronis A. Phenotypic differences in genetically identical organisms: the epigenetic perspective. *Hum Mol Genet*. 2005;14(Spec No 1):R11-R18.
163. World Health O (2005) Preventing chronic diseases: a vital investment.
164. Wu Y, Halverson G, Basir Z, Strawn E, Yan P, Guo SW. Aberrant methylation at HOXA10 may be responsible for its aberrant expression in the endometrium of patients with endometriosis. *Am J Obstet Gynecol*. 2005;193:371-380.
165. Wu Y, Strawn E, Basir Z, Halverson G, Guo SW. Aberrant expression of deoxyribonucleic acid methyltransferases DNMT1, DNMT3A, and DNMT3B in women with endometriosis. *Fertil Steril*. 2007;87:24-32.
166. Xu L, Glass CK, Rosenfeld MG. Coactivator and corepressor complexes in nuclear receptor function. *Curr Opin Genet Dev*. 1999;9:140-147.
167. Yauk C, Polyzos A, Rowan-Carroll A, et al. Germ-line mutations, DNA damage, and global hypermethylation in mice exposed to particulate air pollution in an urban/industrial location. *Proc Natl Acad Sci U S A*. 2008;105: 605-610.

168. Yu J, Zhang H, Gu J, et al. Methylation profiles of thirty four promoter-CpG islands and concordant methylation behaviours of sixteen genes that may contribute to carcinogenesis of astrocytoma. *BMC Cancer*. 2004;4:65.
169. Zammit S, Allebeck P, Dalman C, et al. Paternal age and risk for schizophrenia. *Br J Psychiatry*. 2003;183: 405-408.
170. Zawia NH, Lahiri DK, Cardozo-Pelaez F. Epigenetics, oxidative stress, and Alzheimer disease. *Free Radic Biol Med*. 2009;46:1241-1249.
171. Zhou H, Brockington M, Jungbluth H, et al. Epigenetic allele silencing unveils recessive RYR1 mutations in core myopathies. *Am J Hum Genet*. 2006;79:859-868.

Techniques to Study DNA Methylation and Histone Modification

2

Ester Lara, Vincenzo Calvanese, Agustin F. Fernandez, and Mario F. Fraga

2.1 Introduction

Epigenetics encompasses not only heritable changes in gene activity and expression (in the progeny of cells or individuals) but also stable, long-term alterations in the transcriptional potential of a cell that are not necessarily heritable (a definition from the *NIH Roadmap for Medical Research*). Epigenetic states are not definitive, and changes occur with age in a stochastic manner as well as in response to environmental stimuli.

Epigenetic phenomena are mediated by a variety of molecular mechanisms: DNA methylation, post-translational histone modifications (PTMs), ATP-dependent chromatin remodeling complexes, polycomb/trithorax protein complexes, and noncoding RNAs.

The best-characterized epigenetic mechanisms are DNA methylation and histone modifications. Both have been studied with a variety of techniques developed specifically for the purpose. In recent decades, considerable effort has been devoted to improve these methods, so as to obtain more accurate results in less time and at lower cost. Epigenetic alterations have been identified during differentiation and in the course of normal cell life, as well as in pathological processes such as cancer, diabetes, and Rett syndrome.

The tumorigenic process is characterized by genetic and epigenetic alterations. Genetic modifications include mutations, translocations, deletions, and aberrant amplifications; epigenetic alterations are centered on abnormal patterns of DNA methylation and histone modifications. Tumoral cells often present DNA hypomethylation in repetitive sequences such as centromeres, telomeres, and microsatellites, which are normally hypermethylated in healthy cells to prevent chromosomal instability. Apart from this lack of methylation, the DNA of tumoral cells is usually hypermethylated at promoters of tumor suppressor genes (TSG). This inhibits normal expression which might prevent cancer development. In addition to altered DNA methylation pattern, PTMs play an important role in tumorigenesis, as they are closely associated with the aberrant expression of oncogenes and the abnormal repression of TSG.[20]

Both DNA methylation and histone modifications can be studied at the global and gene-specific levels. Studying DNA methylation at the global level involves measuring the 5-methylcytosine (5MeC) content, while the study of histone modifications at the global level involves determining the overall content of principal histone modifications, methylation, and acetylation in the cell population and in a particular stage. By contrast, epigenetic studies at the gene-specific level involve the detailed characterization of a particular modification that affects the expression of the gene.

2.2 The Study of DNA Methylation

DNA methylation consists of the addition of a methyl group to the fifth carbon of the cytosine pyrimidine ring. In mammals, this occurs mostly in cytosines of the CpG dinucleotides (3–5% of all the cytosine residues within the human genome). It is a dynamic process that takes place in multicellular organisms in the course of development.

M.F. Fraga (✉)
Department of Immunology and Oncology,
National Center for Biotechnology (CNB-CSic),
Madrid, Spain
e-mail: mffraga@ext.cnio.es

H.I. Roach et al. (eds.), *Epigenetic Aspects of Chronic Diseases*,
DOI: 10.1007/978-1-84882-644-1_2,

The study of DNA methylation is facilitated by a variety of techniques that may have been developed for other purposes, but can be applied to the study of epigenetic phenomena. One can determine the methylation status of a gene of interest (locus-specific study) or of a large number of genes (genome-wide study), or study the total DNA methylation content in a cell or tissue under normal or pathological conditions (global study) (Table 2.1). The techniques for studying DNA methylation provide qualitative information concerning the methylation status of a gene or comparative information in paired samples. Quantitative methylation yields information about particular genes or, specifically, a CpG in the region of interest.

The three main approaches for the differential detection of methylated and unmethylated CpGs are the bisulfite modification of DNA, the digestion of DNA with methylation-sensitive restriction enzymes, and the use of anti-5MeC antibody.

The bisulfite-treated DNA is the starting material for many DNA methylation techniques. Most DNA methylation assays involve PCR amplification. However, because DNA polymerases do not recognize the methylated cytosines, the PCR product does not retain the methylation marks of the genomic DNA. For marks to be retained, genomic DNA is treated with sodium bisulfite. In the method, cytosine residues are deaminated to uracils in the presence of NaOH and sodium bisulfite (bisulfite modification of DNA), whereas 5MeC remains unchanged. During PCR amplification, a thymine nucleotide is incorporated in the PCR product for every uracil present in the bisulfite-treated template, with each uracil corresponding to an unmethylated C; a cytosine is introduced for every 5MeC in the template. In this way, the unmethylated and methylated PCR product can be distinguished due to the different content of C or T at CpG sites (Fig. 2.1).

Other techniques are based on the capacity of particular restriction enzymes (methylation-insensitive RE) or the incapacity to cut methylated DNA (methylation-sensitive RE). Among the methylation-insensitive REs are *Bam*HI and *Msp*I. Methylation-sensitive REs, such as *Not*I, *Hha*1, and *Hpa*II, do not cleave methylated DNA.

The production of antibodies that recognize specificity at 5MeC in DNA permits the isolation and identification of 5MeC content by DNA immunoprecipitation or by DNA hybridization on membranes.

2.2.1 Global DNA Methylation

In the human genome, CpGs are asymmetrically distributed into CpG-poor regions and dense regions called "CpG islands," which span the 5′ end of the regulatory region of approximately half of all genes. In normal cells, the CpG islands are usually unmethylated, whereas the sporadic CpG sites in the rest of the genome are generally methylated.[43] 5MeC accounts for ~1% of total DNA bases and consequently affects 70–80% of all CpG dinucleotides in the genome.[19] Repetitive sequences also contain a high percentage of CpG dinucleotides that are normally hypermethylated. DNA methylation is essential for normal development and is associated with a number of key processes including genomic imprinting, X-chromosome inactivation, and suppression of repetitive elements. Hypermethylation of telomeres and centromere sequences is necessary to maintain chromosome stability during normal cell life. Hypermethylation of retrotransposable elements (SINEs – *short interspersed elements* and LINEs – *long interspersed elements*), on the other hand, prevents expression or new insertions that may have fatal consequences for normal gene expression.

Measuring the global content of 5MeC is useful for understanding the interplay between genome-wide alterations in DNA methylation and their effect on genomic stability and gene-specific alterations in epigenetic regulation. In addition, global measurements of DNA methylation are also a proven tool for understanding the molecular pathology of human cancer, for measuring the potential effect of tumor-preventive or tumor-promoting compounds, and for monitoring therapeutic responses to hypomethylating agents.

2.2.1.1 Reverse-Phase High-Performance Liquid Chromatography

This method, developed in the early 1980s, is one of the oldest in methylation analysis.[49] It relies on the total hydrolysis of DNA using DNaseI and nuclease P1, followed by further processing to deoxyribonucleosides by alkaline phosphatase treatment. The free nucleosides are then separated by injection into a column that contains a silica-hydrocarbon stationary phase. Elution of the nucleosides from the column is based on their solubility in the mobile polar phase.

Table 2.1 Study of DNA methylation

	Technique	Type of study	Approach	Time	Cost	Special require-ments	Amount of starting DNA
Global DNA methylation	RP-HPLC	Quantitative	–	X	X	Yes	High
	Immunochemical assay	Quantitative	me5C Ab	XX	XX	No	High
	SssI assay	Quantitative	–	XX	X	No	Small-medium
	Chloroacetaldehyde assay	Quantitative	Bisulfite modification	XXX	XX	No	
	HPCE	Quantitative	–	X	X	Yes	High
	Bisulfite PCR repetitive elements	Quantitative	Bisulfite modification	XXX	XX	Yes	Small
Locus-specific DNA methylation	Bisulfite sequencing	Quantitative	Bisulfite modification	XXX	XXX	Yes	Medium
	MSP	Qualitative	Bisulfite modification	XX	XX	No	Medium
	MALDI-TOF	Quantitative	Bisulfite modification	XX	XX	Yes	Small
	Ms-SNuPE	Quantitative	Bisulfite modification	X	XX	No	Small
	COBRA	Quantitative	RE	XX	XX	No	Medium
	MethyLight	Quantitative	Bisulfite modification	XX	XXX	Yes	Small-medium
	HRM	Quantitative	Bisulfite modification	XX	XX	Yes	Small-medium
	ERMA	Quantitative	Bisulfite modification	XX	XX	Yes	Medium
	Pyrosequencing	Quantitative	Bisulfite modification	XX	XX	Yes	Small
	MethylQuant	Quantitative	Bisulfite modification	XX	XX	Yes	Medium
	Heavy methyl PCR	Qualitative/ quantitative	Bisulfite modification	XX	XX	No	Small
	MS-MLPA	Quantitative	RE	XX	X	Yes	Small
	DNA sequencing	Quantitative	–	X	XX	Yes	Medium
Genome-wide methylation study	RLGS	Qualitative	RE	XXX	XX	Yes	High
	MBD column	Qualitative	–	XX	XX	No	Medium-high
	MCA	Qualitative	RE	XXX	XX	No	Medium
	DMH	Quantitative	RE	XXX	XXX	Yes	Medium-high
	AIMS	Qualitative	RE	XXX	XX	No	Medium-high
	Me-DIP	Qualitative	me5C Ab	XXX	XX	No	High
	MIRA	Qualitative	–	XX	XX	No	High
	MSDK	Quantitative	RE	XXX	XXX	No	Low
	HELP	Quantitative	RE	XXX	XXX	Yes	Medium
	DNA methylation array	Quantitative	Bisulfite modification	XX	XXX	Yes	Low

RE restriction enzymes; *me5C Ab* anti-methylcytosine antibody; Time: fast, average, and labor-intensive (X, XX, and XXX, respectively); Cost: cheap, average, and expensive (X, XX, and XXX, respectively)

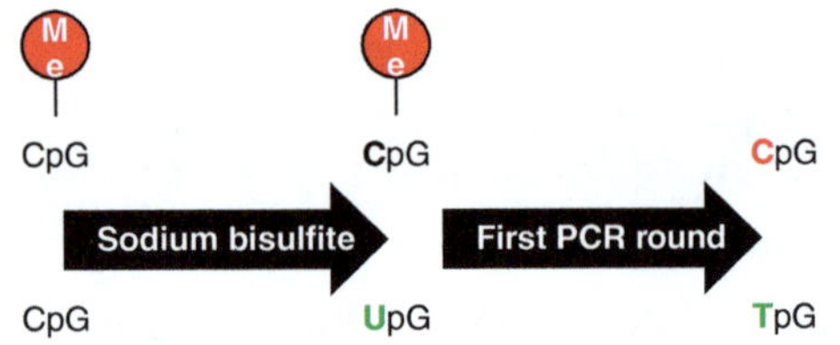

Fig. 2.1 Sodium bisulfite treatment of CpG motifs. Methylated cytosines are protected and remain unchanged, while unmethylated cytosines are deaminated to uracil after treatment with sodium bisulfite

They are detected and quantified by monitoring their UV absorbances at 254 and 280 nm as they exit the column. Greater specificity can be achieved by combining HPLC separation with mass spectrometry (MS), which provides positive identification of the bases. Once the DNA bases have been separated and eluted from the column by changing the concentration of the organic component in the mobile phase, they are introduced into the electrospray system of the mass spectrometer. The MS spectra verify the identity of each HPLC peak for the estimation of DNA methylation; this yields information about the five deoxyribonucleosides (5mdC, dC, dG, dA, and dT). The method provides information about the total 5MeC content of cell lines and tissues, but requires special attention to prevent contamination by RNA; it also requires close attention to the composition of the solutions and the temperature of elution. This technique requires large amounts of DNA and highly specialized equipment that may not be readily available. The combined method, RP/MS, for the quantitative determination of the genomic DNA methylation status is a rapid, selective, sensitive, and accurate method. It is a method that has been used to study the relationship between global DNA methylation levels and different etiological risk factors in the oral cavity and in pharyngeal cancer.[31]

2.2.1.2 Immunochemical Method for Detecting 5MeC in DNA Fragments

This method was primarily developed by Achwal and Chandra[1] and involves spotting DNA on nitrocellulose paper, and incubating the paper strip with the biotinylated anti-5MeC antibody. The paper strip is then incubated with a complex of avidin-DH and biotinylated peroxidase and stained with diaminobenzidine tetrahydrochloride (DABT) and hydrogen peroxide. The reaction produces a colored spot on the paper if the DNA contains 5MeC. The measurement of the staining intensity of these spots gives a quantitative estimate of 5MeC in DNA. In addition the reaction can be visualized using a secondary antibody labeled with fluorescein and detected by immunofluorescence methods. This approach has been used to study the global levels of 5MeC in *Drosophila melanogaster*[2] and to detect levels of urinary 5MeC as a potential biological marker for leukemia.[40]

2.2.1.3 Methyl Group Acceptance Assay

This method is based on isolated DNA accepting radiolabeled methyl groups from *S*-[^{3}H-methyl] adenosylmethionine (SAM), using the bacterial CpG methyltransferase SssI. Given this enzyme methylates all unmethylated CpG dinucleotides in the genome, radiolabeled methyl group acceptance is inversely proportional to the level of preexisting methylation. This method was developed by Balaghi and Wagner[5] and has been used to compare the effects of hypomethylating agents in ovarian cancer cells[6] and to investigate the role of a CpG-binding protein in mouse embryogenesis.[12] The methyl group acceptance assay is one of the most inexpensive and straightforward methods and requires nothing more than a scintillation counter. However, its accuracy depends on knowing the DNA concentration very well. Futhermore obtaining reproducible results may become difficult due to the instability of SAM and SssI.

2.2.1.4 The Chloroacetaldehyde Assay

This is a fluorescent method for detecting DNA methylation levels.[61] Purines are removed from the DNA by treatment with sulfuric acid. The depurinated DNA is modified with sodium bisulfite and incubated with chloroacetaldehyde, forming the intensely fluorescent derivative of 5MeC. Fluorescence can be measured in a fluorimeter; it is proportional to the level of 5MeC in the genome. The method does not require extensive downstream purification, because neither chloroacetaldehyde nor DNA is fluorescent. However, it is time-consuming and the use of a very toxic reagent, chloroacetaldehyde, has limited its use.

2.2.1.5 High-Performance Capillary Electrophoresis

Fraga and collaborators[25] used high-performance capillary electrophoresis (HPCE) to quantify 5MeC content in the genome.[25] In this method, molecules are separated on the basis of size, charge, structure, and hydrophobicity by means of specific and high voltages and the use of a narrow-bore fused-silica capillary. Separation and quantification of cytosine and methylcytosine are by the dodecylsulfate (SDS) micelle system, which is faster than HPLC and is also reasonably inexpensive, but cannot be used for preparative analysis due to the low injection volume. The method utilizes expensive equipment and large amounts of good quality DNA. It has been used to detect differences in global DNA methylation between monozygotic twins, highlighting the importance of epigenetics in individual development.[23]

2.2.1.6 Bisulfite PCR of Repetitive DNA Elements

This method is used to determine global DNA methylation. DNA is treated with bisulfate, and the PCR of multiple DNA repetitive elements is prepared. This includes Alu elements and long interspersed nucleotide elements (LINEs). There are approximately 1.5 million Alu repetitive elements in the human genome and half a million LINE elements that are normally hypermethylated. More than one-third of DNA methylation is estimated to occur in repetitive elements. Therefore, analysis of their methylation is an indirect evaluation of global genomic DNA methylation. Methylation of Alu and LINE PCR products can be measured by bisulfite sequencing pyrosequencing. Kaneda and collaborators[47] first related the 5MeC content quantified by HPLC with the methylation status of the LINE1 element in gastric cancer. This indirect measurement of DNA methylation has subsequently been utilized in other studies, such as pyrosequencing[88] and in studies of tumors, such as leukemia[67] and colon cancer.[68]

2.2.2 Locus-Specific DNA Methylation

DNA methylation can be also evaluated at the locus-specific level to determine whether the methylation at the promoter level represses gene transcription. Numerous TSG are repressed by promoter methylation during tumorigenic processes. This can lead to pro-tumorigenic characteristics or provide an escape pathway for cellular control mechanisms.[14,21,22,79] Information about the specific methylation at gene promoters can be valuable for predicting cancer development.

2.2.2.1 Bisulfite Sequencing

This method, described by Frommer and collaborators in 1992,[28] sequences a specific region in order to obtain quantitative data about the C or T content in a particular CpG. DNA has been converted with bisulfite (Fig. 2.1), the DNA sequence under study is amplified by PCR with a primer that is specific for the bisulfite-converted DNA. The resulting PCR product is cloned, and individual clones are sequenced. The percentage methylation at each CpG position can be derived from the number of sequenced clones that display methylated and unmethylated cytosines. This is an expensive and time-consuming process, and only 10–20 colonies of alleles are usually analysed; this makes it difficult to obtain statistically significant results. The technique has been used to create a methylation profile of chromosomes 6, 20, and 22[18] and also to detect aberrantly methylated genes, as in cancer.[3]

2.2.2.2 Methylation-Specific PCR

First described by Herman in 1996,[35] methylation-specific PCR (MSP) makes possible a qualitative and sensitive analysis of promoter hypermethylation at CpG islands in cell lines and in clinical samples. The method involves the specific amplification by PCR of bisulfite-modified DNA (Fig. 2.1) with two sets of primers, one complementary to the methylated DNA (with unmodified cytosines) and the other complementary to the unmethylated DNA (with cytosines modified to uracils). PCR products are detected by gel electrophoresis without the need for further restriction or sequencing analysis. The MSP assay must include a positive control for the unmethylated primer reaction, a positive control for the methylated reaction (in vitro methylated DNA or IVD) and a negative control for the PCR reaction. Because primer design is critical, dedicated software has been designed expressly for this purpose

(MSPPrimer and MethylPrimer Express). The method has been used extensively to characterize TSG hypermethylated in cancer and other diseases. For instance, in 1997, Wong and collaborators[84] related the hypermethylation of p16 (INK4a) studied by MSP to esophageal adenocarcinomas. Studies have since identified hypermethylation of this gene, analyzed by MSP, in other types of cancer, such as hepatocellular carcinoma[55] and colorectal cancer.[91]

2.2.2.3 Matrix-Assisted Laser Desorption/Ionization Time-of-Flight Mass Spectrometry

First used to determine the methylation specificity of sequence-specific DNA methyltransferases,[74] the technique has more recently been utilized to study the methylation profile in breast cancer.[63] A PCR amplification product from bisulfite-treated DNA is transcribed in vitro into a single-stranded RNA molecule and, subsequently, the base is cleaved by an endoribonuclease. The conversion of unmethylated cytosine to uracil during bisulfite treatment generates base-specific cleavage products that reflect underlying methylation patterns and that can be readily analyzed by matrix-assisted laser desorption/ionization time-of-flight mass spectrometry (MALDI-TOF MS).

For automated, high-performance analysis of nucleic acid mixtures by MALDI-TOF MS, the sample needs to be transferred to a chip array (SpectroCHIP®, SEQUENOM). This yields quantitative results without the need to clone PCR products and is suitable for the analysis of samples obtained from a variety of sources, such as laser capture microdissection, because it does not require large amounts of starting material. The system can detect DNA methylation levels as low as 5%. Its principal disadvantage is the need for sophisticated equipment.

2.2.2.4 Methylation-Sensitive Single Nucleotide Primer Extension

This method uses single nucleotide primer extension to assess DNA methylation at a specific cytosine.[30] The reaction is based on repeated annealing of an oligonucleotide exactly one base pair upstream of a target CpG. The primer is then extended by incorporating a single fluorescent or ^{32}P-dideoxynucleotide. Reaction products are electrophoresed on polyacrylamide gels and quantified by phosphorimage analysis to determine the proportion of the incorporated labeled C or T. MS can also be used for quantification. Nonradioactive labeling and quantification can also be adapted for this assay by using platforms such as Applied Biosystems SNaPshot technology. The methylation status of several CpG sites can be determined simultaneously by using multiple oligonucleotides in a single primer extension reaction. This method requires a small amount of DNA. The PCR product generated before methylation-sensitive single nucleotide primer extension (Ms-SNuPE) analysis is also suitable for bisulfite sequencing. Strategies for Ms-SNuPE primer design are of particular importance, especially to avoid incorporating potentially polymorphic positions into the primer annealing region. The method permits rapid and accurate determination of DNA methylation. Its limitation is that only up to four CpG sites can be measured in a reaction. The method has been used to demonstrate promoter hypermethylation at the *c-abl* gene in chronic myelogenous leukemia.[59]

2.2.2.5 Combined Bisulfite Restriction Analysis

This method[86] exploits the principle that the bisulfite treatment of genomic DNA translates the epigenetic information encoded by cytosine methylation into sequence differences and thus indicates the presence or absence of restriction enzyme recognition sites in a methylation-dependent manner. PCR is performed after enzyme restriction. This step does not discriminate between templates on the basis of their original methylation status. In the mixture of resulting PCR fragments, the fraction that has retained a restriction site or created a new one directly reflects the percentage DNA methylation at that site in the original DNA. A modified protocol for combined bisulfite restriction analysis (COBRA), named Bio-COBRA, incorporates an electrophoresis step in a microfluid chip. This assesses the DNA methylation status of all DNA molecules. Bio-COBRA provides a platform for a rapid, quantitative, sensitive, and reproducible assessment of DNA methylation patterns in large samples. Young and collaborators[90] used this technique to study DNA methylation of HPP1 gene in colorectal cancer.[90]

2.2.2.6 Quantitative Real-Time Methylation-Specific PCR (MethyLight)

This method[17] describes the methylation status of the MLH1 mismatch repair gene in human colorectal tumor specimens. MethyLight combines MSP with methylation-specific detection technology, converting MSP quantitatively. It can detect very low frequencies of hypermethylated alleles and has the quantitative accuracy of real-time PCR. The assay includes primers that can amplify both methylated and unmethylated sequences and specific fluorescent Taqman probes, overlapping two or more CpG sites for both types of sequence. For accuracy, the method requires optimization of PCR and control probes for methylated and unmethylated DNA and for unconverted DNA. However, it does not require large amounts or high-quality DNA, so it is suitable for use with clinical samples. MethyLight has been used to identify prognostic markers in human breast cancer[56] and cervical cancer.[83]

2.2.2.7 Methylation-Sensitive Melting Curve Assay

The use of melting analysis in methylation studies was first reported in 2001.[85] After sodium bisulfite DNA treatment, methylated cytosines are amplified during the PCR as cytosines, whereas unmethylated cytosines are amplified as thymines (Fig. 2.1). The base composition of the PCR product depends on the 5MeC content of the template. The two complementary strands of DNA are held together by hydrogen bonds. Dissociation of double-stranded DNA is known as DNA melting and can be induced either by increased temperature or denaturing chemicals. The dissociation of the triple hydrogen bond between C and G requires more energy than the dissociation of the double hydrogen bond between T and A; therefore, GC-rich sequences melt at relatively higher temperatures than AT-rich sequences. The melting profile of an amplicon can be determined by subjecting it to a gradually increasing temperature in the presence of an intercalating fluorescent dye, which emits fluorescence when intercalated with double-stranded DNA. The fluorescent dye will bind to double-stranded DNA, emitting high levels of fluorescence until the temperature reaches the melting temperature of the PCR product. At the melting temperature, the PCR product dissociates into two single strands and the dye cannot bind and fluoresce; therefore, fluorescence declines sharply. The changes of fluorescence levels along a denaturing gradient describe an amplicon's melting profile. The melting profiles of PCR products originating from methylated and unmethylated variants of the same template are different due to their distinct GC content. Comparison of the melting profile of unknowns with that of control templates yields its methylation status. The proportional amplification of methylated and unmethylated templates is critical, because the preferential amplification of one template can lead to misinterpretation, inasmuch as methylated templates can be underamplified. Current advances in fluorescence detection technology, new algorithms for data calculation, and the use of novel fluorescent dyes have facilitated the development of high-resolution melting (HRM) analysis. The method has been used to analyze the methylation status of a putative TSG, ADAMTS18[51] and the methylation of BRCA1 in breast cancer.[72]

2.2.2.8 Enzymatic Regional Methylation Assay

This method was developed by Galm and collaborators[29] who studied differences in DNA methylation at the p15INK4B promoter in various cell lines after treatment with the demethylating agent 5′-aza-2′-deoxycytidine (DAC). After bisulfite treatment of genomic DNA, the region of interest is amplified by PCR with the use of primers that are specific for bisulfite-converted DNA and are tagged with two GATC sites at their 5′ ends. The resulting product is purified and sequentially incubated with the two bacterial methyltransferases dam and SssI, in combination with the corresponding radiolabeled SAM. The amount of ^{3}H-methyl groups incorporated into the PCR product is directly proportional to the level of DNA methylation. Methylation status is measured by the ^{3}H/^{14}C ratio in the sample and compared to a standard curve. It is a precise method for the quantitative analysis of several CpGs in a specific DNA region.

2.2.2.9 Pyrosequencing

Pyrosequencing is relatively novel.[76] It quantifies the degree of methylation at CpG positions after bisulfite treatment of genomic DNA. It is a sequence-by-synthesis approach based on the luminometric detection of pyrophosphate release following nucleotide incorporation. The procedure can be divided into several steps. The preparative steps include the design of the assay and treatment of the samples of interest with sodium bisulfite (Fig. 2.1). A target region is amplified by PCR using a pair of primers, one of which carries a biotin label at its 5′ end. The strand with the biotinylated primer is captured by streptavidin-coated beads used to purify the PCR product. A pyrosequencing primer complementary to the single-stranded template is then hybridized to the template, and the pyrosequencing reaction is carried out by the addition of single nucleotides in a defined sequence. Only if the added nucleotide is complementary to the template DNA will it be incorporated by a DNA polymerase. At any given time, only one of the four nucleotides is present in the reaction vessel and the biochemical reactions are carried out with a balanced mixture of four enzymes: the Klenow fragment of the DNA polymerase I from *Escherichia coli*, an ATP sulfurylase, the luciferase, and an apyrase. Pyrosequencing offers several advantages, including direct quantitative sequencing without the need for cloning, high reliability, speed, and ease of use. DNA methylation information can be obtained from whole PCR products to obtain statistically relevant information. Pyrosequencing has been used, for example, to detect the heritable germline epimutation of *MSH2* in a family with hereditary nonpolyposis colorectal cancer.[13]

2.2.2.10 MethylQuant: A Real-Time PCR-Based Method to Quantify DNA Methylation at Single Specific Cytosines

MethylQuant is used to quantify the methylation level of a single cytosine by means of real-time amplification of bisulfite-treated DNA.[75] After bisulfite conversion of DNA and amplification by PCR of the region of interest, the PCR product is produced by real-time PCR with the aid of two different sets of primers: a nondiscriminative primer set that allows the PCR product to be quantified, irrespective of the methylation status, and a discriminative primer set that allows quantification of the PCR product corresponding to the specific methylation status of the region of interest. The latter set is designed so that the most 3′ nucleotide is complementary to the position analyzed. Quantification is performed by comparing the PCR products originating from the bisulfite-converted DNA samples to two references: the template that is perfectly matched to the discriminative primer and to the alternative methylation status that is mismatched to the discriminative primer. MethylQuant allows the methylation levels of a single specific cytosine to be quantified. It uses SYBR Green, which eliminates the need for fluorescently labeled probes, thereby reducing the overall cost. MethylQuant is useful for analyzing regions with a lower density of CpG. However, care is required with the primer design and optimization of the PCR if reliable results are to be obtained.

2.2.2.11 Heavy Methyl PCR

This method was developed by Cottrell and collaborators[15] and is useful for the analysis of very low concentrations of methylated DNA. In their study[15] of the methylation status of calcitonin and glutathione-*S*-transferase, as little as 30 and 60 pg of methylated DNA were detected and amplified. In Heavy Methyl PCR, methylation specificity is provided by a nonextendable blocker oligonucleotide, designed to bind to the bisulfite-treated DNA template in a methylation-dependent manner. As the binding sites of the blocker overlap with those of the amplification primer, the primer cannot bind to the unmethylated template and the amplicon is not generated. By contrast, if the DNA is methylated, the blocker cannot bind to the DNA and the amplicon is generated. The method can be adapted to the qualitative and quantitative analyses of DNA methylation. In real-time PCR, this amplification is monitored and can be quantified with a methylation-specific probe. The method can be used as an alternative to MS-PCR, even though it requires more components and more optimization. However, it enables methylated sequences at low concentration in the DNA mixture to be detected with high specificity.

2.2.2.12 Methylation-Specific Multiplex Ligation-Dependent Probe Amplification

In methylation-specific multiplex ligation-dependent probe amplification (MS-MLPA),[60] the ligation of MLPA probe oligonucleotides is combined with digestion of the genomic DNA–probe hybrid complexes with methylation-sensitive endonucleases. In this method, the genomic DNA is hybridized with an MLPA probe that contains two different oligonucleotides, one of which has a methylation-sensitive restriction site. To establish the amount of methylated sequence present after hybridization of the probe mix to the sample DNA, one part is subjected to a single ligation step, whereas in the other, part ligation is combined with methylation-sensitive digestion. Subsequent PCR amplification exponentially amplifies either total DNA or the methylated fraction only. If the CpG site is unmethylated, the genomic DNA–MS–MLPA probe complex is digested and prevents exponential amplification so that no signal can be detected after fragment analysis.

Capillary gel electrophoresis is performed to identify and quantify PCR products of the individual probes. The undigested part shows the same amount of PCR product as the sample containing 100% methylated DNA, whereas the peak height for the probes containing a methylation-specific sensitive digestion site gradually decreases as the amount of methylated sequences decreases.

This method has several advantages: Bisulfite conversion can be omitted, different CpGs can be analyzed simultaneously, and only a small amount of DNA is required. It has been used in the study of DNA methylation at MGMT promoter gene in gliomas[42] or at FANCC or FANCL in leukemia.[36]

2.2.3 *DNA Sequencing*

There are several DNA sequencing platforms available for studying the sequence of bisulfite-modified DNA, but the two most widely used are the 454 GenomeSequencer (Roche Applied Science) and the Illumina (Solexa) Genome Analyzer. A limiting factor is the high cost of generating the sequence with very high-throughput, but even compared with Sanger sequencing (used in the bisulfite sequencing method), the cost per base is lower by several orders of magnitude.

2.2.3.1 454 Genome Sequencing

In this system, DNA fragments are ligated with specific adapters that cause the binding of one fragment to a bead. PCR amplification is necessary to obtain a sufficiently intense light signal. When PCR is completed, and after denaturation, each bead with its one amplified fragment is placed at the top end of an etched fiber in an optical fiber chip. Polymerase enzyme and primers are then added to the beads. Incorporation of a subsequent base by the enzyme in the growing chain releases a pyrophosphate group, which emits light that is detected and analyzed.

2.2.3.2 Illumina Genome Analyzer

In this platform, DNA fragments are ligated to adapters at both ends and, after denaturation, immobilized at one end on a solid support. Each single-stranded fragment creates a bridge by hybridizing with its free end to the complementary adapter on the surface of the support. The adapters on the surface act as primers for the subsequent PCR amplification. After several PCR cycles, random clusters of single-stranded DNA fragments are created on the surface. The platform detects each incorporated nucleotide labeled with a different fluorescent dye. The length of the read sequence is about 35 nucleotides.

2.3 Genome-Wide Methylation Study

Other techniques can be used to study DNA methylation in discrete compartments of the genome, such as CpG islands or repetitive sequences. In addition, a genome-wide methylation study can provide a useful tool for discovering new target genes with abnormal methylation providing, possible prognostic indicators of disease.

2.3.1 *Restriction Landmark Genomic Scanning*

The preferential analysis of DNA methylation in the context of CpG islands can be accomplished with the restriction landmark genomic scanning (RLGS) technique.[37] RLGS adopts a two-dimensional electrophoresis approach. Genomic DNA is primarily digested by a methylation-sensitive restriction enzyme (e.g., *Not*I), thus profiling radiolabeled unmethylated restriction sites. Subsequently, the DNA is subjected to a second restriction digestion to produce smaller pieces which are separated in one dimension by electrophoresis. Once the second digestion is completed, the DNA is separated by two-dimensional electrophoresis, with DNA cut into smaller fragments by a third digestion. The gel is autoradiogrammed and methylation is detected by a change in signal intensity. Amplification or hypomethylation increases signal intensity or causes addition of a new spot. RLGS results are analyzed by a computer program. The technique has been used to identify novel imprinted genes[62] and genes that are frequently hypermethylated in cancer.[70] The method identifies thousands of landmark fragments in a single run and can be applied to any genome without prior knowledge of the DNA sequence. However, high molecular weight DNA is absolutely necessary to prevent nonspecific labeling of degraded fragments. The method also requires an elaborate gel electrophoresis set-up and a powerful image analysis system.

2.3.2 Methylated DNA Binding Column

This method, designed by Cross and collaborators,[16] is used to isolate CpG islands from genomic DNA. It uses an affinity matrix that contains the methyl-CpG (mCpG)-binding domain from the rat protein Mecp2, attached to a solid support. A column contains the matrix-fractionated DNA according to its degree of methylation and sturdily retains those sequences that are highly methylated. Clustering of mCpGs also influences elution profiles, particularly when the DNA fragments are about 200-bp long and contain some 20 mCpGs. Use of methylated DNA binding column (MBD column) chromatography permits distinguishing and separating genomic DNA fragments from methylated and nonmethylated CpG islands. The fragments contained in the eluted fraction can be cloned and used in different downstream methods like detection by using specific probes or specific primers. This technique is useful for obtaining a library of methylated DNA from genomic DNA. For example, Shiraishi and collaborators[69] used it to characterize methylated CpG islands from adenocarcinomas of the lung. The nondestructive nature of this method is of great advantage for the selective isolation of DNA fragments from unmethylated and methylated CpG islands.

2.3.3 Methylation CPG Island Amplification

Methylation CPG island amplification (MCA) is a highly effective method for identifying hundreds of CpG islands in two samples.[77] Genomic DNA of two samples is digested using two types of restriction endonuclease: methylation-sensitive and methylation-insensitive. The basic principle of MCA involves amplification of DNA sequences with closely spaced methylated *Sma*I (methylation-sensitive enzyme) sites, which are frequently encountered in CpG islands. As only short fragments flanked by two *Sma*I sites are later amplified, MCA ensures enrichment of CpG islands. Those methylated fragments are then digested by a methylation-insensitive enzyme. Sticky end fragments are generated that are linked and PCR-amplified. Southern or dot blotting analysis or hybridization in an array is employed to determine whether a candidate CpG island is differentially methylated. This method has been used to determine the methylation profile in clear cell renal carcinoma.[4] Its limitation is the inability to compare more than two samples at a time and to identify aberrantly hypomethylated sequences.

2.3.4 Differential Methylation Hybridization

This high-throughput microarray technique developed by Huang and collaborators[39] identifies changes in DNA methylation patterns. Genomic DNA is digested with *Mse*I, a methylation-insensitive restriction enzyme, then linkers are ligated to the digested DNA, and the ligation product is digested with two methylation-sensitive restriction enzymes, *Bst*UI and *Hpa*II. The products of these two digestions are amplified by PCR with the use of primers complementary to the linker sequence. The result is the amplification of methylated sequences not digested by *Bst*UI and *Hpa*II. The PCR products are then labeled with fluorescent dyes and hybridized to a CpG island array. One important feature of this method is that thousands of CpG islands can be analyzed at one time, but to avoid false-positive results, it depends on the efficient digestion of genomic DNA. The technique has been used to identify epigenetic alterations in breast cancer.[87]

2.3.5 Amplification of Inter-methylated Sites

Amplification of inter-methylated sites (AIMS)[26] is based on the differential cleavage of isoschizomers with distinct methylation sensitivity. The nonmethylated sites are cut in an initial digestion with the methylation-sensitive endonuclease, which leaves blunt ends. A second digestion is performed with an isoschizomer that leaves a CCGG overhang. DNA fragments flanked by two ligated adaptors are amplified by PCR with specific primers that hybridize to the adapter sequence and the restriction site and one or more additional, arbitrarily chosen nucleotides. The PCR products are separated on polyacrylamide urea sequencing gels. Fingerprints consist of multiple anonymous bands, representing DNA sequences flanked by two methylated sites. Bands appearing differentially methylated can be isolated, cloned, and sequenced. The AIMS method has been used for determining methylation profiles in sporadic colorectal carcinoma[27] and in monozygotic twins.[23]

2.3.6 Methyl-DNA Immunoprecipitation

Weber and collaborators[80] developed this method in which a monoclonal antibody against 5MeC is used to purify methylated DNA. Genomic DNA is fragmented by sonication into 300–600-bp fragments. After fragmentation, DNA is denatured and the methylated fragments are immunoprecipitated using a specific antibody against 5MeC (Fig. 2.2). The purified, immunoprecipitated, and methylated DNA is analyzed individually by using specific primers targeting the specified genes; it can also be hybridized in a high-resolution genomic microarray or sequenced by high-throughput sequencing platforms. Methyl-DNA immunoprecipitation (MeDIP) combined with a microarray containing over 13,000 promoters allows the identification of a large number of genes with hypermethylated CpG islands, but the CpG-rich sequences may indicate greater enrichment than methylated CpG-poor sequences (Fig. 2.2). In this sense, the possibility of combining MeDIP with sequencing makes this the preferred strategy for advancing the genome-wide analysis of the DNA. This method has been used for mapping the hypermethylome of cancer DNA.[81]

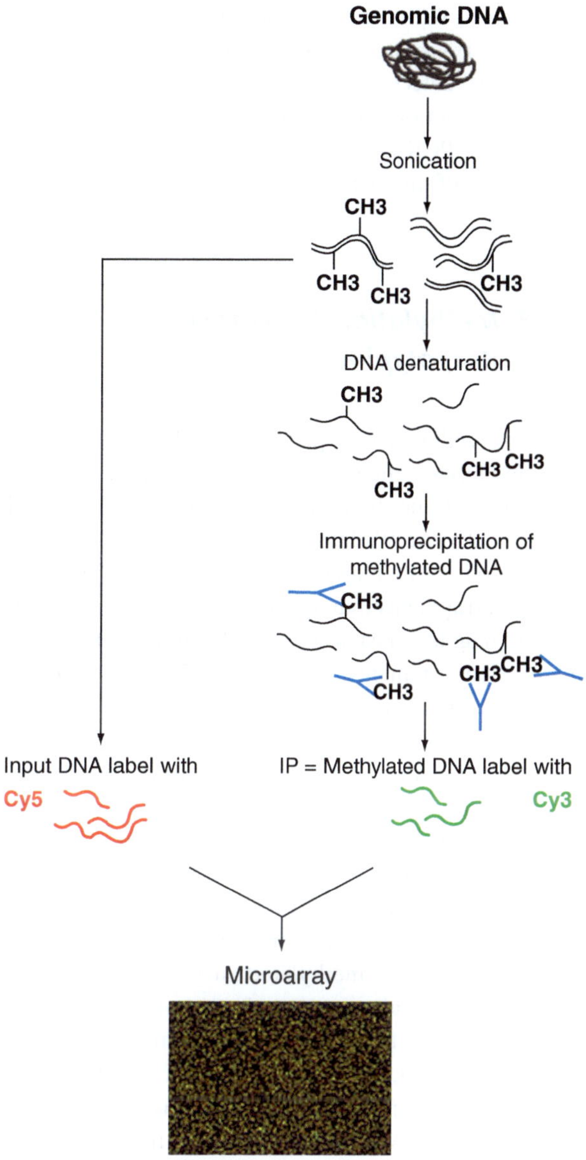

Fig. 2.2 MeDIP assay workflow

2.3.7 Methylated CPG Island Recovery Assay

In this technique, developed in 2005,[64] the methyl-CpG-binding protein MBD2b specifically recognizes methylated CpG dinucleotides, with the interaction strongly enhanced by the MBD3L1 (CpG-binding domain protein 3-like-1). MBD2b protein has a very high affinity for methylated DNA, with little sequence specificity. In this method, fragmented genomic DNA is incubated with the MBD2b/MBD3L1 protein complex. Matrix-assisted binding and simple PCR assays are used to detect methylated DNA sequences in the recovered fraction. The isolated CpG methylated fraction can be PCR-amplified, sequenced, or hybridized in a variety of microarray platforms.

2.3.8 Methylation-Specific Digital Karyotyping

In this procedure, developed by Hu and collaborators,[38] genomic DNA is digested with a methylation-sensitive mapping enzyme, ligated to biotinylated linkers, and fragmented by *Nla*III cleavage. The methylation-sensitive enzyme only cuts unmethylated regions. As a result, the DNA fragments that are bound to streptavidin-conjugated magnetic beads separate the unmethylated and methylated fragments. The bound DNA is digested with other restriction enzymes, giving rise to short sequences that can be cloned for sequencing. The method focuses on unmethylated regions, thus avoiding interference with repetitive sequences (which are normally hypermethylated). The advantages of methylation-specific digital karyotyping (MSDK) are that it is quantitative and requires no prior knowledge of DNA methylation. It therefore can be used to identify new, differentially methylated sites. Given MSDK requires only small amounts of genomic DNA, it can be used to identify methylation sites in clinical samples. By contrast, MSDK is a low-throughput, labor-intensive, and expensive method. It has been used to describe the DNA methylation pattern in distinct subpopulations of mammary epithelial cells[10] and to assess DNA methylation changes in mice.[11]

2.3.9 HpaII Tiny Fragment Enrichment by Ligation-Mediated PCR Assay

In the *Hpa*II tiny fragment enrichment by ligation-mediated PCR (HELP) assay,[48] genomic DNA is digested with the methylation-sensitive restriction enzyme *Hpa*II, which only cuts when the restriction site is unmethylated. Adapters are linked to the fragments and used for ligation-mediated PCR amplification. A second aliquot is digested in parallel with the methylation-insensitive isoschizomer *Msp*I. The two digestion products are labeled with two fluorophores, and DNA methylation can be analyzed by co-hybridization of the two fractions on a microarray.

2.3.10 DNA Methylation Arrays

Illumina® technologies have designed bead arrays to analyze the methylation profile in different samples simultaneously by two principal methylation arrays – Infinium and GoldenGate. The first allows the study of 12 samples in parallel, covering more than 14,000 genes that are not necessarily associated with cancer, whereas the second is suitable for analyzing more than 800 cancer-related genes in 96 samples simultaneously. The starting point in both arrays is bisulfite-treated DNA. The Infinium methylation assay uses two different bead types to detect CpG methylation, one that matches the methylated CpG and the other that matches the unmethylated CpG. When the CpG matches with the specific probe, a single-base extension of the probes incorporates a labeled ddNTP, which is then stained with a fluorescence reagent. The methylation level is calculated from the ratio of methylated to unmethylated fluorescent signals. In the GoldenGate array, four oligos are designed to target a specific CpG dinucleotide. These bind to the bisulfite-treated DNA, become extended, and are linked enzymatically. The linked products are amplified and fluorescently labeled during PCR and finally identified by hybridization to specific sequences on the array. The methylation status of each locus is calculated as the ratio of the signal intensity of the methylated locus to that of the total locus. The GoldenGate array has been used to define a methylation profile in follicular lymphomas.[46] As these arrays require a small amount of DNA, the technique can be used clinically to identify DNA methylation markers of disease.

2.4 DNA Methylation Analysis: Conclusions and Remarks

No single analytical method will be appropriate for every study, and each investigator must select the method that is best suited to their particular research.

Table 2.2 Study of histone modification

	Technique	Time	Cost	Special requirements	Amount of starting DNA
Global histone modification	HPCE	XX	X	Yes	High
	MS	XX	X	Yes	Low-medium
Locus-specific histone modification	ChIP	XXX	XX	No	High
Genome-wide histone modification	ChIP-on-Chip	XXX	XXX	Yes	High
	ChIP-on-PET	XXX	XXX	Yes	High
	ChIP-on-Seq	XXX	XXX	Yes	High

Time: fast, average, and labor-intensive (X, XX, and XXX, respectively); Cost: cheap, average, and expensive (X, XX, and XXX, respectively)

The recent high-throughput methods for DNA methylation analysis at the genome-wide or locus-specific scale are the preferred approaches, but in many cases, they are too expensive or sophisticated to implement in standard laboratories.

The easiest method for specific DNA methylation analysis is MSP. This allows a qualitative and sensitive analyses of many samples at the same time and is a cheap and not too time-consuming technique. It only requires a PCR machine and common reagents for a PCR, thereby making it an ideal technique for beginners.

2.5 Histone Modification Study

Every cell's distinguishing characteristics within a multicellular organism are the result of gene expression at a given time and site. Chromatin regulates gene transcription by the degree of its compaction brought about by PTMs or the association of certain PTMs to promoter regions, these either favoring or inhibiting transcription.

In general, acetylation of histones leads to a relaxed state of the chromatin, allowing access to DNA-modifying enzymes. It is therefore an activating gene transcription mark. Deacetylation of histones, on the other hand, compacts chromatin and represses transcription. Histone methylation affects gene transcription in two ways. High levels of methylated H3K4, H3K36, and H3K79 are associated with relaxed and active chromatin, whereas methylated H3K9, H3K27, and H4K20 are associated with genes that are not transcribed.

Histone modifications can be inferred by studying DNA methylation. Approaches have been developed to study posttranslational modifications (PTMs) at the global, locus-specific, and genome-wide levels (Table 2.2).

2.5.1 Global Histone Modification

Analysis of histone modification generally involves gel systems or specific antibodies. In addition, accurate and specific methods have been developed to detect, separate, and quantify PTMs. Histones have generally been extracted by dissolution in HCl or H_2SO_4, as most nuclear proteins and nucleic acids will precipitate in an acid solution. However, studies centered on the discovery of novel PTMs use high-salt extraction, because some histone structures are acid-labile. High-salt extraction is useful for applications where maintenance of a neutral pH is crucial.

2.5.2 Polyacrylamide Gel Electrophoresis

Electrophoretic separation of histones was introduced in 1959.[57] A variety of polyacrylamide gel electrophoresis (PAGE) have been developed to increase resolution. SDS-PAGE separates the five histone class, namely, H1, H2A, H2B, H3, and H4. Acetic acid-urea PAGE (AU-PAGE) can resolve the five main histones and may also separate some modified forms. Nonionic detergents like Triton X-100 in acetic acid-urea gels (AUT-PAGE) help separate histones and some nonallelic variants.

Resolution of histone separation is improved with two-dimensional gel electrophoresis. This is mainly a combination of SDS in the first dimension and of AU/AUT-PAGE in the second dimension.

AU gels separate histones according to the charges introduced by acetylation and phosphorylation. In AU gel electrophoresis, proteins are denatured by a high concentration of urea, but their charge is unaffected. Acetylation reduces the positive charge of a Lys; phosphorylation introduces an additional negative charge. Two-dimensional electrophoresis has been used to identify H3 phosphorylation at serine 10 as one of the earliest events that occurs after stimulation with serum or phorbol ester.[53]

Antibodies against PTMs can be used in Western blotting to study changes in specific sites. They can also be used in immunofluorescence studies to localize specific modified histones within particular regions of the genome.[78] Although the antibodies are generally very sensitive, some show cross-reactivity due to the great similarity between different modification sites. However, this method does not separate histone isoforms with different acetylation or phosphorylation levels and requires much time and large sample quantities. It is also labor-intensive.

2.5.3 Reversed-Phase High-Performance Liquid Chromatography

This method separates molecules on the basis of hydrophobicity. Histones are separated using an acetonitrile gradient and C18 column. Histones can be previously extracted with acid and high-salt treatments. Gurley and collaborators,[32] who were the first to use this method, obtained several histone fractions within 80 min, but with low recovery. Recovery can be improved somewhat by the use of end-capped column materials and trifluoroacetic acid (TFA) as an ion-pairing reagent. With reversed-phase high-performance liquid chromatography (RP-HPLC), it is laborious and time-consuming. It does not resolve posttranslationally modified histones, but has the advantage that modified forms of one histone protein co-elute in a single fraction that does not contain modified forms of other histones. For this reason, it is frequently used for histone purification that, in conjunction with MS or HPCE, facilitates the study of the various histone modifications.[8,33]

2.5.4 High-Performance Capillary Electrophoresis

Introduced by Jorgenson and Lukacs,[44] HPCE separates solutes by an electric field–induced migration in a fused-silica capillary. The method has several advantages: highly efficient separation, precise quantification, reduced sample volume, and automation. However, because proteins stick to the glass of the capillaries, sensitivity is reduced or resolution is lost. If the capillary surface is coated to reduce interactions between proteins and the surface, results are improved substantially. By coating the capillary surface with hydroxypropylmethylcellulose and using low pH buffers to prevent interactions between the histones and the silica, Fraga and collaborators[24] were able to quantify changes in the modification patterns of histone H4 in tumoral tissues.

2.5.5 Mass Spectrometry

Mass spectrometry provides key techniques for the analysis of PTMs. Histones carry several different modifications localized on a single peptide (at the N-terminus of histones) within a protease digest that can be modified in response to external or internal signals. Since the 1990s, MS has been the best technique to study histone modifications, because every high-resolution MS can analyze the new mass due to a modification of the histone molecule.

MS, however, has several pitfalls. Inasmuch as histones are rich in lysine residues, trypsin cannot be used to digest histones. Instead, other enzymes or a preparative method, such as HPLC, must be used to purify individual histones. Another method that can be used is the derivatization of lysines within the histone molecules with the aid of acid anhydrides.[71] Acid anhydrides react very efficiently with unmodified or monomethylated amino groups of lysine residues. This prevents trypsin from cleaving the modified histone residue, although it can still cleave arginine residues. The resulting peptides are much larger and can be analyzed by MALDI-TOF MS. However, some precautions are necessary. For example, the use of propionic anhydride leads to the propionylation of lysine, which then attains a mass that almost equals the addition of four methyl groups; the same is true for three methyl groups and acetylation. This is important because each of the two modifications has a distinct biological role. The high-resolution TOF mass analyzer can then be used to identify the two modifications. Alternatively, the fragmentation behavior of the modified peptide can be studied in an MS/MS experiment. It is important to determine whether the two modifications are clustered within the same molecules, because the interference between different modifications is often used to integrate signals in a common pathway. A possible solution to this problem is top-down proteomics, which allows sequence determination of intact proteins without prior proteolytic cleavage.

2.5.6 Locus-Specific Histone Modification

2.5.6.1 Chromatin Immunoprecipitation

First described in 1988[73] to demonstrate histone H4 binding to the *Drosophila melanogaster* heat shock protein 70 (*Hsp 70*), the chromatin immunoprecipitation (ChIP) assay indicates what is bound to the chromatin and/or the modification of chromatin in a particular tissue or cell. The ChIP assay can be used to study any target against which an antibody can be raised, including transcription factors, histones, and histone modifications.[54] The key factor for all ChIP-based methods is a high-quality, high-specificity antibody, and its DNA fragmentation efficiency. Assays must include a positive control and a negative control (e.g., a nonspecific IgG) to assess the efficacy of antibody precipitation.

There are two primary types of ChIP assay: conventional and native ChIP. With the *conventional* ChIP, DNA–protein interactions are fixed by the addition of a crosslinking agent, commonly formaldehyde. Cross-linked protein/chromatin is fragmented by sonication or incubation with *Micrococcal nuclease* MNase into approximately 200–800-bp fragments. The protein of interest is then immunoprecipitated from the lysate with the aid of a specific antibody. After the crosslinks are reversed, proteins are removed and the precipitated and purified DNA is used in subsequent analyses, such as Southern blotting, conventional or quantitative real-time PCR, or array hybridization. By contrast, in a *native* ChIP, the crosslinking agent is not used. Instead, chromatin is digested with MNase that cleaves linker regions of chromatin in which DNA is not protected by nucleosomes. The native ChIP is especially useful for the study of proteins such as histones that are tightly bound to DNA, but not for weakly bound proteins such as transcription factors.[58]

DNA bound to an antibody can be analyzed by PCR (semiquantitative-ChIP), quantitative real-time PCR (quantitative-ChIP), flow cytometry (ChIP-on-beads), cloning and sequencing (ChIP-on-PET), or direct sequencing (ChIP-on-Seq). ChIP, a powerful tool for analyzing protein/DNA interactions, is widely used in the epigenetic and other fields[34,89] and forms the basis of numerous genome-wide methods for analyzing PTM profiles.

2.5.7 Genome-Wide Histone Modification

2.5.7.1 ChIP-on-Chip

The traditional Chip is restricted to predetermined target sequences. To get around this problem, the ChIP-on-Chip assay[65] combines a ChIP and a microarray (chip) of genomic DNA fragments. Genomic microarrays consist of selected promoter sets that are a random representation of CpG islands. For ChIP-on-chip, immunoprecipitated DNA is purified and the ends repaired with DNA polymerase to generate blunt ends. A linker is linked to each DNA fragment, thus permitting PCR amplification. A fluorescent label (usually Cy5 for bound DNA and Cy3 for unbound DNA) is incorporated during amplification by a ligation-mediated, polymerase chain reaction. Both DNA fractions are mixed and hybridized onto a microarray imprinted with oligonucleotide probes. A positive signal is established when the signal intensity of the bound (output or ChIP DNA) DNA significantly exceeds that of the unbound (input) DNA on the array.

The method requires large amounts of DNA (approximately 1–5 μg) for DNA microarray hybridization. It is also labor-intensive, quite costly, and generates high proportion of false-positives. In addition, the many steps including fixation, DNA fragmentation, epitope accessibility, and antibody specificity cause relatively high variability. Yet, notwithstanding its limitations, the method yields information that cannot be obtained by other approaches. ChIP-on-chip has been used in studies characterizing genetic regions that are subject to histone and nonhistone protein regulation.[45,88] It can be applied to identify the binding site of a transcription factor,[65] to map genome-wide protein-binding profiles,[41] and to reveal the distribution of local histone modifications.[9]

2.5.7.2 ChIP-PET (Paired-End Tag)

In ChIP-PET, short sequence tags of immunoprecipitated DNA fragments are cloned into a plasmid library and analyzed by sequencing. ChIP-PET is a genome-wide approach and enables the prediction of novel DNA motifs that mediate protein–DNA interactions. It is labor-intensive and expensive because it requires a sufficient quantity of immunoprecipitated fragments. It has been used to generate a global map of *p53* binding sites across the human genome[82] and to establish the transcription regulatory networks of *Oct4* and *Nanog* in mouse embryonic stem cells.[52]

2.5.7.3 ChIP-on-Seq

ChIP-on-Seq couples chromatin immunoprecipitation with ultra-high-throughput DNA sequencing. It relies on the power of new sequencing platforms like Solexa (Illumina) and 454 (Roche). The main objective of the method is to identify genomic regions that are enriched among immunoprecipitated DNA. The method has several advantages over the ChIP-on chip technique, in terms of cost, labor, and sample size. Also, because it does not need an array, it is not restricted to one species. All ChIP-based methods require large quantities of starting material and a specific and validated antibody. ChIP-on-Seq has been used to map genome-wide profiles of transcription factors binding to DNA[66] and histone modifications.[7]

2.6 Histone Modification Analysis: Conclusions and Remarks

As in the case for DNA methylation studies, we should ideally use new technologies for analyzing histone modifications, but sometimes, these require very large equipment that is often not available to all laboratories.

The easiest initial approach to histone modification analysis is to use western blotting techniques to study changes at the global level. Specific antibodies against the specific histone are needed for this. The next step is to use PCR to analyze the DNA bound to an antibody against histone isoforms with different acetylation or phosphorylation levels (ChIPs), although this is a considerably more labor-intensive technique than that previously described.

In conclusion, there are many techniques to analyze epigenetic modifications at different levels. As Laird[50] has noted: "no one technique or general approach is superior, as the competing goals of quantitative accuracy, sensitive detection, and high local or global informational content, compatibility with formalin fixed tissues and compatibility with automation are not found in a single technique."

References

1. Achwal CW, Chandra HS. A sensitive immunochemical method for detecting 5mC in DNA fragments. *FEBS Lett.* 1982;150(2):469-472.
2. Achwal CW, Ganguly P, Chandra HS. Estimation of the amount of 5-methylcytosine in *Drosophila melanogaster* DNA by amplified ELISA and photoacoustic spectroscopy. *EMBO J.* 1984;3(2):263-266.
3. Agrelo R, Setien F, Espada J, et al. Inactivation of the lamin A/C gene by CpG island promoter hypermethylation in hematologic malignancies, and its association with poor survival in nodal diffuse large B-cell lymphoma. *J Clin Oncol.* 2005;23(17):3940-3947.
4. Arai E, Ushijima S, Fujimoto H, et al. Genome-wide DNA methylation profiles in both precancerous conditions and clear cell renal cell carcinomas are correlated with malignant potential and patient outcome. *Carcinogenesis.* 2009;30(2): 214-221.
5. Balaghi M, Wagner C. DNA methylation in folate deficiency: use of CpG methylase. *Biochem Biophys Res Commun.* 1993;193(3):1184-1190.
6. Balch C, Yan P, Craft T, et al. Antimitogenic and chemosensitizing effects of the methylation inhibitor zebularine in ovarian cancer. *Mol Cancer Ther.* 2005;4(10):1505-1514.
7. Barski A, Cuddapah S, Cui K, et al. High-resolution profiling of histone methylations in the human genome. *Cell.* 2007;129(4):823-837.
8. Berger RG, Hoffmann R, Zeppezauer M, et al. Separation and characterisation of bovine histone H1 subtypes by combined ion-exchange and reversed-phase chromatography and mass spectrometry. *J Chromatogr A.* 1995;711(1): 159-165.
9. Bernstein BE, Humphrey EL, Erlich RL, et al. Methylation of histone H3 Lys 4 in coding regions of active genes. *Proc Natl Acad Sci USA.* 2002;99(13):8695-8700.

10. Bloushtain-Qimron N, Yao J, Snyder EL, et al. Cell type-specific DNA methylation patterns in the human breast. *Proc Natl Acad Sci USA*. 2008;105(37):14076-14081.
11. Boon K, Tomfohr JK, Bailey NW, et al. Evaluating genome-wide DNA methylation changes in mice by methylation specific digital karyotyping. *BMC Genomics*. 2008;9:598.
12. Carlone DL, Lee JH, Young SR, et al. Reduced genomic cytosine methylation and defective cellular differentiation in embryonic stem cells lacking CpG binding protein. *Mol Cell Biol*. 2005;25(12):4881-4891.
13. Chan TL, Yuen ST, Kong CK, et al. Heritable germline epimutation of MSH2 in a family with hereditary nonpolyposis colorectal cancer. *Nat Genet*. 2006;38(10):1178-1183.
14. Cheng SH, Ng MH, Lau KM, et al. 4q loss is potentially an important genetic event in MM tumorigenesis: identification of a tumor suppressor gene regulated by promoter methylation at 4q13.3, platelet factor 4. *Blood*. 2007;109(5):2089-2099.
15. Cottrell SE, Distler J, Goodman NS, et al. A real-time PCR assay for DNA-methylation using methylation-specific blockers. *Nucleic Acids Res*. 2004;32(1):e10.
16. Cross SH, Charlton JA, Nan X, Bird AP. Purification of CpG islands using a methylated DNA binding column. *Nat Genet*. 1994;6(3):236-244.
17. Eads CA, Danenberg KD, Kawakami K, et al. MethyLight: a high-throughput assay to measure DNA methylation. *Nucleic Acids Res*. 2000;28(8):E32.
18. Eckhardt F, Lewin J, Cortese R, et al. DNA methylation profiling of human chromosomes 6, 20 and 22. *Nat Genet*. 2006;38(12):1378-1385.
19. Ehrlich M, Gama-Sosa MA, Huang LH, et al. Amount and distribution of 5-methylcytosine in human DNA from different types of tissues of cells. *Nucleic Acids Res*. 1982;10(8):2709-2721.
20. Esteller M. Cancer epigenetics: DNA methylation and chromatin alterations in human cancer. *Adv Exp Med Biol*. 2003;532:39-49.
21. Esteller M, Silva JM, Dominguez G, et al. Promoter hypermethylation and BRCA1 inactivation in sporadic breast and ovarian tumors. *J Natl Cancer Inst*. 2000;92(7):564-569.
22. Esteller M, Toyota M, Sanchez-Cespedes M, et al. Inactivation of the DNA repair gene O6-methylguanine-DNA methyltransferase by promoter hypermethylation is associated with G to A mutations in K-ras in colorectal tumorigenesis. *Cancer Res*. 2000;60(9):2368-2371.
23. Fraga MF, Ballestar E, Paz MF, et al. Epigenetic differences arise during the lifetime of monozygotic twins. *Proc Natl Acad Sci USA*. 2005;102(30):10604-10609.
24. Fraga MF, Ballestar E, Villar-Garea A, et al. Loss of acetylation at Lys16 and trimethylation at Lys20 of histone H4 is a common hallmark of human cancer. *Nat Genet*. 2005;37(4):391-400.
25. Fraga MF, Rodriguez R, Canal MJ. Rapid quantification of DNA methylation by high performance capillary electrophoresis. *Electrophoresis*. 2000;21(14):2990-2994.
26. Frigola J, Ribas M, Risques RA, Peinado MA. Methylome profiling of cancer cells by amplification of inter-methylated sites (AIMS). *Nucleic Acids Res*. 2002;30(7):e28.
27. Frigola J, Sole X, Paz MF, et al. Differential DNA hypermethylation and hypomethylation signatures in colorectal cancer. *Hum Mol Genet*. 2005;14(2):319-326.
28. Frommer M, McDonald LE, Millar DS, et al. A genomic sequencing protocol that yields a positive display of 5-methylcytosine residues in individual DNA strands. *Proc Natl Acad Sci USA*. 1992;89(5):1827-1831.
29. Galm O, Rountree MR, Bachman KE, Jair KW, Baylin SB, Herman JG. Enzymatic regional methylation assay: a novel method to quantify regional CpG methylation density. *Genome Res*. 2002;12(1):153-157.
30. Gonzalgo ML, Jones PA. Rapid quantitation of methylation differences at specific sites using methylation-sensitive single nucleotide primer extension (Ms-SNuPE). *Nucleic Acids Res*. 1997;25(12):2529-2531.
31. Guerrero-Preston R, Baez A, Blanco A, Berdasco M, Fraga M, Esteller M. Global DNA methylation: a common early event in oral cancer cases with exposure to environmental carcinogens or viral agents. *P R Health Sci J*. 2009;28(1):24-29.
32. Gurley LR, Valdez JG, Prentice DA, Spall WD. Histone fractionation by high-performance liquid chromatography. *Anal Biochem*. 1983;129(1):132-144.
33. Hagiwara T, Hidaka Y, Yamada M. Deimination of histone H2A and H4 at arginine 3 in HL-60 granulocytes. *Biochemistry*. 2005;44(15):5827-5834.
34. Heintzman ND, Hon GC, Hawkins RD, et al. Histone modifications at human enhancers reflect global cell-type-specific gene expression. *Nature*. 2009;459(7243):108-112.
35. Herman JG, Graff JR, Myohanen S, Nelkin BD, Baylin SB. Methylation-specific PCR: a novel PCR assay for methylation status of CpG islands. *Proc Natl Acad Sci USA*. 1996;93(18):9821-9826.
36. Hess CJ, Ameziane N, Schuurhuis GJ, et al. Hypermethylation of the FANCC and FANCL promoter regions in sporadic acute leukaemia. *Cell Oncol*. 2008;30(4):299-306.
37. Hirotsune S, Hatada I, Komatsubara H, et al. New approach for detection of amplification in cancer DNA using restriction landmark genomic scanning. *Cancer Res*. 1992;52(13):3642-3647.
38. Hu M, Yao J, Polyak K. Methylation-specific digital karyotyping. *Nat Protoc*. 2006;1(3):1621-1636.
39. Huang TH, Perry MR, Laux DE. Methylation profiling of CpG islands in human breast cancer cells. *Hum Mol Genet*. 1999;8(3):459-470.
40. Itoh K, Aida S, Ishiwata S, Yamaguchi T, Ishida N, Mizugaki M. Immunochemical detection of urinary 5-methyl-2′-deoxycytidine as a potential biologic marker for leukemia. *Clin Chim Acta*. 1995;234(1–2):37-45.
41. Iyer VR, Horak CE, Scafe CS, Botstein D, Snyder M, Brown PO. Genomic binding sites of the yeast cell-cycle transcription factors SBF and MBF. *Nature*. 2001;409(6819):533-538.
42. Jeuken JW, Cornelissen SJ, Vriezen M, et al. MS-MLPA: an attractive alternative laboratory assay for robust, reliable, and semiquantitative detection of MGMT promoter hypermethylation in gliomas. *Lab Invest*. 2007;87(10):1055-1065.
43. Jones PA, Takai D. The role of DNA methylation in mammalian epigenetics. *Science*. 2001;293(5532):1068-1070.
44. Jorgenson JW, Lukacs KD. Capillary zone electrophoresis. *Science*. 1983;222(4621):266-272.
45. Kadota M, Yang HH, Hu N, et al. Allele-specific chromatin immunoprecipitation studies show genetic influence on chromatin state in human genome. *PLoS Genet*. 2007;3(5):e81.

46. Kanduri M, Cahill N, Goransson H, et al. Differential genome-wide array-based methylation profiles in prognostic subsets of chronic lymphocytic leukemia. *Blood*. 2010;115(2):296-305.
47. Kaneda A, Tsukamoto T, Takamura-Enya T, et al. Frequent hypomethylation in multiple promoter CpG islands is associated with global hypomethylation, but not with frequent promoter hypermethylation. *Cancer Sci*. 2004;95(1):58-64.
48. Khulan B, Thompson RF, Ye K, et al. Comparative isoschizomer profiling of cytosine methylation: the HELP assay. *Genome Res*. 2006;16(8):1046-1055.
49. Kuo KC, McCune RA, Gehrke CW, Midgett R, Ehrlich M. Quantitative reversed-phase high performance liquid chromatographic determination of major and modified deoxyribonucleosides in DNA. *Nucleic Acids Res*. 1980;8(20):4763-4776.
50. Laird PW. The power and the promise of DNA methylation markers. *Nat Rev Cancer*. 2003;3(4):253-266.
51. Li Z, Zhang W, Shao Y, et al. High-resolution melting analysis of ADAMTS18 methylation levels in gastric, colorectal and pancreatic cancers. *Med Oncol*. 2010;27(3):998-1004.
52. Loh YH, Wu Q, Chew JL, et al. The Oct4 and Nanog transcription network regulates pluripotency in mouse embryonic stem cells. *Nat Genet*. 2006;38(4):431-440.
53. Mahadevan LC, Willis AC, Barratt MJ. Rapid histone H3 phosphorylation in response to growth factors, phorbol esters, okadaic acid, and protein synthesis inhibitors. *Cell*. 1991;65(5):775-783.
54. Massie CE, Mills IG. ChIPping away at gene regulation. *EMBO Rep*. 2008;9(4):337-343.
55. Matsuda Y, Ichida T, Matsuzawa J, Sugimura K, Asakura H. p16(INK4) is inactivated by extensive CpG methylation in human hepatocellular carcinoma. *Gastroenterology*. 1999;116(2):394-400.
56. Muller HM, Widschwendter A, Fiegl H, et al. DNA methylation in serum of breast cancer patients: an independent prognostic marker. *Cancer Res*. 2003;63(22):7641-7645.
57. Neelin JM, Connell GE. Zone electrophoresis of chicken erythrocyte histone in starch gel. *Biochim Biophys Acta*. 1959;31:539-541.
58. Nguyen TT, Cho K, Stratton SA, Barton MC. Transcription factor interactions and chromatin modifications associated with p53-mediated, developmental repression of the alpha-fetoprotein gene. *Mol Cell Biol*. 2005;25(6):2147-2157.
59. Nguyen TT, Mohrbacher AF, Tsai YC, et al. Quantitative measure of c-abl and p15 methylation in chronic myelogenous leukemia: biological implications. *Blood*. 2000;95(9):2990-2992.
60. Nygren AO, Ameziane N, Duarte HM, et al. Methylation-specific MLPA (MS-MLPA): simultaneous detection of CpG methylation and copy number changes of up to 40 sequences. *Nucleic Acids Res*. 2005;33(14):e128.
61. Oakeley EJ, Schmitt F, Jost JP. Quantification of 5-methylcytosine in DNA by the chloroacetaldehyde reaction. *Biotechniques*. 1999;27(4):744-746. 748–750, 752.
62. Plass C, Shibata H, Kalcheva I, et al. Identification of Grf1 on mouse chromosome 9 as an imprinted gene by RLGS-M. *Nat Genet*. 1996;14(1):106-109.
63. Radpour R, Kohler C, Haghighi MM, Fan AX, Holzgreve W, Zhong XY. Methylation profiles of 22 candidate genes in breast cancer using high-throughput MALDI-TOF mass array. *Oncogene*. 2009;28(33):2969-2978.
64. Rauch T, Pfeifer GP. Methylated-CpG island recovery assay: a new technique for the rapid detection of methylated-CpG islands in cancer. *Lab Invest*. 2005;85(9):1172-1180.
65. Ren B, Robert F, Wyrick JJ, et al. Genome-wide location and function of DNA binding proteins. *Science*. 2000;290(5500):2306-2309.
66. Robertson G, Hirst M, Bainbridge M, et al. Genome-wide profiles of STAT1 DNA association using chromatin immunoprecipitation and massively parallel sequencing. *Nat Methods*. 2007;4(8):651-657.
67. Roman-Gomez J, Jimenez-Velasco A, Barrios M, et al. Poor prognosis in acute lymphoblastic leukemia may relate to promoter hypermethylation of cancer-related genes. *Leuk Lymphoma*. 2007;48(7):1269-1282.
68. Schernhammer ES, Giovannucci E, Kawasaki T, Rosner B, Fuchs C, Ogino S. Dietary folate, alcohol, and B vitamins in relation to LINE-1 hypomethylation in colon cancer. *Gut*. 2010;59(6):794-799.
69. Shiraishi M, Chuu YH, Sekiya T. Isolation of DNA fragments associated with methylated CpG islands in human adenocarcinomas of the lung using a methylated DNA binding column and denaturing gradient gel electrophoresis. *Proc Natl Acad Sci USA*. 1999;96(6):2913-2918.
70. Smiraglia DJ, Smith LT, Lang JC, et al. Differential targets of CpG island hypermethylation in primary and metastatic head and neck squamous cell carcinoma (HNSCC). *J Med Genet*. 2003;40(1):25-33.
71. Smith CM, Gafken PR, Zhang Z, Gottschling DE, Smith JB, Smith DL. Mass spectrometric quantification of acetylation at specific lysines within the amino-terminal tail of histone H4. *Anal Biochem*. 2003;316(1):23-33.
72. Snell C, Krypuy M, Wong EM, Loughrey MB, Dobrovic A. BRCA1 promoter methylation in peripheral blood DNA of mutation negative familial breast cancer patients with a BRCA1 tumour phenotype. *Breast Cancer Res*. 2008;10(1):R12.
73. Solomon MJ, Larsen PL, Varshavsky A. Mapping protein-DNA interactions in vivo with formaldehyde: evidence that histone H4 is retained on a highly transcribed gene. *Cell*. 1988;53(6):937-947.
74. Tamura T, Araki Y, Yamaoka S, Inagaki K, Tanaka H. Determination of methylation specificity of sequence-specific DNA methyltransferases using matrix assisted laser desorption/ionization time-of-flight mass spectrometry. *Nucleic Acids Res*. 1997;25(20):4162-4164.
75. Thomassin H, Kress C, Grange T. MethylQuant: a sensitive method for quantifying methylation of specific cytosines within the genome. *Nucleic Acids Res*. 2004;32(21):e168.
76. Tost J, Dunker J, Gut IG. Analysis and quantification of multiple methylation variable positions in CpG islands by Pyrosequencing. *Biotechniques*. 2003;35(1):152-156.
77. Toyota M, Ho C, Ahuja N, et al. Identification of differentially methylated sequences in colorectal cancer by methylated CpG island amplification. *Cancer Res*. 1999;59(10):2307-2312.
78. Turner BM, Birley AJ, Lavender J. Histone H4 isoforms acetylated at specific lysine residues define individual chromosomes and chromatin domains in Drosophila polytene nuclei. *Cell*. 1992;69(2):375-384.
79. Wagner KJ, Cooper WN, Grundy RG, et al. Frequent RASSF1A tumour suppressor gene promoter methylation in

Wilms' tumour and colorectal cancer. *Oncogene*. 2002;21(47): 7277-7282.
80. Weber M, Davies JJ, Wittig D, et al. Chromosome-wide and promoter-specific analyses identify sites of differential DNA methylation in normal and transformed human cells. *Nat Genet*. 2005;37(8):853-862.
81. Weber M, Hellmann I, Stadler MB, et al. Distribution, silencing potential and evolutionary impact of promoter DNA methylation in the human genome. *Nat Genet*. 2007;39(4): 457-466.
82. Wei CL, Wu Q, Vega VB, et al. A global map of p53 transcription-factor binding sites in the human genome. *Cell*. 2006;124(1):207-219.
83. Widschwendter A, Ivarsson L, Blassnig A, et al. CDH1 and CDH13 methylation in serum is an independent prognostic marker in cervical cancer patients. *Int J Cancer*. 2004;109(2): 163-166.
84. Wong DJ, Barrett MT, Stoger R, Emond MJ, Reid BJ. p16INK4a promoter is hypermethylated at a high frequency in esophageal adenocarcinomas. *Cancer Res*. 1997;57(13): 2619-2622.
85. Worm J, Aggerholm A, Guldberg P. In-tube DNA methylation profiling by fluorescence melting curve analysis. *Clin Chem*. 2001;47(7):1183-1189.
86. Xiong Z, Laird PW. COBRA: a sensitive and quantitative DNA methylation assay. *Nucleic Acids Res*. 1997;25(12): 2532-2534.
87. Yan PS, Chen CM, Shi H, et al. Dissecting complex epigenetic alterations in breast cancer using CpG island microarrays. *Cancer Res*. 2001;61(23):8375-8380.
88. Yang AS, Estecio MR, Doshi K, Kondo Y, Tajara EH, Issa JP. A simple method for estimating global DNA methylation using bisulfite PCR of repetitive DNA elements. *Nucleic Acids Res*. 2004;32(3):e38.
89. Yin W, Barkess G, Fang X, et al. Histone acetylation at the human beta-globin locus changes with developmental age. *Blood*. 2007;110(12):4101-4107.
90. Young J, Biden KG, Simms LA, et al. HPP1: a transmembrane protein-encoding gene commonly methylated in colorectal polyps and cancers. *Proc Natl Acad Sci USA*. 2001;98(1):265-270.
91. Zou HZ, Yu BM, Wang ZW, et al. Detection of aberrant p16 methylation in the serum of colorectal cancer patients. *Clin Cancer Res*. 2002;8(1):188-191.

Mechanisms of Epigenetic Gene Silencing

3

Marie-Pierre Lambert and Zdenko Herceg

3.1 Introduction

Epigenetic mechanisms are now recognized to play a crucial role in the regulation of fundamental cellular processes, and their dysregulation contributes to human diseases, most notably cancer. DNA sequences encode the primary information within the genome, but it is epigenetic modifications that provide a powerful and complex platform for accurate regulation of the genetic information and for integration of external signals. Epigenetics is therefore becoming a major field of interest to elucidate the molecular mechanisms that underlie fundamental cellular processes. At the same time, knowledge of epigenetics helps to understand the development and progression of malignancy. Human cancer has traditionally been considered primarily as a genetic disease, but recent evidence has made clear that epigenetic abnormalities play an important role in most, if not all, human malignancies; this understanding also adds further complexity to the concept of tumor development. The term "epigenetic" refers to all stable changes of phenotype not coded in the DNA sequence.[6,27,101] Epigenetic signals utilize three distinct mechanisms: DNA methylation, histone modifications, and noncoding RNA. Changes in their processes permit stable transmission of gene activity states in the course of cell division. Alteration of epigenetic events may therefore lead to tumor formation by disrupting gene expression.

There is increasing interest in the possibility that epigenetic mechanisms interact with and integrate environmental signals as part of the cellular adaptation response. Epigenetic mechanisms appear to play a key role in the interaction between environmental factors and the genome.[40,44,107] Thus, adverse and prolonged exposure to environmental, physical, chemical, and infectious agents, as well as lifestyle factors, may induce aberrant epigenetic changes that lead to chronic diseases and neoplastic processes.

This chapter will deal with the epigenetic mechanisms that are involved in gene silencing. Epigenetic gene silencing can be defined as a nonmutational gene inactivation that is faithfully transmitted from precursor cells to clones of daughter cells.[114] This is a dynamic, tightly, and constantly regulated process that assures the equilibrium between stable and transient repression.[92] However, how the stable inheritance of chromatin structure is transmitted in cell division remains unclear. Epigenetic silencing of gene transcription is mediated by a complicated series of molecular events that trigger remodeling of chromatin configuration and leads to covalent modification of DNA (DNA methylation).[36] What follows is an overview of epigenetic events that regulate gene silencing, and of the role that these events play in some fundamental cellular processes. We next discuss the role of environmental exposures in dysregulation of epigenetic mechanisms and discuss how dysregulation of epigenetic-mediated gene silencing may lead to cancer.

M.-P. Lambert (✉)
Epigenetics Group, International Agency for Research on Cancer, Lyon, France
e-mail: lambertmp@students.iarc.fr

H.I. Roach et al. (eds.), *Epigenetic Aspects of Chronic Diseases*,
DOI: 10.1007/978-1-84882-644-1_3, © Springer-Verlag London Limited 2011

3.2 Epigenetic Mechanisms Involved in Gene Silencing

3.2.1 Structure of the Chromatin

All levels of epigenetic regulation depend on the structure of chromatin. Chromatin constitutes the functional template on which genes are expressed in accordance with cellular need. The configuration of chromatin is highly dynamic with both active and inactive chromatin co-existing in the genome. Chromatin, in addition to compacting and thus protecting the DNA molecule, makes possible access of transcriptional machinery to the gene promoter region.[15] Consequently, remodeling of chromatin is the starting point in gene regulation. Two conformations make up the structure of chromatin: euchromatin and heterochromatin; the predominance of one or the other structure is a function of epigenetic signals. Heterochromatin is subdivided into constitutive heterochromatin, i.e., chromatin that is condensed, and facultative heterochromatin, a more relaxed structure. Constitutive heterochromatin is found in the chromosomal regions at the centromeres and telomeres, whereas genes that have been silenced in the course of development form facultative heterochromatin that is interspersed in the chromosome arms. In organisms with large genomes, constitutively heterochromatic regions are also found along the chromosome arms. Because of irregular nucleosome spacing, euchromatin is not condensed, it is relatively gene- rich and transcriptionally active.[24] However, these differences are not always universal. Recent analysis of the human genome has shown that some pericentromeric regions are decondensed and that some euchromatic regions are compacted.[31] Chromatin remodeling is a dynamic process with condensation or relaxation of the molecular structure depending on epigenetic signals, leading to gene silencing or activation, respectively. It is therefore the epigenetic modifiers that play the key role in regulating gene expression, particularly gene silencing.

Regulation of the chromatin structure is essential to maintain genomic stability. The genome of higher eukaryotes contains a large number of repetitive sequences (such as Alu, LINEs, and SINEs). Stable inhibition of retrotransposons – transposable elements and non-coding sequences, commonly known as "junk" DNA – assures genome stability and integrity.[24] Permanent silencing of these DNA sequences is mainly due to epigenetic mechanisms, notably DNA methylation, which tightly regulate chromatin. Whereas transposons must be stable and totally silenced to prevent genomic instability, expression of genes involved in development is subject to permissive epigenetic control.[94] These findings highlight the existence of two distinct mechanisms of expression within the genome, each being the result of chromatin modification. Epigenetic mechanisms also play a critical role in other fundamental processes affecting DNA and gene expression, including DNA repair, cellular differentiation, X chromosome inactivation, genomic imprinting, and protection against viral infection.[12,44,115] Interestingly, the different epigenetic mechanisms appear to interact and reinforce one another in response to environmental or endogenous stimuli.[81]

3.2.2 DNA Methylation

Many fundamental cellular events are the result of modification by epigenetic signals of DNA methylation in the genome.[12] Changes in DNA methylation have been extensively studied because of their frequent association with human disease[126] and their role in major cellular processes, including embryonic development, transcription, chromatin structure, X chromosome inactivation, genomic imprinting, and chromosome stability.[7,36] DNA methylation is a chemical modification that results from the transfer of a methyl group from a methyl donor substrate (called *S*-adenosyl-L-methionine, SAM) that affects only the 5′ position of cytosine bases in CpG conformations ("p" indicates that the cytosine and the guanine are linked by a phosphodiester bond[21]). DNA methylation that involves non-CpG sequences, such as CpNpG or CpA and CpT, also occur in the eukaryotic genome,[20] especially in mouse embryonic stem cells (ESCs)[61,91]; the specific functional role of non-CpG methylation is not clear.

DNA methylation is a dynamic process that comprises de novo methylation, demethylation, and the maintenance of existing DNA methylation, all acting on gene expression (Fig. 3.1).[12] DNA methylation patterns are not distributed randomly throughout the genome.[6] In the human genome, they are found more frequently within small regions of DNA called CpG islands.[11] These CpG-rich islands are associated with

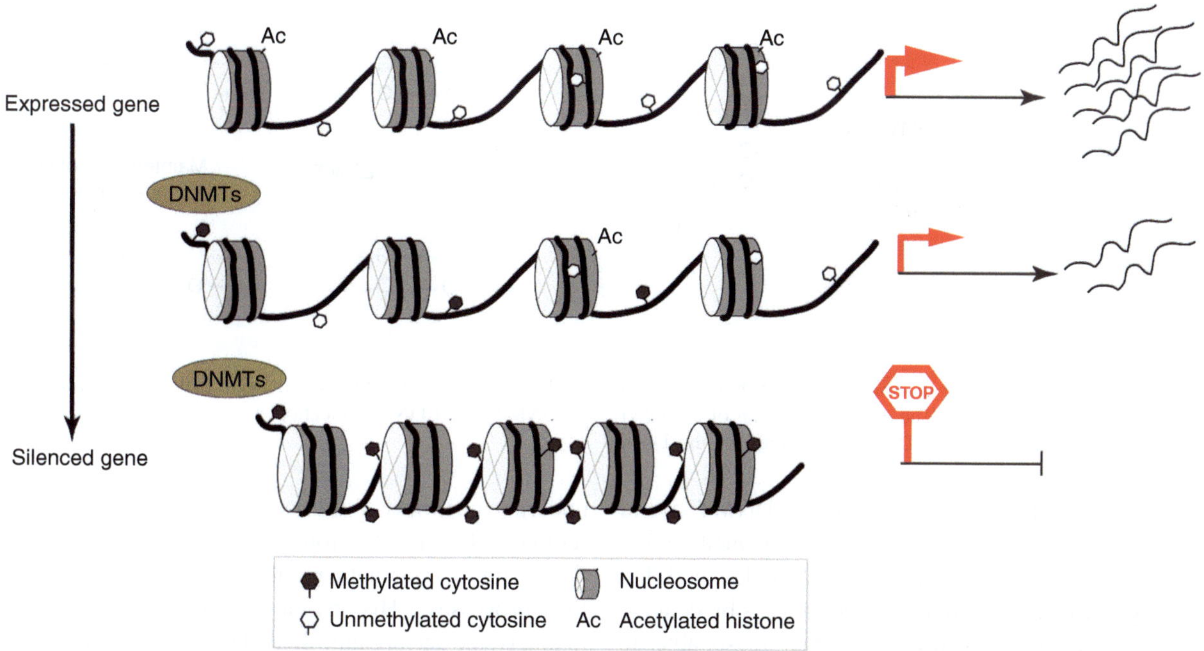

Fig. 3.1 DNA methylation promoter establishment is a progressive process in tumor cells that leads to gene silencing. Promoters of normally expressed genes do not exhibit methylated CpG sites. Progressive hypermethylation of these CpG sites leads to heterochromatin conformation and gene silencing. The level of CpG methylation is inversely correlated to the histone acetylation status

promoter regions and do not exhibit methylation. Aberrant DNA hypermethylation of a CpG island typically involves inhibition of gene transcription and unscheduled gene silencing.[6] Methyl groups are added by enzymes called DNA methyltransferases (DNMTs).[8,9] In mammals, three major DNMTs are identified: DNMT1, DNMT3a, and DNMT3b.[6] DNA methylation maintenance is performed by DNMT1, with hemimethylated DNA as a substrate.[101] Indeed, during replication, DNMT1 transfers the DNA methylation profile from the "parental" to the "daughter" strand; this process maintains accurate DNA methylation pattern over many cell generations.[33] DNMT1 activity is increased in the presence of hemimethylated DNA, compared to activity on unmethylated DNA.[90] The disruption of the DNMT1 gene in the mouse germ line leads to embryonic death.[60] This highlights the crucial role DNA methylation plays in developmental processes.[59] De novo DNA methylation is triggered by the DNA methyltransferases DNMT3a and DNMT3b, which methylate unmethylated CpG sites.[82] These two DNMTs therefore assume a major role in establishing DNA methylation patterns in the course of development. Both DNMT3a and DNMT3b are highly expressed in ESCs and in embryo and germ cells, where de novo DNA methylation takes place. These two methyltransferases are, however, less expressed in somatic cells, where DNMT1 predominates.[83] In contrast to DNMT1, DNMT3a and DNMT3b have no affinity for hemimethylated DNA,[83] but can directly methylate unmethylated CpG sites. DNMT3a and DNMT3b are thus responsible for setting up DNA methylation profiles which are then maintained by DNMT1 in the course of DNA replication (Fig. 3.2). DNMT3a is involved in genomic imprinting during gametogenesis,[50] whereas DNMT3b methylates also repetitive sequences.[36,51] De novo DNA methyltransferase inactivation in embryonic stem (ES) cells results in[9] progressive loss of methylation in various repetitive sequences and single-copy genes.[19] Two additional DNMTs, DNMT2 and DNMT3L, have also been characterized. Even though DNMT2 has been highly conserved throughout evolution,[106] its role is still not understood. The DNA methyltransferase activity of DMNT2 is in fact relatively weak; this suggests a rather different function. Recent studies have described strong RNA methyltransferase activity for DNMT2.[32,49,105] This protein may therefore play a major biological role.[45] DNMT3L, a member of DNMT3 family, does not exhibit catalytic activity,[126] but is involved in genomic imprinting through its

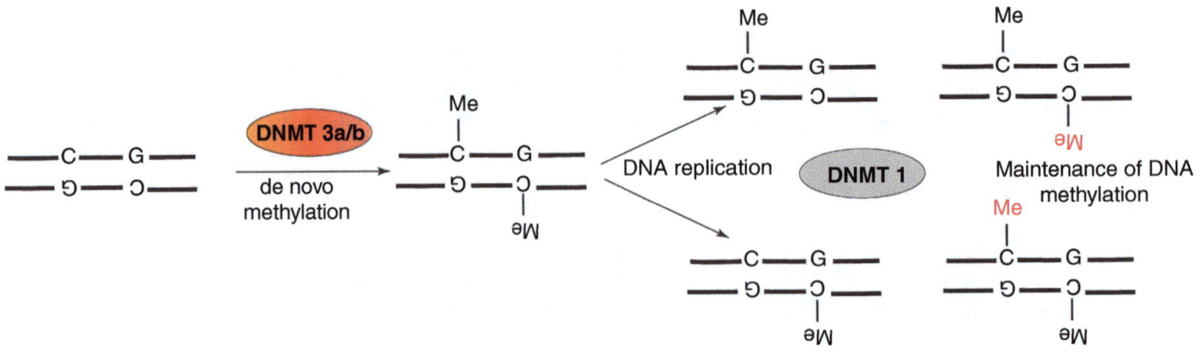

Fig. 3.2 Establishment and transmission of DNA methylation pattern during cell division. The de novo addition of a methyl group to the cytosine base in DNA is catalyzed by the DNA methyltransferases DNMT3a and DNMT3b and is maintained by DNMT1 after DNA replication

interaction with and enhancement of DNMT3a and DNMT3b.[39,109,124] Maintenance of DNA methylation patterns by DNMT1 is relatively well understood, but the significance of methyl cytosine patterns by de novo methylation is not known. It seems reasonable to hypothesize that a given DNA sequence plays a role in targeting DNA methyltransferases. How the primary sequence determines and "guides" cytosine methylation is not known. DNMT3a and DNMT3b enzymes exhibit specificity for de novo methylation of CpG sites, yet cannot discriminate between primary DNA sequences.[19] De novo methylation may therefore be regulated by different mechanisms.

3.2.3 Histone Marks and Gene Silencing

DNA methylation, its association with the DNA sequence notwithstanding, is not sufficient to silence a gene. As discussed above, gene regulation and genomic stability are due to DNA being compacted into chromatin. Regulation of gene expression requires a dynamic equilibrium between the regulatory elements; this involves the promoter regions, chromatin structure, and access for the transcriptional regulator factors.[15] Histones, the principal proteins of chromatin, consist of a highly compact, globular core of alpha-helices arranged in helix-turn-helix motifs that promote oligomerization. Histones are organized within fundamental units of chromatin, nucleosomes, containing 146 bp of DNA wrapped twice around an octamer composed of two copies of each of the following histone proteins: H2A, H2B, H3, and H4.[10] The structure of the nucleosome is well characterized.[18,65,127] Histones undergo chemical modification in the N-terminal region, called the "histone tail"; this constitutes the major site for epigenetic regulation of fundamental cellular processes.[41] Histone modifications include acetylation, lysine and arginine methylation, phosphorylation, ubiquitination, sumoylation, ADP ribosylation, deamination, and proline isomerization, all of which constitute the "histone code." This code may extend the information encoded in the DNA sequence to regulate processes such as transcription, DNA repair, and replication.[10,54] Histone acetylation is largely associated with open chromatin conformation and active transcription. Histone modifications are transmitted to daughter cells and can therefore be considered as a heritable epigenetic mechanism.[104,115]

Histone modifications, as produced by histone acetyltransferases (HATs), histone methyltransferases (HMTs), and protein kinases, are the principal mechanisms assuring dynamic equilibrium of the chromatin structure. Histone modifications thus provide the histone marks that bring about a balance between repressive and active signals and thus assume dose regulation of cellular processes. Here, we will focus on the histone marks commonly associated with gene silencing.

Gene silencing is mainly associated with hypoacetylation and methylation of specific lysines of histones H3 (H3K9, H3K27).[34,97,98,115] Due to its major role in the transcriptional regulation, histone acetylation has been studied extensively.[69,100] Histone acetylation is a covalent modification of specific lysines on the four core histones that is thought to neutralize the positive charge of the target lysine. Histone acetylation is catalyzed by HATs and is reversible via histone deacetyltransferases (HDACs). Thus, HAT activates and HDACs repress gene expression. Histone acetylation is associated with open chromatin structure and an accessible DNA sequence. This open structure makes possible the recruitment of transcriptional activators. In turn, removal of acetyl groups by HDAC activity leads to a closed conformation of the chromatin and to gene silencing.[101]

Gene silencing is associated with methylation of histones H3K9 and H3K27. Histone methylation involves the addition of one, two, or three methyl groups on specific lysine and arginine residues of histones H3 and H4.[10] Addition of each methyl group is mediated by a different HMT. In mammals, H3K9 methylation is mainly catalyzed by SUV39H-1 and -2, SET domain proteins.[93] This in turn induces recruitment of the transcriptional repressor HP1, and results in an inactive chromatin conformation and thus in gene silencing.

EZH2 (enhancer of zeste), another SET domain protein, methylates H3K27. EZH2 is a member of the polycomb repressor complex 2 (PRC2), including EED and SU(Z)12. When PRC2 is recruited via methylated H3K27, gene silencing is initiated. PRC2 is associated with HDACs (via EED)[117] and DNMTs (via EZH2).[118] These interactions suggest that a connection exists between histone acetylation and methylation, on the one hand, and DNA methylation, on the other.

Methylation of H3K4, H3K36, or H3K79 leads to gene activation, whereas methylation of H3K9 and H3K27 or H4K20 leads to gene silencing. This highlights the complexity of the histone code.[116]

3.2.4 Post-transcriptional Silencing: miRNA

The role of microRNAs (miRNAs) in the regulation of gene expression has attracted significant attention. The eukaryotic genome encodes a number of short non-coding transcripts whose precise functions are still unclear. The presence of these post-transcriptional regulators constitutes a new layer of complexity in the regulation of epigenetic gene silencing, but also promises to add to our knowledge of gene silencing mechanisms. Importantly, many chronic diseases, including cancer, have now been associated with dysregulation of miRNA expression.[5,85,87,88,120]

In contrast to DNA methylation or histone modifications, RNA-mediated gene silencing is a post-transcriptional mechanism that reinforces gene silencing. miRNAs are short (20–23 nucleotides) single-stranded RNAs that reduce gene expression.[3] miRNA biogenesis involves multiple steps and several epigenetic factors (Fig. 3.3).[121]

Gene encoding for the miRNA is transcribed by RNA polymerases II or III[14,56] into a precursor RNA (pri-miRNA) with a hairpin conformation. After editing, the pri-miRNA is cleaved by the nuclear microprocessor complex formed by Drosha and DGCR8 (also called Pasha in *Drosophila melanogaster*), a complex with RNAse III activity.[35,37] DGCR8 interacts directly with the pri-miRNA through two double-stranded RNA-binding domains and defines the precise cleavage site. Drosha, on the other hand, cleaves the pri-miRNA at the 5′ and 3′ arms.[38,121] After cleavage, the resulting nuclear pre-miRNA is exported through Exportin-5–Ran-GTP to the cytoplasm, where it is cleaved by the RNase III enzyme Dicer, which forms a complex with the double-stranded RNA-binding protein TRBP. Cleavage generates a miRNA duplex. After degradation of the complementary strand, the functional strand of the mature miRNA is loaded with Argonaute (Ago2) proteins onto the RNA-induced silencing complex (RISC), where it guides RISC to target messenger RNAs and induces gene silencing.[25,72] Once processed, the miRNAs interact directly with the 3′UTR sequence of the targeted mRNA,[4,70] leading to mRNA degradation or translational inhibition. Because of their short length, miRNAs act on many targets genes, including transcription factors downregulating a number of genes. As most miRNAs are expressed in normal cells, they are crucial for the regulation of fundamental cellular processes and their level of expression is tightly regulated. miRNAs are implicated in essential cellular processes, including cell cycle regulation[1,62,123] and apoptosis.[48] In cancers, miRNA expression is globally downregulated.[31] Interestingly, miRNA expression may be downregulated through epigenetic mechanisms.[66] Indeed, miRNA-124a undergoes epigenetic silencing via CpG island DNA hypermethylation in cancer cells; this leads to loss of miRNA-124a expression and activates the oncogenic factor Cyclin D6 kinase.[67] These findings underscore the complexity and interdependence of the epigenetic machineries and their regulation.

3.2.5 Cross talk Between Epigenetic Systems Involved in Gene Silencing

It is now accepted that the equilibrium between active and repressive marks in the chromatin structure is

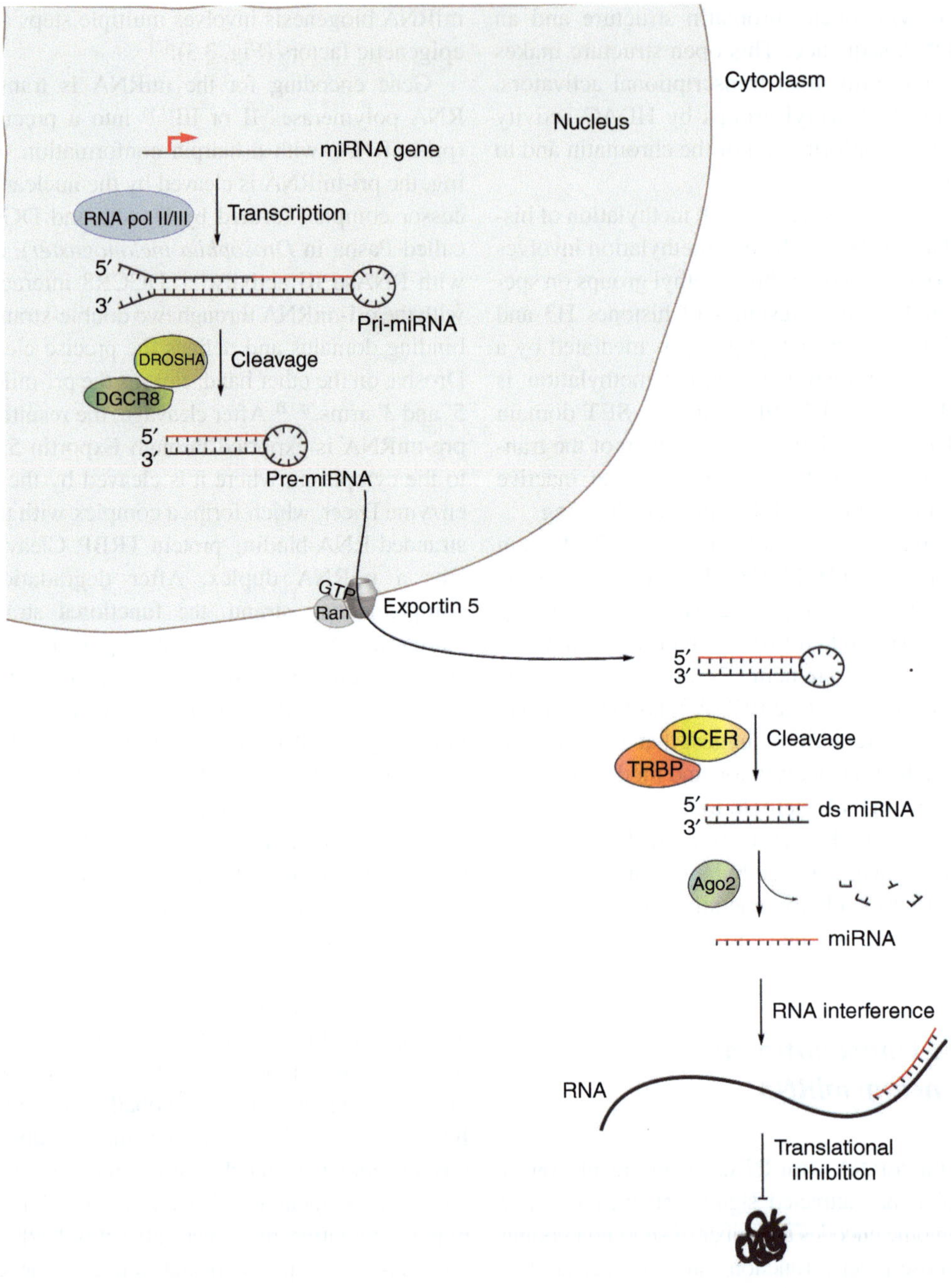

Fig. 3.3 miRNA biogenesis and effect on gene expression. The miRNA machinery is a multistep process. The hairpin primary miRNA (pri-miRNA) is transcribed by polymerases II or III, and cleaved into a precursor miRNA (pre-miRNA) through the Drosha/DGCR8 complex. The pre-miRNA is exported into the cytoplasm via the Exportin5-Ran transporter. The complex DICER/TRBP then cleaves the pre-miRNA into its mature form. The functional strand, together with Argonaute (Ago2), is loaded onto the RNA-induced silencing complex (RISC), with RISC guided to the targeted mRNA to silence the gene

strongly affected by the interplay between epigenetic mechanisms. However, how the different epigenetic signals modulate the balance between stable and transient gene silencing is still subject to debate. Similarly, the order of events that initiate gene silencing remains to be elucidated.

3.2.5.1 Initiation of Gene Silencing: What Is the First Step?

As discussed above, gene silencing is typically associated with global deacetylation and increased DNA methylation. However, little is known about the sequential cascade of events that induce and maintain the repressive marks. DNA methylation was thought to be the initial event for gene silencing either by interfering with the binding of transcriptional factors and/or by inducing inactive chromatin conformation.[13,42] However, current thinking favors the scenario in which loss of histone acetylation acts as the primary step in gene silencing. Histone deacetylation promotes DNA methylation, which is the irrevocable silencing mark.[115] Both hypotheses (Fig. 3.4) have experimental support. This reinforces the possibility that the order of epigenetic events during gene silencing is context-dependent. Several studies have shown that the DNA CpG hypermethylation occurs first (hypothesis 1), leading to recruitment of methyl CpG binding proteins (MeCP1 and 2) at the promoter region.[71,77] MeCP2 is associated with catalytically active deacetylase complexes, including HDAC1 and HDAC2, which in turn are recruited to the promoter. This leads to histone deacetylation.[47,78] Histone methyltransferases (SUV39H1), also recruited to the promoter, initiate H3K9 methylation and HP1 recruitment, to silence the targeted gene.[28,29]

Histone acetylation can also be viewed as a mechanism of protection against aberrant DNA methylation (hypothesis 2). Several studies support the idea that histone acetylation on a transcriptionally active region, by inhibiting DNMTs, prevents DNA methylation.[22,75] In other words, histone acetylation defects may precede and induce DNA methylation during gene silencing. Many studies in different models support this concept.[57,76,111] For instance, silencing of the RASSF1A gene starts with deacetylation of histone H3 and methylation of H3K9, with both events preceding DNA methylation.[108] Further studies are needed, in as much as the relationships between DNA methylation and histone modification bear on our understanding of normal development and of somatic cell reprogramming and tumorigenesis.[97]

3.2.5.2 Connection Between Epigenetic Players: Gordian Knot?

Direct and indirect experimental evidence suggests that DNA methylation and histone marks can crosstalk and interact during gene silencing, non-coding RNAs impacting on these epigenetic events. For example, the mir-29 family interacts with the de novo methyltransferases DNMT3a and DNMT3b through the 3′-UTR sequence, inducing downregulation of de novo DNA methylation activity.[26] The downregulation of the mir-29 family expression in lung cancer causes upregulation of DNMT3a and DNMT3b and an aberrant DNA methylation profile.[26]

Ling and colleagues[61] have shown that another mechanism involving sumoylation (SUMO-1) modifies the interaction of DNMT3a with histone deacetylases (HDACs), thus altering its repression of transcription.[60] Because, as discussed above, miRNA gene expression is also regulated via promoter DNA methylation,[66] there may exist a functional connection between miRNAs and histones. As Barski et al.[2] recently reported, the profile of histone marks in miRNA genes is similar to that of protein-coding genes; miRNA expression may therefore be modulated by histone marks.[2]

3.3 Gene Silencing in Fundamental Cellular Processes

3.3.1 Epigenetic Plasticity During Developmental Processes

Development has by definition a strong epigenetic component because identical pluripotent cells differentiate into different tissues and organs. Indeed, embryonic development is a clear example of how genes are activated and silenced with time and how, at the same time a balance is achieved between transient and stable gene repression. In early development, pluripotency-related genes are associated with active histone marks, whereas genes that are involved in the differentiation process are transiently repressed through repressive histone marks until differentiation is initiated.[94] The pluripotent and indefinite self-renewal of ESCs are regulated by specific genetic factors, including Nanog, Oct4 or Sox2,[84] and chromatin remodeling factors.[73,74] The chromatin structure of ESCs as such allows increased access to "self-renewal" factors. The change from self-renewal to differentiation parallels changes in epigenetic chromatin marks due to silencing-associated histone H3K9 dimethylation and trimethylation (H3K9Me2/Me3) marks and a diminution in levels of acetylated histones H3 and H4.[52,55] Loh et al.[64] have shown that Jmjd1a and Jmjd2c histone H3 Lys 9 demethylases are important for self-renewal. Remarkably, both are upregulated in mouse ESCs, thus preventing HP1, the transcriptional repressor, from binding to the

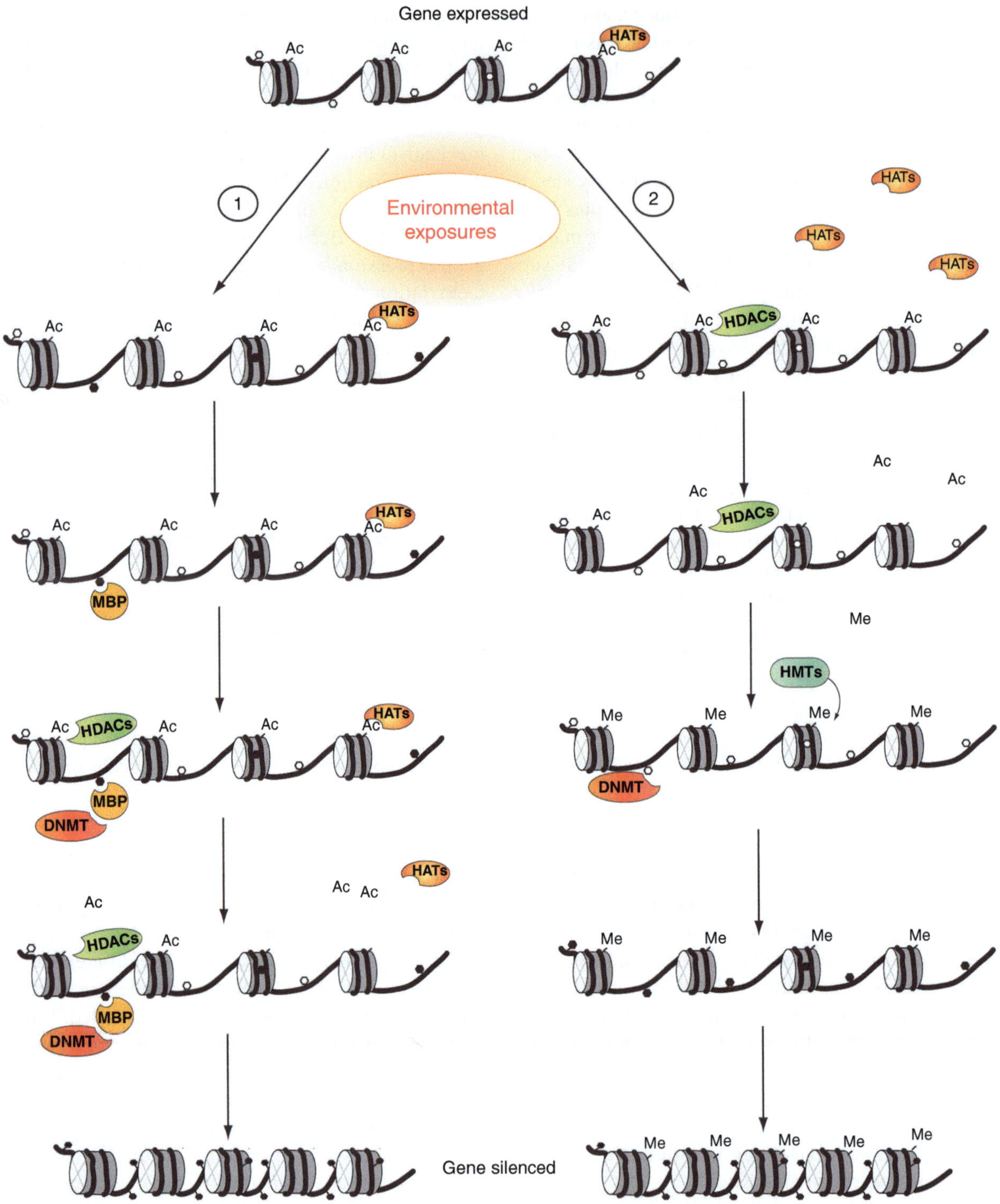

Fig. 3.4 Initiation of gene silencing is due to one of two possible sequences of events. (1) Environmental signals may trigger gene silencing by inducing partial, then total methylation of the DNA at specific CpG sites through DNA methyltransferase proteins (DNMTs). The binding of methyl-binding proteins (MBP) to the methylated CpG sites will trigger histone deacetylase recruitment (HDAC) which will make gene silencing permanent. (2) In an alternate hypothesis, gene silencing is induced through loss of histone acetylation, which triggers DNA methylation. The hypoacetylated chromatin is recognized by DNMTs which methylate CpG sites, with gene silencing the result

Nanog promoter region.[63] When differentiation has occurred, the chromatin architecture becomes less accessible to self-renewal factors, with chromatin remodeling favoring the expression of developmental genes.[94]

3.3.2 Genomic Imprinting

Genomic imprinting is defined as an epigenetic mechanism that permits parental-specific gene expression.[89,96] It depends on an epigenetic marking of parental alleles during gametogenesis and is largely dependent on DNA methylation marks,[58] which are established during embryological development of germ cells. Expression of only one allele guarantees the proper levels of the proteins encoded by the imprinted genes. This is important for embryonic and placental development, and for the metabolism.[95] Inadequately regulated expression results in developmental abnormalities, exemplified by hereditary overgrowth syndromes, as in the Beckwith–Wiedemann or the Prader–Willy syndrome. Given DNMT1 has an affinity for hemimethylated DNA, the maintenance of differentially methylated pattern is likely to be based on a specific mechanism. Imprinting therefore is interesting because it demonstrates how epigenetic events affect normal Mendelian events. The effects of imprinting, beyond causing asymmetric expression, are found throughout clusters.[23] This suggests that imprinted genes within a cluster are subject to the same regulatory elements. Indeed, imprinted genes are regulated through differentially methylated control regions called "imprinting control regions" (ICR) that affect gene expression in cis.[23] The partial silencing of these imprinted genes is not only based on DNA methylation pattern. Long non-coding RNAs that lead to gene silencing have been described for several loci.[102,103,112] For instance, the Kcnq1 ICR, located in intron 10 of the Kcnq1 gene, is unmethylated on the paternal chromosome and methylated on the maternal chromosome. The unmethylated Kcnq1 ICR allows expression of an antisense RNA, Kcnq1ot1, which overlaps the Kcnq1 coding region; this leads to its bidirectional silencing.[112] Both alleles often do not exhibit the same histone modification pattern.[53] Acetylated histone H3 and H4 and methylated lysine 4 of H3 (H3-K4Me) are associated with transcriptionally active alleles, whereas tri-methylated lysine 9 of H3 (H3-K9Me3) marks are localized on silenced alleles.[119]

3.3.3 X Chromosome Inactivation

Epigenetic mechanisms also play a major role in genetic sex determination, an elaborate mechanism based on the precise assignment of X-linked genes to males and females.[68] For this to occur, one randomly selected female X chromosome is silenced in the cells that have originated from the postimplantation epiblast. This process involves counting, choice, and mutually exclusive silencing and is regulated by a genomic locus termed the "X-inactivation center" (XIC), a locus that contains multiple non-coding genes, including Xist, Tsix, and Xite.[86] Interestingly, Xist RNA upregulation is quickly followed by several chromatin modifications, including the recruitment of Polycomb Repressive Complexes 1 (PRC1) and 2 (PRC2) to the inactivated X chromosome.[64,79,122] Although an association between genomic imprinting and X inactivation has not been reported, imprinting of the paternal X chromosome has been shown.[113] More studies are needed to elucidate the interaction with epigenetic systems.

3.4 Gene Silencing in Cancer

Epigenetic silencing mechanisms play a major role in multiple cellular processes, and their dysregulation can lead to tumor development. Epigenetic changes are gradual and progressive and may constitute the mechanism by which environmental exposure in early life causes changes that lead to chronic disease susceptibility, especially cancer.[40,46] Cancer cells have defects in DNA methylation, histone modification, or miRNA expression and are characterized by global hypomethylation and promoter-specific hypermethylation. Indeed, CpG islands that are typically unmethylated in normal cells are frequently hypermethylated in cancer cells, with abnormal silencing of tumor suppressor and other cancer-associated genes.[6,16,43,110,125] Dysregulation of histone acetylation has also been reported in a variety of cancers.[17,22,99] Increasingly, miRNAs have become associated with cancer, as they can act as tumor suppressors (by inhibiting oncogenes) or as oncogenes (by inhibiting tumor suppressor genes, oncomiRs).[30,80] Thus, epigenetic changes are seen to constitute early events in carcinogenesis, which can promote later genetic changes. For example, aberrant DNA hypermethylation and histone deacetylation of DNA repair genes can induce genomic instability and increase mutation rates. As dysregulation of epigenetic events in cancer becomes better understood, opportunities may arise to improve cancer detection and to develop therapeutic strategies, based on the reversible nature of epigenetic changes.

3.5 Summary and Conclusions

Gene transcription is a chromatin-based process that is modulated not only by DNA and DNA-associated proteins, but also by epigenetic mechanisms. Chromatin is a dynamic structure that depends on a combination of specific epigenetic factors to maintain the proper pattern of active and silenced genes. Inhibition of gene expression depends on signaling by specific factors, including non-coding RNAs. It is therefore obvious that gene silencing is not a random process, but one that is closely regulated.

Gene silencing involves a number of epigenetic players, including histone acetylases, methyltransferases, DNA methyltransferases and their antagonists, histone deacetylases, demethylases. These induce specific activating and repressive marks, thereby assuring the coexistence in the genome of expressed and repressed genes. Specific epigenetic mechanisms have been well characterized, but their interaction and hierarchical are far from being understood. Moreover, epigenetic networks may undergo change in the course of life and be modulated by environmental factors. Exposure to environmental stimuli may modify the epigenetic infrastructure to a point where the resulting genetic and epigenetic defects bring about disease, cancer included.

In conclusion, because gene silencing plays a fundamental role in cellular processes and because unscheduled gene silencing results in human diseases, further studies on epigenetic mechanisms that underlie normal and aberrant gene silencing will advance our knowledge of the molecular mechanisms of human diseases and provide information for the development of therapeutic strategies.

References

1. Bandi N, Zbinden S, Gugger M, et al. miR-15a and miR-16 are implicated in cell cycle regulation in a Rb-dependent manner and are frequently deleted or down-regulated in non-small cell lung cancer. *Cancer Res*. 2009;69:5553-5559.
2. Barski A, Jothi R, Cuddapah S, et al. Chromatin poises miRNA- and protein-coding genes for expression. *Genome Res*. 2009;19:1742-1751.
3. Bartel DP. MicroRNAs: genomics, biogenesis, mechanism, and function. *Cell*. 2004;116:281-297.
4. Bartel DP. MicroRNAs: target recognition and regulatory functions. *Cell*. 2009;136:215-233.
5. Bartels CL, Tsongalis GJ. MicroRNAs: novel biomarkers for human cancer. *Clin Chem*. 2009;55:623-631.
6. Baylin SB. DNA methylation and gene silencing in cancer. *Nat Clin Pract Oncol*. 2005;2(Suppl 1):S4-S11.
7. Baylin SB, Esteller M, Rountree MR, Bachman KE, Schuebel K, Herman JG. Aberrant patterns of DNA methylation, chromatin formation and gene expression in cancer. *Hum Mol Genet*. 2001;10:687-692.
8. Bestor TH. Cloning of a mammalian DNA methyltransferase. *Gene*. 1988;74:9-12.
9. Bestor TH. The DNA methyltransferases of mammals. *Hum Mol Genet*. 2000;9:2395-2402.
10. Bhaumik SR, Smith E, Shilatifard A. Covalent modifications of histones during development and disease pathogenesis. *Nat Struct Mol Biol*. 2007;14:1008-1016.
11. Bird AP. CpG-rich islands and the function of DNA methylation. *Nature*. 1986;321:209-213.
12. Bird A. DNA methylation patterns and epigenetic memory. *Genes Dev*. 2002;16:6-21.
13. Bird AP, Wolffe AP. Methylation-induced repression – belts, braces, and chromatin. *Cell*. 1999;99:451-454.
14. Borchert GM, Lanier W, Davidson BL. RNA polymerase III transcribes human microRNAs. *Nat Struct Mol Biol*. 2006;13:1097-1101.
15. Cairns BR. The logic of chromatin architecture and remodelling at promoters. *Nature*. 2009;461:193-198.
16. Calvisi DF, Ladu S, Gorden A, et al. Mechanistic and prognostic significance of aberrant methylation in the molecular pathogenesis of human hepatocellular carcinoma. *J Clin Invest*. 2007;117:2713-2722.
17. Cang S, Feng J, Konno S, et al. Deficient histone acetylation and excessive deacetylase activity as epigenomic marks of prostate cancer cells. *Int J Oncol*. 2009;35:1417-1422.
18. Caterino TL, Hayes JJ. Chromatin structure depends on what's in the nucleosome's pocket. *Nat Struct Mol Biol*. 2007;14:1056-1058.
19. Chen T, Ueda Y, Dodge JE, Wang Z, Li E. Establishment and maintenance of genomic methylation patterns in mouse embryonic stem cells by Dnmt3a and Dnmt3b. *Mol Cell Biol*. 2003;23:5594-5605.

20. Clark SJ, Harrison J, Frommer M. CpNpG methylation in mammalian cells. *Nat Genet*. 1995;10:20-27.
21. Doerfler W. DNA methylation and gene activity. *Annu Rev Biochem*. 1983;52:93-124.
22. Dong E, Guidotti A, Grayson DR, Costa E. Histone hyperacetylation induces demethylation of reelin and 67-kDa glutamic acid decarboxylase promoters. *Proc Natl Acad Sci USA*. 2007;104:4676-4681.
23. Edwards CA, Ferguson-Smith AC. Mechanisms regulating imprinted genes in clusters. *Curr Opin Cell Biol*. 2007;19: 281-289.
24. Elgin SC, Grewal SI. Heterochromatin: silence is golden. *Curr Biol*. 2003;13:R895-R898.
25. Eulalio A, Huntzinger E, Izaurralde E. Getting to the root of miRNA-mediated gene silencing. *Cell*. 2008;132:9-14.
26. Fabbri M, Garzon R, Cimmino A, et al. MicroRNA-29 family reverts aberrant methylation in lung cancer by targeting DNA methyltransferases 3A and 3B. *Proc Natl Acad Sci USA*. 2007;104:15805-15810.
27. Feinberg AP, Ohlsson R, Henikoff S. The epigenetic progenitor origin of human cancer. *Nat Rev Genet*. 2006;7:21-33.
28. Fuks F, Hurd PJ, Deplus R, Kouzarides T. The DNA methyltransferases associate with HP1 and the SUV39H1 histone methyltransferase. *Nucleic Acids Res*. 2003;31:2305-2312.
29. Fuks F, Hurd PJ, Wolf D, Nan X, Bird AP, Kouzarides T. The methyl-CpG-binding protein MeCP2 links DNA methylation to histone methylation. *J Biol Chem*. 2003;278:4035-4040.
30. Gartel AL, Kandel ES. RNA interference in cancer. *Biomol Eng*. 2006;23:17-34.
31. Gilbert N, Boyle S, Fiegler H, Woodfine K, Carter NP, Bickmore WA. Chromatin architecture of the human genome: gene-rich domains are enriched in open chromatin fibers. *Cell*. 2004;118:555-566.
32. Goll MG, Kirpekar F, Maggert KA, et al. Methylation of tRNAAsp by the DNA methyltransferase homolog Dnmt2. *Science*. 2006;311:395-398.
33. Goyal R, Reinhardt R, Jeltsch A. Accuracy of DNA methylation pattern preservation by the Dnmt1 methyltransferase. *Nucleic Acids Res*. 2006;34:1182-1188.
34. Grandjean V, O'Neill L, Sado T, Turner B, Ferguson-Smith A. Relationship between DNA methylation, histone H4 acetylation and gene expression in the mouse imprinted Igf2-H19 domain. *FEBS Lett*. 2001;488:165-169.
35. Gregory RI, Yan KP, Amuthan G, et al. The microprocessor complex mediates the genesis of microRNAs. *Nature*. 2004;432:235-240.
36. Gronbaek K, Hother C, Jones PA. Epigenetic changes in cancer. *APMIS*. 2007;115:1039-1059.
37. Han J, Lee Y, Yeom KH, Kim YK, Jin H, Kim VN. The Drosha–DGCR8 complex in primary microRNA processing. *Genes Dev*. 2004;18:3016-3027.
38. Han J, Lee Y, Yeom KH, et al. Molecular basis for the recognition of primary microRNAs by the Drosha–DGCR8 complex. *Cell*. 2006;125:887-901.
39. Hata K, Okano M, Lei H, Li E. Dnmt3L cooperates with the Dnmt3 family of de novo DNA methyltransferases to establish maternal imprints in mice. *Development*. 2002;129: 1983-1993.
40. Herceg Z. Epigenetics and cancer: towards an evaluation of the impact of environmental and dietary factors. *Mutagenesis*. 2007;22:91-103.
41. Herceg Z, Hainaut P. Genetic and epigenetic alterations as biomarkers for cancer detection, diagnosis and prognosis. *Mol Oncol*. 2007;1:26-41.
42. Iguchi-Ariga SM, Schaffner W. CpG methylation of the cAMP-responsive enhancer/promoter sequence TGACGTCA abolishes specific factor binding as well as transcriptional activation. *Genes Dev*. 1989;3:612-619.
43. Issa JP. CpG island methylator phenotype in cancer. *Nat Rev Cancer*. 2004;4:988-993.
44. Jaenisch R, Bird A. Epigenetic regulation of gene expression: how the genome integrates intrinsic and environmental signals. *Nat Genet*. 2003;33(Suppl):245-254.
45. Jeltsch A, Nellen W, Lyko F. Two substrates are better than one: dual specificities for Dnmt2 methyltransferases. *Trends Biochem Sci*. 2006;31:306-308.
46. Jirtle RL, Skinner MK. Environmental epigenomics and disease susceptibility. *Nat Rev Genet*. 2007;8:253-262.
47. Jones PL, Veenstra GJ, Wade PA, et al. Methylated DNA and MeCP2 recruit histone deacetylase to repress transcription. *Nat Genet*. 1998;19:187-191.
48. Jovanovic M, Hengartner MO. miRNAs and apoptosis: RNAs to die for. *Oncogene*. 2006;25:6176-6187.
49. Jurkowski TP, Meusburger M, Phalke S, et al. Human DNMT2 methylates tRNA(Asp) molecules using a DNA methyltransferase-like catalytic mechanism. *RNA*. 2008;14: 1663-1670.
50. Kaneda M, Okano M, Hata K, et al. Essential role for de novo DNA methyltransferase Dnmt3a in paternal and maternal imprinting. *Nature*. 2004;429:900-903.
51. Kato Y, Kaneda M, Hata K, et al. Role of the Dnmt3 family in de novo methylation of imprinted and repetitive sequences during male germ cell development in the mouse. *Hum Mol Genet*. 2007;16:2272-2280.
52. Keohane AM, O'Neill LP, Belyaev ND, Lavender JS, Turner BM. X-Inactivation and histone H4 acetylation in embryonic stem cells. *Dev Biol*. 1996;180:618-630.
53. Koerner MV, Pauler FM, Huang R, Barlow DP. The function of non-coding RNAs in genomic imprinting. *Development*. 2009;136:1771-1783.
54. Kouzarides T. Chromatin modifications and their function. *Cell*. 2007;128:693-705.
55. Lee JH, Hart SR, Skalnik DG. Histone deacetylase activity is required for embryonic stem cell differentiation. *Genesis*. 2004;38:32-38.
56. Lee Y, Kim M, Han J, et al. MicroRNA genes are transcribed by RNA polymerase II. *EMBO J*. 2004;23:4051-4060.
57. Lehnertz B, Ueda Y, Derijck AA, et al. Suv39h-mediated histone H3 lysine 9 methylation directs DNA methylation to major satellite repeats at pericentric heterochromatin. *Curr Biol*. 2003;13:1192-1200.
58. Li E, Beard C, Jaenisch R. Role for DNA methylation in genomic imprinting. *Nature*. 1993;366:362-365.
59. Li E, Bestor TH, Jaenisch R. Targeted mutation of the DNA methyltransferase gene results in embryonic lethality. *Cell*. 1992;69:915-926.
60. Ling Y, Sankpal UT, Robertson AK, McNally JG, Karpova T, Robertson KD. Modification of de novo DNA methyltransferase 3a (Dnmt3a) by SUMO-1 modulates its interaction with histone deacetylases (HDACs) and its capacity to repress transcription. *Nucleic Acids Res*. 2004;32: 598-610.

61. Lister R, Pelizzola M, Dowen RH, et al. Human DNA methylomes at base resolution show widespread epigenomic differences. *Nature*. 2009;462(7271):315-322.
62. Liu M, Wu H, Liu T, et al. Regulation of the cell cycle gene, BTG2, by miR-21 in human laryngeal carcinoma. *Cell Res*. 2009;19:828-837.
63. Loh YH, Zhang W, Chen X, George J, Ng HH. Jmjd1a and Jmjd2c histone H3 Lys 9 demethylases regulate self-renewal in embryonic stem cells. *Genes Dev*. 2007;21:2545-2557.
64. Lucchesi JC, Kelly WG, Panning B. Chromatin remodeling in dosage compensation. *Annu Rev Genet*. 2005;39: 615-651.
65. Luger K, Mader AW, Richmond RK, Sargent DF, Richmond TJ. Crystal structure of the nucleosome core particle at 2.8 A resolution. *Nature*. 1997;389:251-260.
66. Lujambio A, Esteller M. CpG island hypermethylation of tumor suppressor microRNAs in human cancer. *Cell Cycle*. 2007;6:1455-1459.
67. Lujambio A, Ropero S, Ballestar E, et al. Genetic unmasking of an epigenetically silenced microRNA in human cancer cells. *Cancer Res*. 2007;67:1424-1429.
68. Lyon MF. Gene action in the X-chromosome of the mouse (*Mus musculus* L.). *Nature*. 1961;190:372-373.
69. Marmorstein R, Roth SY. Histone acetyltransferases: function, structure, and catalysis. *Curr Opin Genet Dev*. 2001; 11:155-161.
70. Mayr C, Bartel DP. Widespread shortening of 3′UTRs by alternative cleavage and polyadenylation activates oncogenes in cancer cells. *Cell*. 2009;138:673-684.
71. Meehan RR, Lewis JD, Bird AP. Characterization of MeCP2, a vertebrate DNA binding protein with affinity for methylated DNA. *Nucleic Acids Res*. 1992;20:5085-5092.
72. Meister G, Landthaler M, Patkaniowska A, Dorsett Y, Teng G, Tuschl T. Human Argonaute2 mediates RNA cleavage targeted by miRNAs and siRNAs. *Mol Cell*. 2004;15: 185-197.
73. Meshorer E, Misteli T. Chromatin in pluripotent embryonic stem cells and differentiation. *Nat Rev Mol Cell Biol*. 2006;7:540-546.
74. Meshorer E, Yellajoshula D, George E, Scambler PJ, Brown DT, Misteli T. Hyperdynamic plasticity of chromatin proteins in pluripotent embryonic stem cells. *Dev Cell*. 2006;10:105-116.
75. Mutskov VJ, Farrell CM, Wade PA, Wolffe AP, Felsenfeld G. The barrier function of an insulator couples high histone acetylation levels with specific protection of promoter DNA from methylation. *Genes Dev*. 2002;16:1540-1554.
76. Mutskov V, Felsenfeld G. Silencing of transgene transcription precedes methylation of promoter DNA and histone H3 lysine 9. *EMBO J*. 2004;23:138-149.
77. Nan X, Cross S, Bird A. Gene silencing by methyl-CpG-binding proteins. *Novartis Found Symp*. 1998;214:6-16. discussion 16–21, 46–50.
78. Nan X, Ng HH, Johnson CA, et al. Transcriptional repression by the methyl-CpG-binding protein MeCP2 involves a histone deacetylase complex. *Nature*. 1998;393:386-389.
79. Ng K, Pullirsch D, Leeb M, Wutz A. Xist and the order of silencing. *EMBO Rep*. 2007;8:34-39.
80. Nimmo RA, Slack FJ. An elegant miRror: microRNAs in stem cells, developmental timing and cancer. *Chromosoma*. 2009;118(4):405-418.
81. Niwa H. Open conformation chromatin and pluripotency. *Genes Dev*. 2007;21:2671-2676.
82. Okano M, Bell DW, Haber DA, Li E. DNA methyltransferases Dnmt3a and Dnmt3b are essential for de novo methylation and mammalian development. *Cell*. 1999;99: 247-257.
83. Okano M, Xie S, Li E. Cloning and characterization of a family of novel mammalian DNA (cytosine-5) methyltransferases. *Nat Genet*. 1998;19:219-220.
84. Pan G, Thomson JA. Nanog and transcriptional networks in embryonic stem cell pluripotency. *Cell Res*. 2007;17: 42-49.
85. Pandey AK, Agarwal P, Kaur K, Datta M. MicroRNAs in diabetes: tiny players in big disease. *Cell Physiol Biochem*. 2009;23:221-232.
86. Payer B, Lee JT. X chromosome dosage compensation: how mammals keep the balance. *Annu Rev Genet*. 2008; 42:733-772.
87. Pekarsky Y, Santanam U, Cimmino A, et al. Tcl1 expression in chronic lymphocytic leukemia is regulated by miR-29 and miR-181. *Cancer Res*. 2006;66:11590-11593.
88. Perry MM, Moschos SA, Williams AE, Shepherd NJ, Larner-Svensson HM, Lindsay MA. Rapid changes in microRNA-146a expression negatively regulate the IL-1beta-induced inflammatory response in human lung alveolar epithelial cells. *J Immunol*. 2008;180:5689-5698.
89. Pfeifer K. Mechanisms of genomic imprinting. *Am J Hum Genet*. 2000;67:777-787.
90. Pradhan S, Bacolla A, Wells RD, Roberts RJ. Recombinant human DNA (cytosine-5) methyltransferase. I. Expression, purification, and comparison of de novo and maintenance methylation. *J Biol Chem*. 1999;274:33002-33010.
91. Ramsahoye BH, Biniszkiewicz D, Lyko F, Clark V, Bird AP, Jaenisch R. Non-CpG methylation is prevalent in embryonic stem cells and may be mediated by DNA methyltransferase 3a. *Proc Natl Acad Sci USA*. 2000;97:5237-5242.
92. Razin A. CpG methylation, chromatin structure and gene silencing – a three-way connection. *EMBO J*. 1998;17: 4905-4908.
93. Rea S, Eisenhaber F, O'Carroll D, et al. Regulation of chromatin structure by site-specific histone H3 methyltransferases. *Nature*. 2000;406:593-599.
94. Reik W. Stability and flexibility of epigenetic gene regulation in mammalian development. *Nature*. 2007;447: 425-432.
95. Reik W, Dean W, Walter J. Epigenetic reprogramming in mammalian development. *Science*. 2001;293:1089-1093.
96. Reik W, Walter J. Genomic imprinting: parental influence on the genome. *Nat Rev Genet*. 2001;2:21-32.
97. Rice JC, Futscher BW. Transcriptional repression of BRCA1 by aberrant cytosine methylation, histone hypoacetylation and chromatin condensation of the BRCA1 promoter. *Nucleic Acids Res*. 2000;28:3233-3239.
98. Richards EJ, Elgin SC. Epigenetic codes for heterochromatin formation and silencing: rounding up the usual suspects. *Cell*. 2002;108:489-500.
99. Ropero S, Esteller M. The role of histone deacetylases (HDACs) in human cancer. *Mol Oncol*. 2007;1:19-25.
100. Roth SY, Denu JM, Allis CD. Histone acetyltransferases. *Annu Rev Biochem*. 2001;70:81-120.

101. Rountree MR, Bachman KE, Herman JG, Baylin SB. DNA methylation, chromatin inheritance, and cancer. *Oncogene*. 2001;20:3156-3165.
102. Royo H, Bortolin ML, Seitz H, Cavaille J. Small non-coding RNAs and genomic imprinting. *Cytogenet Genome Res*. 2006;113:99-108.
103. Royo H, Cavaille J. Non-coding RNAs in imprinted gene clusters. *Biol Cell*. 2008;100:149-166.
104. Sawan C, Vaissiere T, Murr R, Herceg Z. Epigenetic drivers and genetic passengers on the road to cancer. *Mutat Res*. 2008;642:1-13.
105. Schaefer M, Hagemann S, Hanna K, Lyko F. Azacytidine inhibits RNA methylation at DNMT2 target sites in human cancer cell lines. *Cancer Res*. 2009;69:8127-8132.
106. Schaefer M, Lyko F. Solving the Dnmt2 enigma. *Chromosoma*. 2009;119:35-40.
107. Shen L, Ahuja N, Shen Y, et al. DNA methylation and environmental exposures in human hepatocellular carcinoma. *J Natl Cancer Inst*. 2002;94:755-761.
108. Strunnikova M, Schagdarsurengin U, Kehlen A, Garbe JC, Stampfer MR, Dammann R. Chromatin inactivation precedes de novo DNA methylation during the progressive epigenetic silencing of the RASSF1A promoter. *Mol Cell Biol*. 2005;25:3923-3933.
109. Suetake I, Shinozaki F, Miyagawa J, Takeshima H, Tajima S. DNMT3L stimulates the DNA methylation activity of Dnmt3a and Dnmt3b through a direct interaction. *J Biol Chem*. 2004;279:27816-27823.
110. Suzuki K, Suzuki I, Leodolter A, et al. Global DNA demethylation in gastrointestinal cancer is age dependent and precedes genomic damage. *Cancer Cell*. 2006;9:199-207.
111. Tamaru H, Selker EU. A histone H3 methyltransferase controls DNA methylation in *Neurospora crassa*. *Nature*. 2001;414:277-283.
112. Thakur N, Tiwari VK, Thomassin H, et al. An antisense RNA regulates the bidirectional silencing property of the Kcnq1 imprinting control region. *Mol Cell Biol*. 2004;24:7855-7862.
113. Thornhill AR, Burgoyne PS. A paternally imprinted X chromosome retards the development of the early mouse embryo. *Development*. 1993;118:171-174.
114. Tycko B. Epigenetic gene silencing in cancer. *J Clin Invest*. 2000;105:401-407.
115. Vaissiere T, Sawan C, Herceg Z. Epigenetic interplay between histone modifications and DNA methylation in gene silencing. *Mutat Res*. 2008;659:40-48.
116. Vakoc CR, Sachdeva MM, Wang H, Blobel GA. Profile of histone lysine methylation across transcribed mammalian chromatin. *Mol Cell Biol*. 2006;26:9185-9195.
117. van der Vlag J, Otte AP. Transcriptional repression mediated by the human polycomb-group protein EED involves histone deacetylation. *Nat Genet*. 1999;23:474-478.
118. Vire E, Brenner C, Deplus R, et al. The Polycomb group protein EZH2 directly controls DNA methylation. *Nature*. 2006;439:871-874.
119. Vu TH, Jirtle RL, Hoffman AR. Cross-species clues of an epigenetic imprinting regulatory code for the IGF2R gene. *Cytogenet Genome Res*. 2006;113:202-208.
120. Williams AE, Larner-Svensson H, Perry MM, et al. MicroRNA expression profiling in mild asthmatic human airways and effect of corticosteroid therapy. *PLoS One*. 2009;4:e5889.
121. Winter J, Jung S, Keller S, Gregory RI, Diederichs S. Many roads to maturity: microRNA biogenesis pathways and their regulation. *Nat Cell Biol*. 2009;11:228-234.
122. Wutz A, Gribnau J. X inactivation Xplained. *Curr Opin Genet Dev*. 2007;17:387-393.
123. Xia H, Qi Y, Ng SS, et al. MicroRNA-15b regulates cell cycle progression by targeting cyclins in glioma cells. *Biochem Biophys Res Commun*. 2009;380:205-210.
124. Xie ZH, Huang YN, Chen ZX, et al. Mutations in DNA methyltransferase DNMT3B in ICF syndrome affect its regulation by DNMT3L. *Hum Mol Genet*. 2006;15:1375-1385.
125. Yang B, Guo M, Herman JG, Clark DP. Aberrant promoter methylation profiles of tumor suppressor genes in hepatocellular carcinoma. *Am J Pathol*. 2003;163:1101-1107.
126. Zardo G, Fazi F, Travaglini L, Nervi C. Dynamic and reversibility of heterochromatic gene silencing in human disease. *Cell Res*. 2005;15:679-690.
127. Zhou J, Fan JY, Rangasamy D, Tremethick DJ. The nucleosome surface regulates chromatin compaction and couples it with transcriptional repression. *Nat Struct Mol Biol*. 2007;14:1070-1076.

4 Mechanisms of Epigenetic Gene Activation in Disease: Dynamics of DNA Methylation and Demethylation

Thierry Grange and Edio Eligio Lourenço

4.1 Introduction

In vertebrates, methylation at the carbon-5 position of cytosine is found essentially within the dinucleotide CpG.[9] The DNMT3A/B and DNMT1 families of DNA cytosine-5 methyltransferases (DNMT) are responsible for the establishment and maintenance of methylation patterns, respectively.[41] The methylation patterns are correlated with gene expression and constitute relatively stable epigenetic marks that are transmitted through DNA replication and cellular division. Cytosine methylation exerts its effects on genome activity by preventing regulators from binding to their target, and/or by favoring the formation of inactive chromatin through the recruitment of repressive complexes.[9,67] It participates in the control of several aspects of mammalian development, including X chromosome inactivation, parental imprinting, and tissue-specific expression of genes and is involved in the silencing of transposable elements.[9,110,117] Abnormalities in DNA methylation have been observed in a number of pathological situations and have often been proposed to cause the pathology, although this has not been demonstrated unambiguously. Cancer is the disease where dysfunction of methylation regulation is most widely believed to play a role.[58] Many changes in DNA methylation levels have been observed in tumors involving both global hypomethylation and local hypermethylation. Hypermethylation of the promoters of tumor-suppressor genes is associated with epigenetic inactivation in some tumors.[58]

Hypomethylation of DNA repeats may cause genetic instability: Hypomethylation of tandem repeats, especially in the vicinity of the centromere, may favor chromosomal rearrangements or interfere with proper chromosome segregation, whereas hypomethylation of interspersed repeats may enhance DNA rearrangements.[33] Drugs that target the DNMTs have shown promising results in the treatment of some tumors, but whether the effects are due to demethylation and reactivation of tumor-suppressor genes or to induction of a cytotoxic DNA-damage response is far from clear.[101] Abnormal DNA methylation levels in key promoters have also been associated with other diseases, including imprinting disorders,[111] cognitive disorders like Alzheimer's disease and schizophrenia[42,103] and, atherosclerosis[103] among others. The difficulty to establish unambiguously that DNA methylation plays a causal role in a normal or pathological situation stems from the fact that it is involved in complex epigenetic feedback loops that also involve chromatin modifications and transcription regulators. It is therefore unclear whether modified DNA methylation patterns are directly responsible for changes in gene expression or merely constitute convenient markers. To fully appreciate the importance of epigenetic modifications in the etiology of diseases and to develop appropriate therapeutic strategies, it is essential to understand how DNA methylation patterns are established, maintained, and modified. The multiplicity of the pathways controlling epigenetic memory and the intricacy of the regulatory networks interconnecting them make it difficult to draw a simple unequivocal picture of the mechanisms involved. This is particularly true when it comes to the mechanisms of DNA demethylation, for which contradictory findings and interpretations have led some scientists to doubt the existence of active DNA demethylation in mammals.[97]

T. Grange (✉)
Department of Genomes and Epigenomes, Institut Jacques Monod, 15 rue Helene Brion, Paris, France
e-mail: thierry.grange@univ-paris-diderot.fr

H.I. Roach et al. (eds.), *Epigenetic Aspects of Chronic Diseases*,
DOI: 10.1007/978-1-84882-644-1_4, © Springer-Verlag London Limited 2011

4.2 Dynamics of the Methylation Landscape

Recently developed methodologies have led to large-scale and genome-scale methylation maps of the mouse and human genomes.[131] Oligonucleotide tiling microarrays have been used to analyze methylated or unmethylated DNA, enriched with antibodies or MeCpG-binding proteins,[32,108,130,144] or differentially cleaved with methylation-sensitive restriction enzymes.[55] Large-scale sequencing of bisulfite-treated DNA has allowed precise partial or complete genome-scale views of the methylation landscape.[1,16,71,79,84,150] These studies have made it possible to integrate previously fragmented knowledge, but due to the heterogeneity of the methods and the experimental systems, global-scale analyses have not yet led to a unified, coherent picture. The analyses have revealed that 70–80% of the cytosines at CpG dinucleotides are methylated and distributed bio-modally, with 55–80% of the CpG sites frequently methylated (80–100%) and some 20% largely unmethylated.[71,79,84] CpGs in regions of high CpG density are unmethylated, except for 0.3%, whereas some 90% of CpG in low-density regions tend to be methylated, with some 10% unmethylated.[84]

Based on their CpG contents, promoters have been divided into three classes, high-CpG density promoters (HCP) that correspond approximately to the strict definition of CpG islands (CGI),[9] the intermediate-CpG density promoters (ICP) that correspond roughly to the less-stringent definition, and the low-CpG density promoters (LCP) that correspond to the non-CpG island promoters (Fig. 4.1).[144] HCP are associated with the ubiquitous "housekeeping" genes and "key developmental" genes, whereas the LCP are generally associated with tissue-specific genes. HCP are largely unmethylated in normal, pluripotent cells, irrespective of gene expression, but a small percentage can become methylated when cultured in vitro.[84] As cells differentiate, CpG methylation can either increase or decrease. A gain in methylation has been observed for both HCP and LCP; methylation losses essentially affect LCP and a subset of the ICP (Fig. 4.1).[16,84] A much higher proportion of methylated CpG islands (16%) has been observed in some tissues.[130] It has proved possible to identify nucleotide sequence features in the unmethylated status and to develop an algorithm that, based only on the sequence, predicts the unmethylated regions.[130] Another microarray-based study has identified tissue-specific differentially methylated regions (T-DMR) many of which are localized near CpG islands in regions with an intermediate CpG density; these have been termed CpG shores (Fig. 4.1).[55,123] Strikingly, the same study,[55] searching for DNA methylation differences between normal and colon cancer cells, found that cancer-specific DMR (C-DMR) were enriched in CpG shores and that about half of the C-DMRs were T-DMRs. Finally, in a follow-up study, a search for DMR between fibroblasts and reprogrammed pluripotent stem cells (R-DMR) revealed a significant overlap between the three classes of DMRs.[30] Overall, these studies reveal that changes in the DNA methylation pattern that are observed both during cellular differentiation and in pathological

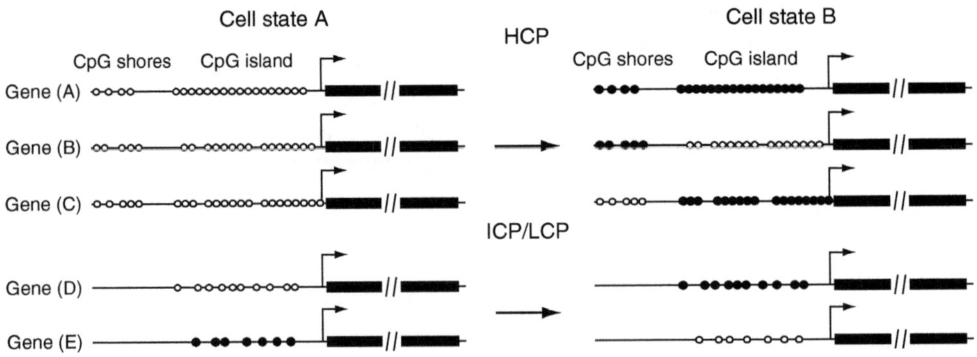

Fig. 4.1 Schematization of typical changes of methylation in the regulatory regions that differ in CpG density, following cell differentiation or transition from a normal to a pathological situation. The scheme depicts typical changes that affect the promoters of high CpG density (HCP), or low density (LCP) and a subset with intermediate density (ICP). CpG islands and CpG shores are indicated for HCP, with *filled circles* representing methylated and *empty circles* non-methylated CpGs. Cell state A is a precursor of cell state B, during development or differentiation, or in a change from a normal to a pathological situation. See text for details

situations occur more readily in gene-proximal regions with intermediate and low CpG density (Fig. 4.1).

DNA methylation is not only involved in the regulation of promoter activity, but also in the regulation of more distal regulatory elements. Using a new strategy to capture CGIs with a non-methylated CpG-binding protein, Bird and colleagues[54] have reported that about half of the CGIs from this more comprehensive set did not overlap with an annotated promoter. Six to eight percent of these CGIs showed methylation in one or more tissues, with inter- and intragenic CGIs being preferentially susceptible to methylation. Indeed, a search for DMRs within a selection of genomic regions in T cells revealed that the majority of about one hundred of the identified DMRs were located at promoter-distal sites, many clearly corresponding to enhancers.[119] Most of these distal DMRs had a low level of CpG, a property shared with promoter-proximal DMRs.

In most of the gene-proximal DMRs, DNA methylation is inversely related to expression of the nearby gene. In contrast, gene-body methylation is directly correlated to gene expression, with highly expressed genes showing high gene-body methylation.[1,71,131] Intragenic methylation may prevent spurious transcription initiation events within active transcription units that would otherwise be favored by transcription-induced chromatin disruption.[131] Finally, non-CpG methylation, a well-known phenomenon in plants, has been shown to occur also in embryonic stem (ES) cells, comprising 20–25% of methylated cytosines (MeC).[71,79,107] In a study by Laurin and colleagues, this non-CpG methylation affected 85% of the CHG and CHH sites (H=A, C or T), and each site showed intermediate methylation levels (10–40%).[79] In another study, a marked preference of non-CpG methylation at CpA sites was observed.[71] This methylation also occurred more frequently within the body of the genes and was correlated with gene expression. Interestingly, non-CpG methylation was decreased upon ES cell differentiation.[71,79]

4.3 Establishment and Maintenance of the Methylation Landscape: The Link to Chromatin

Even though DNA methylation may have a function very early as an epigenetic mark, it has been difficult to demonstrate unambiguously that it plays a causal role in the regulation of gene expression in the course of development. This is most likely due to the fact that DNA methylation is part of a complex interrelated network of epigenetic modifications that involve many overlapping, interdependent, and redundant contributions. These have evolved to provide robustness to spurious and insignificant regulatory fluctuations and enable appropriate responses to specific signals. DNA methylation is linked in many ways to both chromatin modifications and RNA production, even though the exact relevance of each individual link has been difficult to demonstrate.

Establishment and maintenance of DNA methylation patterns are ensured by DNMT3A/B and DNMT1, respectively.[41] DNMT3A/B act on unmethylated DNA sites and have been identified as de novo DNMTs. DNMT1 acts preferentially on the hemimethylated DNA that is produced upon DNA replication, thus ensuring the maintenance of the methylation patterns throughout cell division. DNMT1 action is likely to be favored both by its interaction with the DNA polymerase processivity factor (PCNA) that allows its recruitment at the replication fork, as well as by the UHRF1 factor that binds hemimethylated DNA and DNTM1.[13,23,124] The action of DNMT1 on hemimethylated DNA by itself may not be sufficient to maintain the DNA methylation pattern faithfully at every cell cycle. Hairpin bisulfite experiments that allow simultaneous analysis of both DNA strands of the same DNA molecule have revealed that a small percentage of the methylated CpGs remain hemimethylated or are fully demethylated just before the subsequent DNA replication.[69,70] Despite this, the overall pattern is maintained and does not drift upon successive cell divisions. This provides evidence for the existence of an additional proofreading mechanism that does not depend on DNMT1 acting on the hemimethylated substrate. Furthermore, cells that lack DNMT3A/B progressively lose the methylation of both repetitive and single-copy sequences. This shows that DNMT1 alone is not sufficient and must work with de novo DNMTs to maintain the methylation patterns faithfully.[20] De novo methylation may therefore function as a proofreading device to ensure fidelity of the methylation patterns at the hemimethylated CpG sites missed by DNMT1.

De novo DNMT methylation and its maintenance require specific targeting mechanisms. In plants, small RNAs appear to participate in both aspects.[82,136] In mammalian cells, it has been possible to induce de novo methylation with the aid of siRNAs targeting promoters.[92] RNA-mediated mechanisms for targeting de novo methylation may therefore also be active in mammalian cells. In support of this concept, transcription elongation through a maternally imprinted DMR has been shown to be required for targeting DNA methylation in the mouse.[22] Other mechanisms have been proposed to target DNA methylation to previously unmethylated regions in tumor cells. Targeting could involve recruitment of DNMT3 to promoters by sequence-specific transcription factors. For example, the PML-RAR oncogenic fusion protein promotes DNA methylation of the RAR target genes.[28]

Targeting of DNA methylation could also involve specific chromatin modifications and chromatin-modifying complexes because histone and DNA methylation are linked. One link involves the Polycomb Repressive Complex (PRC2) that is responsible for the establishment and maintenance of a repressive chromatin mark, the trimethylation of K27 of histone H3. The K27 histone methyltransferase EZH2 interacts with both de novo and maintenance DNMT and facilitates DNA methylation at target genes.[142] In agreement with such a link, it has been observed that the genes marked by trimethylated K27 in stem cells may also become hypermethylated in the course of differentiation and in cancer cells.[89,95,118] However, the causal relationship between K27 trimethylation and DNA methylation is subtle. Thus, trimethylation of K27 may increase the probability of de novo methylation only slightly. Indeed, in normal cells, no correlation has been observed between these two epigenetic marks at the genome-wide level.[71]

A clearer link between trimethylation of K9 of histone H3 and DNA methylation has been demonstrated in filamentous fungi and plants. In *Neurospora*, H3K9 trimethylation is essential to establish DNA methylation.[135] In *Arabidopsis*, the DNMT Chromomethylase 3 is recruited to target sequences through its chromodomain that recognizes the dual methylation of both K9 and K27 of histone H3.[78] In mouse cells, interspersed repeat families that are enriched in H3K9 trimethylation, like long terminal repeats (LTRs) and long interspersed nuclear elements (LINEs), are also enriched in DNA methylation. Such enrichment is not seen in short interspersed nuclear elements (SINEs) that are not enriched in H3K9 trimethylation.[84] In the mouse, DNMT3B is recruited at heterochromatic foci marked by H3K9 trimethylation. DNA methylation of the pericentric satellite repeats is impaired, following inactivation of the gene that codes for the methyltransferase Suv39h which mediates trimethylation of H3 K9.[74] Finally, the methyltransferase G9a mediating dimethylation of H3-K9 in euchromatic regions can recruit DNMT3A/B directly to promote de novo DNA methylation of target genes in a manner that appears independent of its histone methyltransferase activity.[31,35,133]

The recruitment of de novo DNMT3A/B appears to be also directly dependent on the methylation status of the histone tails. DNMT3A/B and the DNMT3 positive regulator DNMT3L contain a domain similar to the plant homeodomain (PHD) finger called the ADD domain.[98,100] The ADD domain of DNMT3A and DNMT3L binds the first seven amino acids of the tail of histone H3, with binding decreased by di- or trimethylation of K4.[98,100] H3K4 methylation may therefore downregulate DNA methylation by DNMT3A/B. Indeed, trimethylation of K4 of histone H3 is highly enriched at active promoters,[46] sites where CpGs are unmethylated.[71] A repressive mark, the symmetrical dimethylation of R3 of histone H4, deposited by the arginine methyltransferase PRMT5 may in contrast favor recruitment of DNMT3A through recognition of the modified tail by the ADD domain.[152] There may be additional chromatin modifications or chromatin-binding proteins that favor the recruitment of DNMTs, because DNMT3A/B are strongly anchored to a subset of nucleosomes that by presently unknown interactions are enriched in methylated DNA.[56]

Both the effects of chromatin modifications on DNA methylation and of methylated DNA on chromatin have been described. Methylated DNA transferred to vertebrate cells is assembled into a transcriptionally repressive chromatin structure,[51,65] with both the extent and robustness of transcriptional repression related to the density of methylated CpGs.[14,50] Methylated DNA recruits several families of proteins, including proteins with methyl-binding domains (MBD) that mediate, at least partially, the assembly of the repressive chromatin structures.[10,67,115] MBD proteins recruit histone-modifying enzymes that help form repressive chromatin. MeCP2 interacts with the Sin3 complex that contains the HDAC1 and HDAC2 deacetylases.[94] MBD2 and MBD3 are components of two distinct nucleosome remodeling and deacetylase complexes (NuRD), both containing the HDAC1 and HDAC2 deacetylases.[72] The MBD2/NuRD complex contains the PRMT5 methyltransferase that methylates R3 of histone H4,[72] whereas the MBD3/NuRD complex contains the LSD1 H3-K4 demethylase.[143] MBD1 forms a stable complex with the H3-K9 methyltransferase SETDB1/ESET and recruits ESET to the chromatin assembly factor CAF-1 to allow trimethylation of H3-K9 during replication-coupled chromatin assembly.[114] Methylated DNA participates in the formation of repressive chromatin, whereas non-methylated CpGs participate in the formation of active chromatin. Non-methylated CpG islands are recognized by proteins with zinc finger domains of the CXXC type, in particular MLL1, a H3-K4 methyltransferase; KDM2a, a H3-K36 demethylase; and Cfp1, a transcriptional activator.[11,24,138] Cfp1 interacts with the H3-K4 methyltransferase SetD1 and thereby mediates H3-K4 trimethylation at the non-methylated CpG islands.[138]

The interaction of DNMT with histone-modifying enzymes also functions in the reciprocal coupling between DNA methylation and chromatin modifications. DNMT1 interacts with the DNA replication processivity factor PCNA and recruits the H3K9 methyltransferase G9a responsible for K9 dimethylation at the replication foci; this coordinates the maintenance of DNA and K9 methylation after chromatin replication.[36] Thus, G9a may be recruited by the maintenance DNMT and, as mentioned earlier, may in turn recruit de novo DNMTs. G9a thus participates in the proofreading mechanism that ensures the fidelity of DNA methylation of repressed genes.

In summary, DNA methylation and the DNA methylation machineries participate in an intricate network that regulates chromatin and the epigenetic state of genes plus repeated sequences. Numerous positive feedback loops are likely responsible for the bistability of the epigenetic state of genes. This in turn ensures the robustness of the expression status toward noise which results from fluctuations of regulatory signals and transcription factors.[29] The frequency of switching between alternate epigenetic states varies widely between genes and cell types, with stem cells being more plastic than differentiated cells. Among the parameters that influence gene to gene variability is the stability of the repression exerted by DNA methylation. Methylated CpG density appears to be an important parameter, presumably because methylation density modulates recruitment of complexes that mediate repressive chromatin structure. Accordingly, regions that are differentially methylated in the course of development and differentiation mostly have a low or intermediate CpG content.

4.4 Epigenetic Reprogramming by DNA Demethylation

The identification of the DNA methylating machinery has led to better understanding of the de novo methylation events that occur at DMR during development and differentiation. The targeting mechanisms are not yet fully understood, but because the identity of the machinery is clearly established, it has been possible to identify interacting partners. This in turn makes it possible to test specific hypotheses.

Unfortunately, the situation is much less clear with respect to DNA demethylation. The field is full of controversies and recurrently published data that seem spectacular have not been reproduced or independently tested (see Kress et al.[68] and Ooi and Bestor[97] for discussion). Part of the confusion may be due to the use of different experimental systems, some with perhaps as yet unidentified systematic flaws. Nevertheless, it appears likely that a context-dependent biological diversity underlies the situation. The resulting complexity may due to one or more of these: (1) multiple pathways under independent regulatory controls, (2) pathways of different evolutionary origins, (3) a balance between epigenetic control and maintenance of genomic integrity that was established in the course of evolution.

Two major DNA demethylation events that occur during development may or may not be based on the same mechanisms: (1) global demethylation events, affecting most genes and interspersed repeats, that occur at major developmental transitions; (2) site-specific demethylation events that affect specific regulatory sequences, targeted by site-specific transcriptional regulators. These occur throughout development and ultimately in differentiated cells.[68,110] Demethylation mechanisms can be active or passive. Passive demethylation occurs if methylation is not maintained following DNA replication. Methylation then becomes diluted after several replication rounds. Active demethylation implies that one or several enzymes are involved in removing the methyl groups, the methylated bases, or nucleotides. Both active and passive demethylation mechanisms seem to be involved in global and site-specific demethylation (Table 4.1).

Table 4.1 Summary of the various DNA demethylation pathways used

		Mechanisms	Pathways	References
Global demethylation	Early embryogenesis Zygote	Passive (female)	Cytoplasmic sequestration of DNMT1	18
		Active (male)	Elongator complex	83, 96, 99
	Early embryogenesis reprogramming of PGC	Passive	Cytoplasmic sequestration of DNMT1	122
		Active	AID (BER?)	104
	Mouse myoblast differentiation	Active	BER	60
Site-specific demethylation triggered by transcriptional factors	OriP of Epstein Barr virus by EBNA	Passive	Protection from methylation	52
	Vitellogenin gene by ER	Active	N.D.	112
	pS2 gene by ER	Active	BER	85
	TAT gene by GR	Active	BER	69, 137
	RARβ2 gene by RAR	Active	NER	73
	CYP27B1 gene by parathyroid hormone	Active	BER	66
	rDNA	Active	Direct removal of Me group by Mbd3	15
			NER	120

N.D. Not determined

4.4.1 Global DNA Demethylation

In mammals, global DNA demethylation is transiently observed during early embryogenesis and in some differentiating cells, at stages when important genome reprogramming events take place.[60,109,110] During early mammalian development, two waves of global DNA demethylation act on the zygote and primordial germ cells (PGC).[110] Upon fertilization, the paternal genome undergoes active genome-wide DNA demethylation.[83,99] From the zygote to the morula stage, cytoplasmic sequestration of most of DNMT1 causes passive replication-dependent demethylation of both the paternal and maternal genomes except for the imprinted genes.[18,48] Thereafter, in the pluripotent epiblast, DNMT3A and 3B become active and the methylation levels in the epiblast cells increase.[75] PGCs are derived from some epiblast cells that migrate and colonize the genital ridge, where they undergo epigenetic reprogramming between embryonic days (E) 7 and 12.5 of mouse development.[122] Around E8, at a stage when DNMT3A/B is either absent or expressed at low levels in PGCs, DNMT1 is lost from PGC nuclei for one cell cycle; this presumably triggers global DNA methylation loss that lasts until E13.5.[104,122] This passive global demethylation may not be sufficient to cause the observed global demethylation, because absence of maintenance methylation for one replication should produce hemimethylated DNA that leads to remethylation of half of the daughter molecules at the next round of replication, when maintenance methylation is restored. Indeed, recent findings imply that active DNA demethylation also occurs around this developmental period.[104] In conclusion, major reprogramming events associated with global DNA demethylation seem to result from a combination of passive and active mechanisms.

4.4.2 Local DNA Demethylation Triggered by Transcription Factors

Site-specific CpG demethylation triggered by transcription regulatory factors has long been known.[145,148] Various transcription factors can induce local DNA demethylation. Some studies, aimed at analyzing this phenomenon with the aid of transfected methylated templates, yielded contradictory results, with demethylation occurring by both active and passive mechanisms.[52,81,102] It is uncertain whether these differences are due to different mechanisms or to some bias resulting from the artificial nature of the systems. This is not true for events that take place on endogenous genes in the native chromatin context. A better characterization of the mechanisms and consequences of local DNA demethylation has emerged from the study of transcription activation by nuclear receptors, in part because their ligand-induced activation has allowed kinetic analyses.

During activation of the chicken vitellogenin gene by estrogen, site-specific demethylation occurs in the vicinity of the estrogen receptor (ER)–binding site within the promoter.[145] Demethylation was not prevented by DNA replication inhibitors; this suggests that it was due to an active mechanism, with no contribution by a passive step.[146] Four CpGs are demethylated progressively over the course of 2–3 days of induction, and demethylation is much faster on one of the two DNA strands.[112]

During activation of the tyrosine aminotransferase gene *Tat* by glucocorticoids in hepatoma cells, four CpGs within a glucocorticoid-responsive enhancer 2.5 kb upstream from the promoter are demethylated progressively over the course of 2–3 days of induction.[137] Demethylation of these four CpGs occurred with similar kinetics on the two DNA strands.[69,137] Quantification in the course of the cell cycle revealed that DNA demethylation can occur in G1 with the templates fully methylated. It is therefore due to an active mechanism with no contribution by a passive step.[69] The four CpGs are located in the vicinity of two glucocorticoid receptor (GR)-binding sites, are included within a region where chromatin remodeling is triggered by the GR, and overlap two binding sites of other transcription factors.[37,43,137] The kinetics of the recruitment of these factors parallels that of demethylation. In fact, in vitro binding of one of these factors is impaired by DNA methylation.[137] Strikingly, DNA demethylation is associated with the memorization of glucocorticoid activation. After 3 days of induction, when the four CpGs are fully demethylated, hormone withdrawal causes cessation of transcriptional activation, chromatin remodeling, and transcription factor recruitment, but the four CpGs remain demethylated for at least 3 months.[137] If a second induction occurs, GR-induced chromatin remodeling causes rapid recruitment of transcription factors that were recruited slowly during the first induction. It also leads to more rapid kinetics of the transcriptional activation of the *Tat* gene.[137] Thus, targeted DNA demethylation regulates the recruitment of transcription factors and provides gene-specific memory of a previous regulatory event. During development, this demethylation occurs prenatally in the liver when there is a peak of glucocorticoids. It also prepares the enhancer to fully induce the *Tat* gene at birth in response to hypoglycemia.[43,116,137]

Nuclear receptors do not always induce DNA demethylation, but can induce DNA methylation at some target genes. The vitamin D receptor (VDR) that represses the cytochrome p450 27B1 (*CYP27B1*) gene directly recruits DNMT1 and DNMT3B to its promoter. These in turn methylate up to 12 CpGs.[66] The methylated *CYP27B1* promoter is reactivated and demethylated within an hour when the parathyroid hormone activates the protein kinases PKA and PKC.[66] No study of the long-term stability of these modifications has been reported, and it is not clear if these modifications contribute to some epigenetic memory. DNA methylation switching can therefore be seen to participate in hormonal regulation of transcription.

Finally, studies of nuclear receptors have for the first time provided clear evidence that DNA methylation changes can fluctuate very dynamically at several regulatory regions. Activation of the trefoil factor I (*TFF1* or *pS2*) gene by estrogen induces cyclical and ordered recruitment of cofactors to the promoter,[86] involving multiple cofactors and chromatin modifications, all of which have mostly short cycles that last between 20 and 40 min at the cell population level.[86] Binding of ERα itself cycles and peaks every 40 min at the *pS2* promoter. The methylation state of about eight CpGs within the *pS2* promoter also cycles, but out of phase.[85] The CpGs are demethylated during the first two rounds of ER binding, remethylated during the following period, when ER is not bound, and demethylated again at the third cycle of ER binding.[85] Not all CpGs cycle in the same manner. Most CpGs affected by the cycling are demethylated on only one DNA strand, which is the same for all CpGs. Under different cell synchronization conditions, ERα induced methylation cycles on the *pS2* gene have a different period, 100 min.[64] Cycles of CpG methylation and demethylation have also been reported for four other

ERα target genes.[64] Cycles of DNA methylation and transcription factor binding can be detected only if most cells within a population respond synchronously to the inducer. This synchrony has been attained under specific conditions, including nuclear receptor overexpression and/or drug treatment.[64,85,86] Conceivably, such cycles may also occur asynchronously for the previously described nuclear receptor target genes, but, because of asynchrony, may not have been observed.

DNA methylation cycling implies that DNA methylation is not necessarily a stable epigenetic mark, even if so in some situations, as is the case for the *Tat* gene. Stability of epigenetic DNA methylation marks may depend on how they are interconnected with other epigenetic modifications or factors. In other words, chemical stability of modifications is not the key property; rather it seems to be the structure of the network of which they are part. These networks are likely to differ from gene to gene, or as a result of changes in the duration or intensity of activation or repression. Cycles between alternate DNA methylation states could occur during short-term activation events, but when activation is prolonged, the gene might switch to a more stable epigenetic state that does not cycle so readily. Unfortunately, no data are presently available to support this conjecture, because genes with rapid changes of DNA methylation states have not been analyzed after long-term activation, nor have genes with long-term changes been analyzed under conditions where rapid DNA methylation can be observed. Finally, one should note that the DNA demethylation events reviewed here affect regulatory regions with low or intermediate CpG density, which might be more prone to switching between alternate epigenetic states, because the methylation density is not high enough to efficiently elicit a stable repressive chromatin structure.

4.5 The Mechanisms of DNA Demethylation

A number of active DNA demethylation mechanisms in vertebrates have been proposed. As yet it is not clear whether there is a single, uncontroversial active mechanism. However, it is difficult to believe that all DNA demethylation events in mammals are due to passive demethylation.[97] As discussed earlier, passive global DNA demethylation does not involve complex mechanisms: It is sufficient to inactivate the maintenance methylation DNMT. Nuclear exclusion of DNMT1 has been documented for the two major global DNA demethylation events that take place during early development.[18,122] With respect to targeted local DNA demethylation, a passive mechanism resulting from site-specific protection from maintenance methylation by stably bound transcription factors has been reasonably well documented.[52] However, for such a passive mechanism to occur, a saturating concentration of transcription factors is needed for the sites to be fully occupied during the post-replicative maintenance methylation phase.[77] Active mechanisms are needed, however, for demethylation events to occur outside of the S phase. Indeed, the existence of active demethylation mechanisms has been firmly established in mammals in a variety of situations.

Currently described active DNA demethylation mechanisms are akin to one of the DNA repair mechanisms and may involve recruitment of a component of the DNA repair machinery. In principle, the simplest way to demethylate a cytosine in DNA is to remove the methyl group at the base (for review see Kress et al.[68]). The alkylated bases 1-meA and 3-meC have been shown to be repaired directly by oxidative demethylation catalyzed by the AlkB dioxygenase.[121] The oxidative demethylation in question involves a N–C bond, a thermodynamic process that differs from the demethylation of the carbon-5 of cytosine. Interestingly, histone lysine demethylation by the demethylase LSD1 involves oxidative demethylation of another N–C bond.[126] Two years before the oxidative demethylation by AlkB was elucidated, it was proposed that the methyl-binding protein MBD2b could remove the methyl group from 5-meC in DNA by transposing it to a water molecule and producing methanol.[7,106] This mechanism was severely criticized, because the biochemistry of purification and the thermodynamics of the reaction were unlikely, and because other laboratories were unable to demonstrate DNA demethylation catalyzed by the MBD2 protein.[147] The Szyf laboratory has nevertheless published many related reports, showing recently that MBD3, a protein related to MBD2b, demethylated the rDNA promoter.[15] In the absence of independent confirmation and given that the reagents are widely available, it is difficult to give full credit to the MBD pathway of demethylation, yet it is also difficult to disregard this pathway, given the perseverance shown by the Szyf laboratory.

All the other proposed DNA demethylation pathways rely on more conventional DNA repair mechanisms involving removal of the methylated cytosine and, in some cases, of the neighboring nucleotides.

4.5.1 Demethylation by a Base Excision Repair Pathway

The proposal that in vertebrates, demethylation occurs by a base excision repair (BER) pathway has been controversial for many years, even though this mechanism has been accepted for plants.[39,40,90] The BER pathway is initiated by a DNA glycosylase that cleaves the *N*-glycosidic bond linking the N1 of cytosine to the C1′ of deoxyribose; the cleavage is followed by repair of the abasic site (Fig. 4.2). There are two classes of glycosylases: monofunctional and bifunctional. Monofunctional glycosylases have only glycosylase activity, whereas bifunctional glycosylase/lyases couple base excision with 3′-phosphodiester bond breakage (AP lyase activity) (compare the middle and lower pathways of Fig. 4.2). In plants, the Demeter and Ros1 family of DNA glycosylase/lyases are involved in removing the epigenetic DNA methylation marks responsible for imprinting the maternal alleles of several genes in the endosperm.[39,90] Demeter is also involved in genome-wide demethylation in the endosperm with effects on genes and repetitive sequences.[38,53] In vertebrates, the proposal of demethylation by a DNA glycosylase was based on in vitro experiments that employed crude nuclear extracts from HeLa cells.[141] This finding was criticized 2 years later, on the basis of studies using a similar in vitro system.[128] On the other hand, Jost and colleagues confirmed the earlier finding using a partially purified nuclear fraction from a chick embryo.[62] Subsequently, Jost and colleagues, having refined the in vitro characterization of its demethylating activity, found that the enzyme acted preferentially on hemimethylated substrates and that RNA molecules were targeting the demethylation activity to the complementary DNA strand.[59,62] They identified a DEAD box RNA helicase, p68, that is tightly associated with the protein–RNA

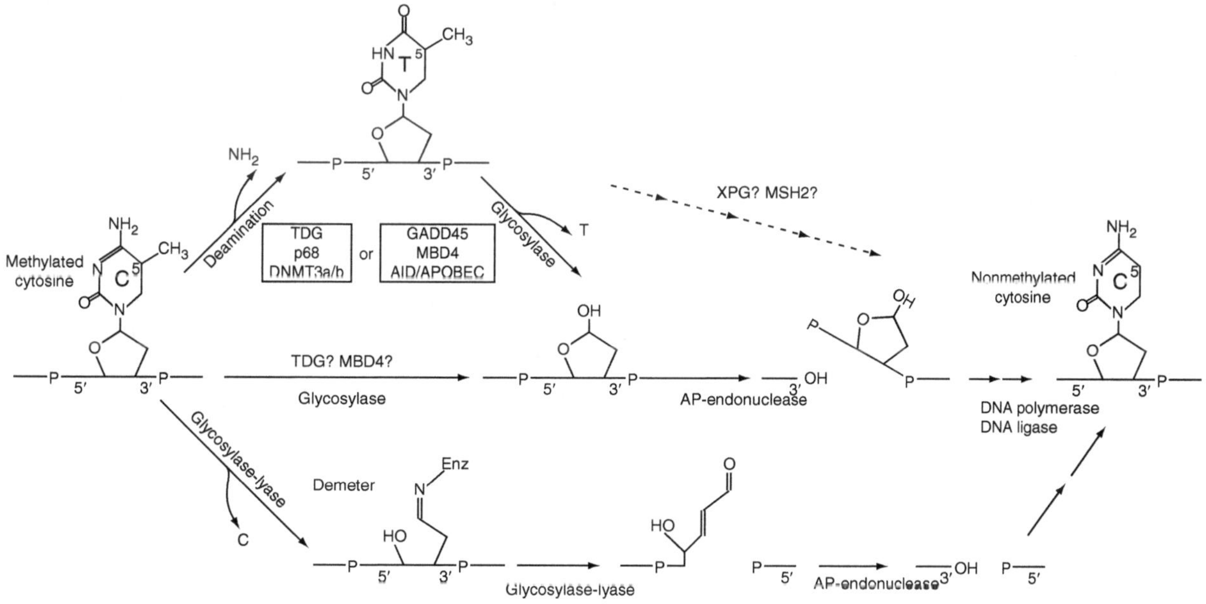

Fig. 4.2 MeC demethylation mechanisms following the BER pathway. The various pathways that allow transition from a MeC (*left*) to a non-methylated C (*right*) are shown. The lower pathway corresponds to the BER pathway, initiated by a bifunctional glycosylase-lyase, as for the plant enzymes Demeter and Ros1. The central pathway corresponds to the BER pathway initiated by a monofunctional glycosylase, such as TDG and MBD4/MED1. The latter corresponds to the hypothesis proposed by Jost and colleagues (see text), where the glycosylase acts directly on MeC. The upper pathway represents the initiation of the process by MeC-deamination that gives rise to a T, and thus a G/T mismatch. The possibility that the deaminated MeC could initiate an alternative NER or MMR pathway is represented by a *dotted line* (multiple steps not shown). See text for details

complex.[61] Finally, they proposed that the G/T mismatch DNA glycosylase TDG was responsible for the 5-methylcytosine (MeC)-glycosylase activity observed in vitro.[155] The authors also proposed that the G/T mismatch MBD-glycosylase MBD4/MED1 was active on MeC.[154] Simple functional assays to assess the relevance of these enzymes for DNA demethylation in cell culture have shown that a morpholino antisense oligonucleotide complementary to the TDG sequence decreased the genome-wide demethylation in differentiating myoblasts.[60] Massive overexpression of TDG has led to the demethylation of a methylated construct transfected in cultured cells.[153] The involvement of TDG and MBD4/MED1 in DNA demethylation failed to gain wide recognition because the in vitro activity was observed in extreme conditions. Thus, even with a fivefold molar excess of enzyme over substrate, only a small percentage of the MeC was removed after 1-h incubation.[155] Furthermore, the overexpression and morpholino experiments lacked specificity controls and the data were obtained in artificial settings and not monitored with reliable techniques.[60,153] In addition, it was shown that TDG behaved as a coactivator and was interacting with ERα and the coactivator p300.[19,139] Interestingly, the coactivator function is independent of the glycosylase activity. This contrasts with the CpG demethylase Demeter in plants, where the activating function requires glycosylase activity.[21] TDG function in association with p300 and ERα may represent a more general coupling between DNA repair and transcriptional regulation, inasmuch as several other DNA glycosylases, each with different substrate specificities, interact with p300 or ERα.[5,6,76] In contrast to the findings of the Jost's laboratory, other reports on the in vitro activity of TDG and MBD4/MED1 have failed to show activity toward methylated cytosines when the enzymes were fully active on T/G mismatches.[25,44,47,149] These findings made it less likely that TDG and MBD4/MED1 are responsible for DNA demethylation. Possibly demethylation by TDG and MBD4/MED1 was the result of a base modification like deamination that converts MeC into a T. However, this pathway was not actively pursued for several years after JP Jost retired.

Evidence of demethylation by a BER pathway has come from a study of the GR-mediated demethylation of an enhancer of the *Tat* gene.[69] As mentioned earlier, this active demethylation can occur in G1 on fully methylated CpGs. Simultaneous analysis of both strands of single molecules using the hairpin bisulfite method was utilized to analyze the enhancer in the course of its demethylation. Demethylation was revealed to be distributive: Each MeC is replaced one by one by a slow process and prior demethylation has little effect on the probability of subsequent demethylation.[69] In addition, demethylation intermediates were detected in cultured cells: Specific cleavage of the phosphodiester backbone was induced between the MeC undergoing demethylation and its neighboring G in the CpG dinucleotide, leaving the phosphate on the 5′ end of the guanosine.[69] As shown in Fig. 4.2, a glycosylase-driven BER pathway produces this type of reaction intermediate, which would be more readily detected if a glycosylase/lyase is acting (compare the lower and middle pathways, Fig. 4.2). TDG and MBD4/MED1 are glycosylases devoid of lyase activity, and so may not be involved, but the responsible enzymes have not yet been identified.

Even though many laboratories have been unable to show significant glycosylase activity of TDG and MBD4 toward MeC, this concept was recently revived when recruitment of MBD4/MED1 to the promoter undergoing demethylation was observed in the course of demethylation of the *CYP27B1* gene following activation of PKC by parathyroid hormone.[66] The BER pathway is thought to be involved in demethylation because the AP endonuclease APE1 is recruited alongside MBD4 and required for the recruitment of DNA polymerase b and DNA ligase I.[66] In vitro glycosylase assays performed with bacterially produced MBD4/MED1 have revealed a slightly higher activity toward MeC-containing oligonucleotides following phosphorylation of the enzyme by PKC.[66] These findings[66] contradict the data published 10 years earlier that showed that MBD4/MED1 was not active on MeC-containing DNA.[47] It should be noted, however, that in the 1999 studies, the molar ratio of enzyme to substrate was one, whereas in 2009, the enzyme to substrate ratio was at least 100, a ratio that is not functionally meaningful. Therefore, it is unlikely that the PKC-activated MBD4/MED1 is solely responsible for the enzymatic removal of MeC. It is playing a key role, however, since siRNA-mediated MBD4/MED1 knock down inhibits demethylation of the *CYP27B1* promoter, and MBD4/MED1 knock-out impairs reactivation by parathyroid hormone of the vitamin D–repressed gene in mouse kidney.[66] The effect on gene expression of MBD4/MED1 knock-out is subtle though, because it does not lead to an obvious phenotypic change.[88] Furthermore, the developmental induction and demethylation of the *CYP27B1* promoter and the expression of the gene in the absence of hormonal treatment are not significantly impaired in the mutant mice.[66] In other words, demethylation of the *CYP27B1* promoter either affects only the kinetics of gene induction, as when the *Tat* gene is induced by glucocorticoids,[69] or the kinetics of demethylation are slower because other pathways rescue the MBD4/MED1 deficiency. If direct removal of MeC by MBD4/MED1 does not occur, the enzyme may act after MeC has first been modified.

The cyclical ERα triggered demethylation of the *TFF1/pS2* promoter is accompanied by cyclical interactions of TDG and of several components of the machinery of the BER pathway, namely, the DEAD box p68 helicase, the AP endonuclease REF1, DNA polymerase β and DNA ligase I.[85] The cycle is not perfectly synchronous with the demethylation cycle, with the first demethylation event preceding the recruitment of the repair machinery.[85] Available evidence does not show that recruitment of TDG plays a role in the demethylation process. As MeC is not thought to be a preferred substrate for TDG, the authors hypothesized that a MeC-deaminase generates a G/T mismatch and proposed that the DNMTs act as the deaminases; this reasoning is based on the fact that in the absence of the methyl group-donor S-adenosyl methionine (SAM), bacterial DNMTs exhibit cytosine deaminase activity and because DNMT1 and DNMT3A/B are recruited cyclically just before recruitment of the TDG and BER proteins.[85] Even though Metivier et al.[85] show that the eukaryotic DNMTs exhibit C- and MeC-deaminase activities in the absence of SAM, some of their data are inconsistent with that role of DNMTs. Indeed, a competitive inhibitor of DNMTs (RG108) does not inhibit the first demethylation, even if added at the time of induction.[85] Only upon prolonged incubation can it inhibit this demethylation under conditions where indirect effects become a major concern. A deaminase function of DNMTs to promote DNA demethylation is a concept that has encountered skepticism based on three arguments: (1) it appears difficult to imagine that the SAM pool is synchronously regulated with the cycling of various genes; (2) the SAM depletion that promotes DNMT deamination is incompatible with cell metabolism; and (3) the efficiency of the reaction is too low and therefore incompatible with the rapidity of cycling.[97] Demethylation of the *TFF1/pS2* promoter by a BER pathway also raises the problem of the mechanism that would ensure the strand specificity of the demethylation. The BER pathway is usually distributive, but strand specificity implies a processive pathway that keeps track of the DNA strand; this is difficult to conceive if only proteins are involved. In contrast, an RNA molecule produced by transcription through the demethylated region could provide strand specificity, either because an R-loop would favor methylated cytosine deamination of the single-strand region[3,125] or because the RNA molecule would target TDG to the hybridized strand, as proposed by Jost.[59] Alternatively, another mechanism other than BER could provide strand specificity.

AID and Apobec1 can deaminate MeC in DNA.[91] The concept that deamination of MeC in DNA could create the substrate for the G/T mismatch glycosylase gained ground from a zebrafish study.[105] The injection of massive quantities of methylated DNA into the embryo (10^8 molecules per cell) induced the expression of three cytidine deaminases: activation-induced deaminase (AID), apolipoprotein B-editing catalytic subunit (Apobec) 2a, and Apobec 2b. This could constitute an innate defense mechanism against foreign DNA since cytidine deaminases are involved in mutagenesis or degradation of intruding DNA.[45,129] Demethylation by cytidine deaminase and MBD4/MED1 was coordinated by the growth arrest and DNA-damage-inducible protein 45 alpha (GADD45α) that promotes their interaction.[105] GADD45α is involved in the maintenance of genome integrity through participation in the control of a cell cycle checkpoint, several DNA repair pathways, and signal transduction.[151] GADD45α/β were observed previously to be associated with a MeC glycosylase activity in human cells.[140] In developing zebrafish embryos, antisense oligonucleotides against AID, GADD45α or MBD4 impaired neuronal development and the induction and demethylation of neuronal genes.[105] GADD45α does not play such a key function in vertebrates, inasmuch as GADD45α deficient mice show marked genome instability, but no clear developmental defects and no detectable change in DNA methylation.[34,49] GADD45β, the major GADD45 member expressed at that stage, was found dispensable for paternal demethylation in the zygote.[96] GADD45β deficient mice, however, even though devoid of major developmental defects, showed subtle deficits in neural activity-induced proliferation of neural progenitors of hippocampal neurons that could be linked to impaired demethylation of specific promoters of genes critical for adult neurogenesis.[80] Finally, when both GADD45α and β were knocked down, the full differentiation of epidermal progenitor cells was impaired coincident with the demethylation of the promoter of some of the genes that are normally demethylated upon differentiation.[123]

Mice deficient in either AID, Apobec2, or Apobec3 have no general developmental defects.[87,93] AID does however participate in two epigenetic reprogramming events in mammals. The global DNA demethylation observed in mouse PGCs at E13.5 is lower in AID –/– mice than in the wild type.[104] The effect of AID deficiency is more striking in the female than in male PGCs, but overall the effects are weak; this indicates that AID contribution is marginal. Indeed, median methylation levels in WT PGCs are about 16.3% in the male and 7.8% in the female, whereas in the AID –/– mice, the levels rise to about 22% and 20%, respectively.[104] These differences are significant, but are much lower than in other tissues where median values range from 73.2% to 85%. It is therefore evident that most of the demethylation that takes place in the PGCs is by an AID-independent process. In the other tissues, global DNA methylation levels did not change due to AID deficiency.[104] The involvement of AID in local gene-specific demethylation during reprogramming of somatic cell nuclei toward pluripotency has been reported as involving two key regulators of induced pluripotent stem cells (iPS), Oct4 and Nanog, that undergo active DNA demethylation upon reprogramming.[8] Several siRNAs against AID inhibit almost all demethylation events at these two promoters; they also inhibit induction of these two genes, an indication of the key role played by AID.[8]

The generation of 5-hydroxymethylcytosine when MeC is modified by oxidation may also play a role in DNA demethylation. This base modification is catalyzed by several enzymes, Tet1–3, members of the 2-oxoglutarate (2OG)- and Fe(II)-dependent enzyme family.[134] Tet1 is a fusion partner of the H3-K4 histone methyltranferase MLL in acute myeloid leukemia, and both MLL and Tet are found mutated in various tumors.[134] It is believed that 5-hydroxymethylcytosine may constitute an intermediate in the DNA demethylation process either by direct conversion to cytosine or by removal by a glycosylase that is present in mammalian nuclear extracts.[17]

In conclusion, in contrast to the rather simple situation in plants where several DNA glycosylases act directly on MeC, the situation is not so simple in vertebrates. After more than 15 years of controversy, a DNA demethylation mechanism in mammals that involves a DNA repair pathway involving a DNA glycosylase is becoming increasingly accepted, even though the experimental findings are limited. DNA demethylation is now believed to involve a multistep process, including MeC modification and recognition by a glycosylase involved in the repair of MeC DNA damage. DNA glycosylases are generally associated with the BER pathway that in some instances is involved in DNA demethylation.

4.5.2 Demethylation by the Nucleotide Excision Repair Pathways

Several reports propose that the nucleotide excision repair (NER) pathway is responsible for DNA demethylation. This DNA repair pathway involves many proteins, including the TFIIH transcription and repair complex and the 5′ and 3′ endonucleases XPF/ERCC1 and XPG, respectively, and is responsible for the recognition of mostly bulky lesions (Fig. 4.3). Following recognition of the lesion and recruitment of the various members of the pathway, a 24–32-base single-strand fragment is excised on both sides of the lesion and replaced by DNA polymerase δ– and ε–mediated DNA synthesis.[113] GADD45α has also been linked to DNA demethylation through the NER pathway. It interacts with XPG and promotes reactivation of a methylation-silenced gene in a XPB (a TFIIH subunit)- and XPG-dependent manner.[2] It can be recruited by transcription factors at promoters where it is involved in DNA demethylation.[73,120] It is recruited by the transcriptional cofactor Taf12 to the rDNA promoter along with the NER proteins XPA, XPF, and XPG[120] and by the retinoic acid receptor (RAR) to the RARβ2 promoter, along with TFIIH and various proteins of the NER pathway.[73] Proteins of the NER pathway are required, and the hierarchy of protein recruitment to the RARβ2 promoter is known[73] (Fig. 4.3). RAR recruits TFIIH and RNA polymerase II, XPC is simultaneously recruited and required for the recruitment of all the other NER proteins, XPA and presumably RPA are recruited and required for the

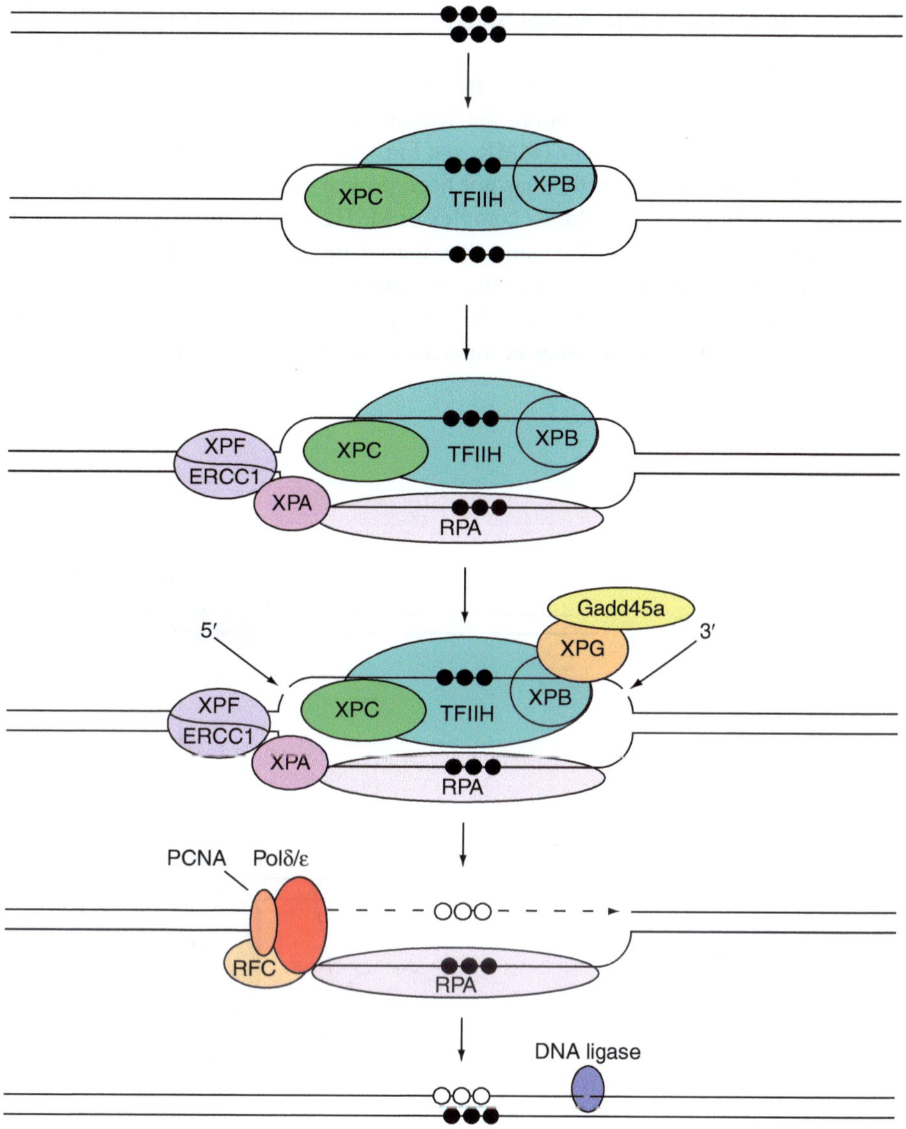

Fig. 4.3 MeC demethylation by the NER pathway. The scheme depicts proteins of the NER pathway[113] and their effect on DNA demethylation as proposed by LeMay et al.[73] The *filled black circles* represent MeCs and the *empty circles* non-methylated Cs

recruitment of the 5′ and 3′ endonuclease XPF/ERCC1 and XPG along with GADD45α[73] (Fig. 4.3). GADD45α however is not always involved in targeted DNA demethylation.[57]

In conclusion, several DNA demethylation events are performed by the BER or NER pathways, with GADD45α involved in both. GADD45α therefore may play a role in the coordination or selection of the pathways. Demethylation by way of the NER pathway allows simultaneous demethylation of neighboring MeCs on the same DNA strand within a 24–32-base segment. It can therefore be excluded for some DNA demethylation events, like that of the GR-activated *Tat* gene. Since the NER pathway is preferentially involved in the repair of bulky lesions and a methyl group is not at all bulky, it is conceivable that the MeC is not recognized as such and that the NER pathway functions in transcriptional activation irrespective of DNA demethylation. In particular, the single-strand DNA breaks that arise in the course of the repair may be needed for a chromatin remodeling event, as is true for the double-strand break induced by Topoisomerase II during nuclear receptor–mediated activation.[63] DNA demethylation may therefore constitute a collateral effect and not represent the objective of the pathway reactions.

4.5.3 The Links to Other Pathways

Given that three possible DNA repair pathways have been linked to DNA demethylation, the fourth DNA repair pathway, mismatch repair (MMR, see Fig. 4.4), may also be involved. So far this has not been proven to be the case. However, the MutSα complex of MSH2 and MSH6 can recognize G/T mismatches that result from MeC-deamination.[4] Because the glycosylase MBD4/MED1 interacts with the MMR protein MLH1,[26] the MMR pathway may be linked to repair damage in methylated DNA regions and eventually, in some instances, to DNA demethylation initiated by a deamination event.

Finally, Okada and colleagues have reported that the elongator complex has a role in zygotic paternal genome demethylation.[96] One of the elongator subunits, Elp3, possesses histone acetyltransferase (HAT) and radical S-adenosyl methionine (SAM) domains. Demethylation is impaired upon either knock down of Elp3 or overexpression of a dominant negative radical SAM domain mutant of Elp3.[96] In contrast, the HAT

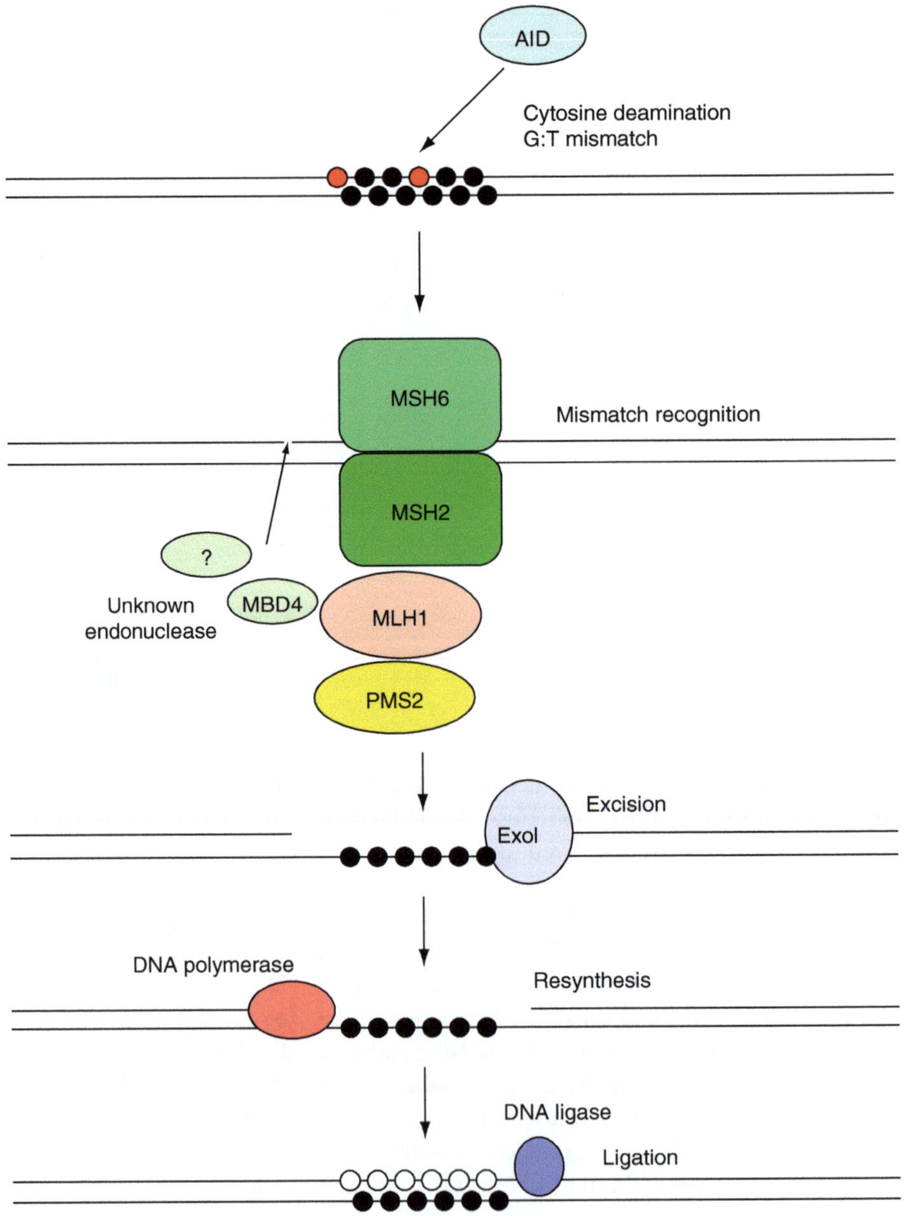

Fig. 4.4 CpG demethylation by the MMR pathway. The diagram suggests that MBD4/MED1 is the yet unidentified endonuclease that initiates DNA synthesis and replacement to effect DNA demethylation. The *filled black circles* represent MeCs, the *filled red circles* deaminated methylated Cs (i.e., Ts); *empty circles* represent the non-methylated Cs

activity of Elp3 appeared dispensable.[96] The elongator complex is involved in RNA polymerase II–driven transcriptional elongation, but also has other functions such as regulation of microtubule dynamics and intracellular trafficking, cytoplasmic kinase signaling, exocytosis, and tRNA modification.[27,132] It is not clear whether it is the transcriptional elongation or one of its other functions that is involved in DNA demethylation, nor is it clear which demethylating enzymes are involved and how they are regulated by the elongator function. Conceivably, the radical SAM domain could play a role in the removal of the methyl group from the MeC in DNA.[12,127]

4.6 Conclusions

It is possible that multiple pathways are involved in demethylating DNA either under specific conditions or to ensure robustness through redundancy. Given many factors that have been tentatively associated with DNA demethylation are also involved in DNA repair and in maintaining genome integrity, tight regulation of the epigenetic and genome surveillance functions of these factors is bound to be needed. This may explain why it has proven difficult to identify the factors' function. The inactivation of genes coding for the DNA demethylation factors has had a negligible effect on development and only a marginal effect on gene expression, perhaps because many proteins implicated in demethylation belong to families with one or more members. Alternatively, DNA demethylation may regulate gene expression by kinetic modulation of the responses rather than playing an all-or-none role. Indeed, the vast majority of demethylation events affect genes whose CpG density is low or intermediate, presumably to allow epigenetic flexibility. Demethylation may therefore function in the recruitment of methylation-sensitive transcription factors, rather than in regulating chromatin switches. Transcription factor recruitment may provide more subtle variations in expression levels and in the ability to integrate different signals in a gene-specific manner. It may also help memorize previous regulatory events, as for the *Tat* gene.[137] Conceivably these type of differences, even though important for human health, may be more difficult to identify in animal models.

Acknowledgments The DNA demethylation work from our team is supported in part by the CNRS and by grants from the Association de Recherche sur le Cancer and La Ligue Nationale contre le Cancer. We thank Eva-Maria Geigl, Peter Brooks, Gisele Picchi, Sophie Laget, and Pierre-Antoine Defossez for critical reading of the manuscript.

References

1. Ball MP, Li JB, Gao Y, et al. Targeted and genome-scale strategies reveal gene-body methylation signatures in human cells. *Nat Biotechnol*. 2009;27:361-368.
2. Barreto G. Gadd45a promotes epigenetic gene activation by repair-mediated DNA demethylation. *Nature*. 2007;445:671-675.
3. Beletskii A, Bhagwat AS. Transcription-induced mutations: increase in C to T mutations in the nontranscribed strand during transcription in *Escherichia coli*. *Proc Natl Acad Sci USA*. 1996;93:13919-13924.
4. Bellacosa A. Functional interactions and signaling properties of mammalian DNA mismatch repair proteins. *Cell Death Differ*. 2001;8:1076-1092.
5. Bhakat KK, Hazra TK, Mitra S. Acetylation of the human DNA glycosylase NEIL2 and inhibition of its activity. *Nucleic Acids Res*. 2004;32:3033-3039.
6. Bhakat KK, Mokkapati SK, Boldogh I, Hazra TK, Mitra S. Acetylation of human 8-oxoguanine-DNA glycosylase by p300 and its role in 8-oxoguanine repair in vivo. *Mol Cell Biol*. 2006;26:1654-1665.
7. Bhattacharya SK, Ramchandani S, Cervoni N, Szyf M. A mammalian protein with specific demethylase activity for mCpG DNA. *Nature*. 1999;397:579-583.
8. Bhutani N, Brady JJ, Damian M, Sacco A, Corbel SY, Blau HM. Reprogramming towards pluripotency requires AID-dependent DNA demethylation. *Nature*. 2010;463:1042-1047.
9. Bird A. DNA methylation patterns and epigenetic memory. *Genes Dev*. 2002;16:6-21.
10. Bird AP, Wolffe AP. Methylation-induced repression – belts, braces, and chromatin. *Cell*. 1999;99:451-454.
11. Blackledge NP, Zhou JC, Tolstorukov MY, Farcas AM, Park PJ, Klose RJ. CpG islands recruit a histone H3 lysine 36 demethylase. *Mol Cell*. 2010;38:179-190.
12. Booker SJ. Anaerobic functionalization of unactivated C–H bonds. *Curr Opin Chem Biol*. 2009;13:58-73.
13. Bostick M, Kim JK, Esteve PO, Clark A, Pradhan S, Jacobsen SE. UHRF1 plays a role in maintaining DNA methylation in mammalian cells. *Science*. 2007;317:1760-1764.
14. Boyes J, Bird A. Repression of genes by DNA methylation depends on CpG density and promoter strength: evidence for involvement of a methyl-CpG binding protein. *EMBO J*. 1992;11:327-333.
15. Brown SE, Szyf M. Epigenetic programming of the rRNA promoter by MBD3. *Mol Cell Biol*. 2007;27:4938-4952.
16. Brunner AL, Johnson DS, Kim SW, et al. Distinct DNA methylation patterns characterize differentiated human embryonic stem cells and developing human fetal liver. *Genome Res*. 2009;19:1044-1056.

17. Cannon SV, Cummings A, Teebor GW. 5-Hydroxymethylcytosine DNA glycosylase activity in mammalian tissue. *Biochem Biophys Res Commun*. 1988;151:1173-1179.
18. Cardoso MC, Leonhardt H. DNA methyltransferase is actively retained in the cytoplasm during early development. *J Cell Biol*. 1999;147:25-32.
19. Chen D, Lucey MJ, Phoenix F, et al. T:G mismatch-specific thymine-DNA glycosylase potentiates transcription of estrogen-regulated genes through direct interaction with estrogen receptor alpha. *J Biol Chem*. 2003;278:38586-38592.
20. Chen T, Ueda Y, Dodge JE, Wang Z, Li E. Establishment and maintenance of genomic methylation patterns in mouse embryonic stem cells by Dnmt3a and Dnmt3b. *Mol Cell Biol*. 2003;23:5594-5605.
21. Choi Y, Harada JJ, Goldberg RB, Fischer RL. An invariant aspartic acid in the DNA glycosylase domain of DEMETER is necessary for transcriptional activation of the imprinted MEDEA gene. *Proc Natl Acad Sci USA*. 2004;101:7481-7486.
22. Chotalia M, Smallwood SA, Ruf N, et al. Transcription is required for establishment of germline methylation marks at imprinted genes. *Genes Dev*. 2009;23:105-117.
23. Chuang LS, Ian HI, Koh TW, Ng HH, Xu G, Li BF. Human DNA-(cytosine-5) methyltransferase-PCNA complex as a target for p21WAF1. *Science*. 1997;277:1996-2000.
24. Cierpicki T, Risner LE, Grembecka J, et al. Structure of the MLL CXXC domain-DNA complex and its functional role in MLL-AF9 leukemia. *Nat Struct Mol Biol*. 2010;17:62-68.
25. Cortazar D, Kunz C, Saito Y, Steinacher R, Schar P. The enigmatic thymine DNA glycosylase. *DNA Repair (Amst)*. 2007;6:489-504.
26. Cortellino S, Turner D, Masciullo V, et al. The base excision repair enzyme MED1 mediates DNA damage response to antitumor drugs and is associated with mismatch repair system integrity. *Proc Natl Acad Sci USA*. 2003;100:15071-15076.
27. Creppe C, Malinouskaya L, Volvert ML, et al. Elongator controls the migration and differentiation of cortical neurons through acetylation of alpha-tubulin. *Cell*. 2009;136:551-564.
28. Di Croce L, Raker VA, Corsaro M, et al. Methyltransferase recruitment and DNA hypermethylation of target promoters by an oncogenic transcription factor. *Science*. 2002;295:1079-1082.
29. Dodd IB, Micheelsen MA, Sneppen K, Thon G. Theoretical analysis of epigenetic cell memory by nucleosome modification. *Cell*. 2007;129:813-822.
30. Doi A, Park IH, Wen B, et al. Differential methylation of tissue- and cancer-specific CpG island shores distinguishes human induced pluripotent stem cells, embryonic stem cells and fibroblasts. *Nat Genet*. 2009;41:1350-1353.
31. Dong KB, Maksakova IA, Mohn F, et al. DNA methylation in ES cells requires the lysine methyltransferase G9a but not its catalytic activity. *EMBO J*. 2008;27:2691-2701.
32. Eckhardt F, Lewin J, Cortese R, et al. DNA methylation profiling of human chromosomes 6, 20 and 22. *Nat Genet*. 2006;38:1378-1385.
33. Ehrlich M. DNA methylation and cancer-associated genetic instability. *Adv Exp Med Biol*. 2005;570:363-392.
34. Engel N, Tront JS, Erinle T, et al. Conserved DNA methylation in Gadd45a(−/−) mice. *Epigenetics*. 2009;4:98-99.
35. Epsztejn-Litman S, Feldman N, Abu-Remaileh M, et al. De novo DNA methylation promoted by G9a prevents reprogramming of embryonically silenced genes. *Nat Struct Mol Biol*. 2008;15:1176-1183.
36. Esteve PO, Chin HG, Smallwood A, et al. Direct interaction between DNMT1 and G9a coordinates DNA and histone methylation during replication. *Genes Dev*. 2006;20:3089-3103.
37. Flavin M, Cappabianca L, Kress C, Thomassin H, Grange T. Nature of the accessible chromatin at a glucocorticoid-responsive enhancer. *Mol Cell Biol*. 2004;24:7891-7901.
38. Gehring M, Bubb KL, Henikoff S. Extensive demethylation of repetitive elements during seed development underlies gene imprinting. *Science*. 2009;324:1447-1451.
39. Gehring M, Huh JH, Hsieh TF, et al. DEMETER DNA glycosylase establishes MEDEA polycomb gene self-imprinting by allele-specific demethylation. *Cell*. 2006;124:495-506.
40. Gehring M, Reik W, Henikoff S. DNA demethylation by DNA repair. *Trends Genet*. 2009;25:82-90.
41. Goll MG, Bestor TH. Eukaryotic cytosine methyltransferases. *Annu Rev Biochem*. 2005;74:481-514.
42. Graff J, Mansuy IM. Epigenetic dysregulation in cognitive disorders. *Eur J Neurosci*. 2009;30:1-8.
43. Grange T, Cappabianca L, Flavin M, Sassi H, Thomassin H. In vivo analysis of the model tyrosine aminotransferase gene reveals multiple sequential steps in glucocorticoid receptor action. *Oncogene*. 2001;20:3028-3038.
44. Hardeland U, Bentele M, Jiricny J, Schar P. The versatile thymine DNA-glycosylase: a comparative characterization of the human, Drosophila and fission yeast orthologs. *Nucleic Acids Res*. 2003;31:2261-2271.
45. Harris RS, Bishop KN, Sheehy AM, et al. DNA deamination mediates innate immunity to retroviral infection. *Cell*. 2003;113:803-809.
46. Heintzman ND, Stuart RK, Hon G, et al. Distinct and predictive chromatin signatures of transcriptional promoters and enhancers in the human genome. *Nat Genet*. 2007;39:311-318.
47. Hendrich B, Hardeland U, Ng HH, Jiricny J, Bird A. The thymine glycosylase MBD4 can bind to the product of deamination at methylated CpG sites. *Nature*. 1999;401:301-304.
48. Hirasawa R, Chiba H, Kaneda M, et al. Maternal and zygotic Dnmt1 are necessary and sufficient for the maintenance of DNA methylation imprints during preimplantation development. *Genes Dev*. 2008;22:1607-1616.
49. Hollander MC, Sheikh MS, Bulavin DV, et al. Genomic instability in Gadd45a-deficient mice. *Nat Genet*. 1999;23:176-184.
50. Hsieh CL. Dependence of transcriptional repression on CpG methylation density. *Mol Cell Biol*. 1994;14:5487-5494.
51. Hsieh CL. Stability of patch methylation and its impact in regions of transcriptional initiation and elongation. *Mol Cell Biol*. 1997;17:5897-5904.
52. Hsieh CL. Evidence that protein binding specifies sites of DNA demethylation. *Mol Cell Biol*. 1999;19:46-56.
53. Hsieh TF, Ibarra CA, Silva P, et al. Genome-wide demethylation of Arabidopsis endosperm. *Science*. 2009;324:1451-1454.
54. Illingworth R, Kerr A, Desousa D, et al. A novel CpG island set identifies tissue-specific methylation at developmental gene loci. *PLoS Biol*. 2008;6:e22.
55. Irizarry RA, Ladd-Acosta C, Wen B, et al. The human colon cancer methylome shows similar hypo- and hypermethyla-

tion at conserved tissue-specific CpG island shores. *Nat Genet.* 2009;41:178-186.
56. Jeong S, Liang G, Sharma S, et al. Selective anchoring of DNA methyltransferases 3A and 3B to nucleosomes containing methylated DNA. *Mol Cell Biol.* 2009;29:5366-5376.
57. Jin SG, Guo C, Pfeifer GP. GADD45A does not promote DNA demethylation. *PLoS Genet.* 2008;4:e1000013.
58. Jones PA, Baylin SB. The epigenomics of cancer. *Cell.* 2007;128:683-692.
59. Jost JP, Fremont M, Siegmann M, Hofsteenge J. The RNA moiety of chick embryo 5-methylcytosine-DNA glycosylase targets DNA demethylation. *Nucleic Acids Res.* 1997;25:4545-4550.
60. Jost JP, Oakeley EJ, Zhu B, et al. 5-Methylcytosine DNA glycosylase participates in the genome-wide loss of DNA methylation occurring during mouse myoblast differentiation. *Nucleic Acids Res.* 2001;29:4452-4461.
61. Jost JP, Schwarz S, Hess D, et al. A chicken embryo protein related to the mammalian DEAD box protein p68 is tightly associated with the highly purified protein-RNA complex of 5-MeC-DNA glycosylase. *Nucleic Acids Res.* 1999;27:3245-3252.
62. Jost JP, Siegmann M, Sun L, Leung R. Mechanisms of DNA demethylation in chicken embryos. Purification and properties of a 5-methylcytosine-DNA glycosylase. *J Biol Chem.* 1995;270:9734-9739.
63. Ju BG, Lunyak VV, Perissi V, et al. A topoisomerase IIbeta-mediated dsDNA break required for regulated transcription. *Science.* 2006;312:1798-1802.
64. Kangaspeska S, Stride B, Metivier R, et al. Transient cyclical methylation of promoter DNA. *Nature.* 2008;452:112-115.
65. Kass SU, Landsberger N, Wolffe AP. DNA methylation directs a time-dependent repression of transcription initiation. *Curr Biol.* 1997;7:157-165.
66. Kim MS, Kondo T, Takada I, et al. DNA demethylation in hormone-induced transcriptional derepression. *Nature.* 2009;461:1007-1012.
67. Klose RJ, Bird AP. Genomic DNA methylation: the mark and its mediators. *Trends Biochem Sci.* 2006;31:89-97.
68. Kress C, Thomassin H, Grange T. Local DNA demethylation in vertebrates: how could it be performed and targeted? *FEBS Lett.* 2001;494:135-140.
69. Kress C, Thomassin H, Grange T. Active cytosine demethylation triggered by a nuclear receptor involves DNA strand breaks. *Proc Natl Acad Sci USA.* 2006;103:11112-11117.
70. Laird CD, Pleasant ND, Clark AD, et al. Hairpin-bisulfite PCR: assessing epigenetic methylation patterns on complementary strands of individual DNA molecules. *Proc Natl Acad Sci USA.* 2004;101:204-209.
71. Laurent L, Wong E, Li G, et al. Dynamic changes in the human methylome during differentiation. *Genome Res.* 2010;20:320-331.
72. Le Guezennec X, Vermeulen M, Brinkman AB, et al. MBD2/NuRD and MBD3/NuRD, two distinct complexes with different biochemical and functional properties. *Mol Cell Biol.* 2006;26:843-851.
73. Le May N, Mota-Fernandes D, Velez-Cruz R, Iltis I, Biard D, Egly JM. NER factors are recruited to active promoters and facilitate chromatin modification for transcription in the absence of exogenous genotoxic attack. *Mol Cell.* 2010;38:54-66.
74. Lehnertz B, Ueda Y, Derijck AA, et al. Suv39h-mediated histone H3 lysine 9 methylation directs DNA methylation to major satellite repeats at pericentric heterochromatin. *Curr Biol.* 2003;13:1192-1200.
75. Li E. Chromatin modification and epigenetic reprogramming in mammalian development. *Nat Rev Genet.* 2002;3:662-673.
76. Likhite VS, Cass EI, Anderson SD, Yates JR, Nardulli AM. Interaction of estrogen receptor alpha with 3-methyladenine DNA glycosylase modulates transcription and DNA repair. *J Biol Chem.* 2004;279:16875-16882.
77. Lin IG, Tomzynski TJ, Ou Q, Hsieh CL. Modulation of DNA binding protein affinity directly affects target site demethylation. *Mol Cell Biol.* 2000;20:2343-2349.
78. Lindroth AM, Shultis D, Jasencakova Z, et al. Dual histone H3 methylation marks at lysines 9 and 27 required for interaction with CHROMOMETHYLASE3. *EMBO J.* 2004;23:4286-4296.
79. Lister R, Pelizzola M, Dowen RH, et al. Human DNA methylomes at base resolution show widespread epigenomic differences. *Nature.* 2009;462:315-322.
80. Ma DK, Jang MH, Guo JU, et al. Neuronal activity-induced Gadd45b promotes epigenetic DNA demethylation and adult neurogenesis. *Science.* 2009;323:1074-1077.
81. Matsuo K, Silke J, Georgiev O, Marti P, Giovannini N, Rungger D. An embryonic demethylation mechanism involving binding of transcription factors to replicating DNA. *EMBO J.* 1998;17:1446-1453.
82. Matzke M, Kanno T, Huettel B, Daxinger L, Matzke AJ. Targets of RNA-directed DNA methylation. *Curr Opin Plant Biol.* 2007;10:512-519.
83. Mayer W, Niveleau A, Walter J, Fundele R, Haaf T. Demethylation of the zygotic paternal genome. *Nature.* 2000;403:501-502.
84. Meissner A, Mikkelsen TS, Gu H, et al. Genome-scale DNA methylation maps of pluripotent and differentiated cells. *Nature.* 2008;454:766-770.
85. Metivier R, Gallais R, Tiffoche C, et al. Cyclical DNA methylation of a transcriptionally active promoter. *Nature.* 2008;452:45-50.
86. Metivier R, Penot G, Hubner MR, et al. Estrogen receptor-alpha directs ordered, cyclical, and combinatorial recruitment of cofactors on a natural target promoter. *Cell.* 2003;115:751-763.
87. Mikl MC, Watt IN, Lu M, et al. Mice deficient in APOBEC2 and APOBEC3. *Mol Cell Biol.* 2005;25:7270-7277.
88. Millar CB, Guy J, Sansom OJ, et al. Enhanced CpG mutability and tumorigenesis in MBD4-deficient mice. *Science.* 2002;297:403-405.
89. Mohn F, Weber M, Rebhan M, et al. Lineage-specific polycomb targets and de novo DNA methylation define restriction and potential of neuronal progenitors. *Mol Cell.* 2008;30:755-766.
90. Morales-Ruiz T, Ortega-Galisteo AP, Ponferrada-Marin MI, Martinez-Macias MI, Ariza RR, Roldan-Arjona T. DEMETER and REPRESSOR OF SILENCING 1 encode 5-methylcytosine DNA glycosylases. *Proc Natl Acad Sci USA.* 2006;103:6853-6858.
91. Morgan HD, Dean W, Coker HA, Reik W, Petersen-Mahrt SK. Activation-induced cytidine deaminase deaminates 5-methylcytosine in DNA and is expressed in pluripotent tissues: implications for epigenetic reprogramming. *J Biol Chem.* 2004;279:52353-52360.

92. Morris KV, Chan SW, Jacobsen SE, Looney DJ. Small interfering RNA-induced transcriptional gene silencing in human cells. *Science*. 2004;305:1289-1292.
93. Muramatsu M, Kinoshita K, Fagarasan S, Yamada S, Shinkai Y, Honjo T. Class switch recombination and hypermutation require activation-induced cytidine deaminase (AID), a potential RNA editing enzyme. *Cell*. 2000;102: 553-563.
94. Nan X, Ng HH, Johnson CA, et al. Transcriptional repression by the methyl-CpG-binding protein MeCP2 involves a histone deacetylase complex. *Nature*. 1998;393:386-389.
95. Ohm JE, McGarvey KM, Yu X, et al. A stem cell-like chromatin pattern may predispose tumor suppressor genes to DNA hypermethylation and heritable silencing. *Nat Genet*. 2007;39:237-242.
96. Okada Y, Yamagata K, Hong K, Wakayama T, Zhang Y. A role for the elongator complex in zygotic paternal genome demethylation. *Nature*. 2010;463:554-558.
97. Ooi SK, Bestor TH. The colorful history of active DNA demethylation. *Cell*. 2008;133:1145-1148.
98. Ooi SK, Qiu C, Bernstein E, et al. DNMT3L connects unmethylated lysine 4 of histone H3 to de novo methylation of DNA. *Nature*. 2007;448:714-717.
99. Oswald J. Active demethylation of the paternal genome in the mouse zygote. *Curr Biol*. 2000;10:475-478.
100. Otani J, Nankumo T, Arita K, Inamoto S, Ariyoshi M, Shirakawa M. Structural basis for recognition of H3K4 methylation status by the DNA methyltransferase 3A ATRX-DNMT3-DNMT3L domain. *EMBO Rep*. 2009;10: 1235-1241.
101. Palii SS, Van Emburgh BO, Sankpal UT, Brown KD, Robertson KD. DNA methylation inhibitor 5-Aza-2′-deoxycytidine induces reversible genome-wide DNA damage that is distinctly influenced by DNA methyltransferases 1 and 3B. *Mol Cell Biol*. 2008;28:752-771.
102. Paroush Z, Keshet I, Yisraeli J, Cedar H. Dynamics of demethylation and activation of the alpha-actin gene in myoblasts. *Cell*. 1990;63:1229-1237.
103. Pogribny IP, Beland FA. DNA hypomethylation in the origin and pathogenesis of human diseases. *Cell Mol Life Sci*. 2009;66:2249-2261.
104. Popp C, Dean W, Feng S, et al. Genome-wide erasure of DNA methylation in mouse primordial germ cells is affected by AID deficiency. *Nature*. 2010;463:1101-1105.
105. Rai K, Huggins IJ, James SR, Karpf AR, Jones DA, Cairns BR. DNA demethylation in zebrafish involves the coupling of a deaminase, a glycosylase, and gadd45. *Cell*. 2008;135:1201-1212.
106. Ramchandani S, Bhattacharya SK, Cervoni N, Szyf M. DNA methylation is a reversible biological signal. *Proc Natl Acad Sci USA*. 1999;96:6107-6112.
107. Ramsahoye BH, Biniszkiewicz D, Lyko F, Clark V, Bird AP, Jaenisch R. Non-CpG methylation is prevalent in embryonic stem cells and may be mediated by DNA methyltransferase 3a. *Proc Natl Acad Sci USA*. 2000;97: 5237-5242.
108. Rauch TA, Wu X, Zhong X, Riggs AD, Pfeifer GP. A human B cell methylome at 100-base pair resolution. *Proc Natl Acad Sci USA*. 2009;106:671-678.
109. Razin A, Szyf M, Kafri T, et al. Replacement of 5-methylcytosine by cytosine: a possible mechanism for transient DNA demethylation during differentiation. *Proc Natl Acad Sci USA*. 1986;83:2827-2831.
110. Reik W. Stability and flexibility of epigenetic gene regulation in mammalian development. *Nature*. 2007;447: 425-432.
111. Robertson KD. DNA methylation and human disease. *Nat Rev Genet*. 2005;6:597-610.
112. Saluz HP, Jiricny J, Jost JP. Genomic sequencing reveals a positive correlation between the kinetics of strand-specific DNA demethylation of the overlapping estradiol/glucocorticoid-receptor binding sites and the rate of avian vitellogenin mRNA synthesis. *Proc Natl Acad Sci USA*. 1986;83:7167-7171.
113. Sancar A. DNA excision repair. *Annu Rev Biochem*. 1996;65:43-81.
114. Sarraf SA, Stancheva I. Methyl-CpG binding protein MBD1 couples histone H3 methylation at lysine 9 by SETDB1 to DNA replication and chromatin assembly. *Mol Cell*. 2004;15:595-605.
115. Sasai N, Defossez PA. Many paths to one goal? The proteins that recognize methylated DNA in eukaryotes. *Int J Dev Biol*. 2009;53:323-334.
116. Sassi H, Pictet R, Grange T. Glucocorticoids are insufficient for neonatal gene induction in the liver. *Proc Natl Acad Sci USA*. 1998;95:5621-5625.
117. Schaefer CB, Ooi SK, Bestor TH, Bourc'his D. Epigenetic decisions in mammalian germ cells. *Science*. 2007;316: 398-399.
118. Schlesinger Y, Straussman R, Keshet I, et al. Polycomb-mediated methylation on Lys27 of histone H3 pre-marks genes for de novo methylation in cancer. *Nat Genet*. 2007;39:232-236.
119. Schmidl C, Klug M, Boeld TJ, et al. Lineage-specific DNA methylation in T cells correlates with histone methylation and enhancer activity. *Genome Res*. 2009;19:1165-1174.
120. Schmitz KM, Schmitt N, Hoffmann-Rohrer U, Schafer A, Grummt I, Mayer C. TAF12 recruits Gadd45a and the nucleotide excision repair complex to the promoter of rRNA genes leading to active DNA demethylation. *Mol Cell*. 2009;33:344-353.
121. Sedgwick B. Repairing DNA-methylation damage. *Nat Rev Mol Cell Biol*. 2004;5:148-157.
122. Seki Y, Hayashi K, Itoh K, Mizugaki M, Saitou M, Matsui Y. Extensive and orderly reprogramming of genome-wide chromatin modifications associated with specification and early development of germ cells in mice. *Dev Biol*. 2005;278:440-458.
123. Sen GL, Reuter JA, Webster DE, Zhu L, Khavari PA. DNMT1 maintains progenitor function in self-renewing somatic tissue. *Nature*. 2010;463:563-567.
124. Sharif J, Muto M, Takebayashi S, et al. The SRA protein Np95 mediates epigenetic inheritance by recruiting Dnmt1 to methylated DNA. *Nature*. 2007;450:908-912.
125. Shen JC, Rideout WM 3rd, Jones PA. The rate of hydrolytic deamination of 5-methylcytosine in double-stranded DNA. *Nucleic Acids Res*. 1994;22:972-976.
126. Shi Y, Lan F, Matson C, et al. Histone demethylation mediated by the nuclear amine oxidase homolog LSD1. *Cell*. 2004;119:941-953.
127. Sofia HJ, Chen G, Hetzler BG, Reyes-Spindola JF, Miller NE. Radical SAM, a novel protein superfamily linking unresolved steps in familiar biosynthetic pathways with

radical mechanisms: functional characterization using new analysis and information visualization methods. *Nucleic Acids Res*. 2001;29:1097-1106.
128. Steinberg RA. Enzymic removal of 5-methylcytosine from poly(dG-5-methyl-dC) by HeLa cell nuclear extracts is not by a DNA glycosylase. *Nucleic Acids Res*. 1995;23: 1621-1624.
129. Stenglein MD, Burns MB, Li M, Lengyel J, Harris RS. APOBEC3 proteins mediate the clearance of foreign DNA from human cells. *Nat Struct Mol Biol*. 2010;17: 222-229.
130. Straussman R, Nejman D, Roberts D, et al. Developmental programming of CpG island methylation profiles in the human genome. *Nat Struct Mol Biol*. 2009;16:564-571.
131. Suzuki MM, Bird A. DNA methylation landscapes: provocative insights from epigenomics. *Nat Rev Genet*. 2008;9:465-476.
132. Svejstrup JQ. Elongator complex: how many roles does it play? *Curr Opin Cell Biol*. 2007;19:331-336.
133. Tachibana M, Matsumura Y, Fukuda M, Kimura H, Shinkai Y. G9a/GLP complexes independently mediate H3K9 and DNA methylation to silence transcription. *EMBO J*. 2008;27:2681-2690.
134. Tahiliani M, Koh KP, Shen Y, et al. Conversion of 5-methylcytosine to 5-hydroxymethylcytosine in mammalian DNA by MLL partner TET1. *Science*. 2009;324:930-935.
135. Tamaru H, Zhang X, McMillen D, et al. Trimethylated lysine 9 of histone H3 is a mark for DNA methylation in Neurospora crassa. *Nat Genet*. 2003;34:75-79.
136. Teixeira FK, Heredia F, Sarazin A, et al. A role for RNAi in the selective correction of DNA methylation defects. *Science*. 2009;323:1600-1604.
137. Thomassin H, Flavin M, Espinas ML, Grange T. Glucocorticoid-induced DNA demethylation and gene memory during development. *EMBO J*. 2001;20:1974-1983.
138. Thomson JP, Skene PJ, Selfridge J, et al. CpG islands influence chromatin structure via the CpG-binding protein Cfp1. *Nature*. 2010;464:1082-1086.
139. Tini M, Benecke A, Um SJ, Torchia J, Evans RM, Chambon P. Association of CBP/p300 acetylase and thymine DNA glycosylase links DNA repair and transcription. *Mol Cell*. 2002;9:265-277.
140. Vairapandi M. Characterization of DNA demethylation in normal and cancerous cell lines and the regulatory role of cell cycle proteins in human DNA demethylase activity. *J Cell Biochem*. 2004;91:572-583.
141. Vairapandi M, Duker NJ. Enzymic removal of 5-methylcytosine from DNA by a human DNA-glycosylase. *Nucleic Acids Res*. 1993;21:5323-5327.
142. Vire E, Brenner C, Deplus R, et al. The Polycomb group protein EZH2 directly controls DNA methylation. *Nature*. 2006;439:871-874.
143. Wang Y, Zhang H, Chen Y, et al. LSD1 is a subunit of the NuRD complex and targets the metastasis programs in breast cancer. *Cell*. 2009;138:660-672.
144. Weber M, Hellmann I, Stadler MB, et al. Distribution, silencing potential and evolutionary impact of promoter DNA methylation in the human genome. *Nat Genet*. 2007;39:457-466.
145. Wilks AF, Cozens PJ, Mattaj IW, Jost JP. Estrogen induces a demethylation at the 5′ end region of the chicken vitellogenin gene. *Proc Natl Acad Sci USA*. 1982;79:4252-4255.
146. Wilks A, Seldran M, Jost JP. An estrogen-dependent demethylation at the 5′ end of the chicken vitellogenin gene is independent of DNA synthesis. *Nucleic Acids Res*. 1984;12:1163-1177.
147. Wolffe AP, Jones PL, Wade PA. DNA demethylation. *Proc Natl Acad Sci USA*. 1999;96:5894-5896.
148. Yisraeli J, Adelstein RS, Melloul D, Nudel U, Yaffe D, Cedar H. Muscle-specific activation of a methylated chimeric actin gene. *Cell*. 1986;46:409-416.
149. Yoon JH, Iwai S, O'Connor TR, Pfeifer GP. Human thymine DNA glycosylase (TDG) and methyl-CpG-binding protein 4 (MBD4) excise thymine glycol (Tg) from a Tg:G mispair. *Nucleic Acids Res*. 2003;31:5399-5404.
150. Zemach A, McDaniel IE, Silva P, Zilberman D. Genome-wide evolutionary analysis of eukaryotic DNA methylation. *Science*. 2010;328:916-919.
151. Zhan Q. Gadd45a, a p53- and BRCA1-regulated stress protein, in cellular response to DNA damage. *Mutat Res*. 2005;569:133-143.
152. Zhao Q, Rank G, Tan YT, et al. PRMT5-mediated methylation of histone H4R3 recruits DNMT3A, coupling histone and DNA methylation in gene silencing. *Nat Struct Mol Biol*. 2009;16:304-311.
153. Zhu B, Benjamin D, Zheng Y, et al. Overexpression of 5-methylcytosine DNA glycosylase in human embryonic kidney cells EcR293 demethylates the promoter of a hormone-regulated reporter gene. *Proc Natl Acad Sci USA*. 2001;98:5031-5036.
154. Zhu B, Zheng Y, Angliker H, et al. 5-Methylcytosine DNA glycosylase activity is also present in the human MBD4 (G/T mismatch glycosylase) and in a related avian sequence. *Nucleic Acids Res*. 2000;28:4157-4165.
155. Zhu B, Zheng Y, Hess D, et al. 5-Methylcytosine-DNA glycosylase activity is present in a cloned G/T mismatch DNA glycosylase associated with the chicken embryo DNA demethylation complex. *Proc Natl Acad Sci USA*. 2000;97: 5135-5139.

The Role of Histone Demethylases in Disease

5

Paul Cloos

5.1 Introduction

Epigenetic factors modulate cellular differentiation by altering histone structure and changing DNA methylation.[9,97] These processes package genes into "open and accessible" or "closed and inaccessible" chromatin states in a cell type–specific manner.[9,52] Chromatin structure shapes gene-expression programs in a variety of cell types by regulating access of transcription factors to specific regions. This permits different cells to maintain characteristic transcriptional programs that lead to specialized functions, even though all cells contain the same DNA-code.

Histones are subject to a variety of post-translational modifications, primarily at their N-termini. These modifications cause remodeling of the chromatin structure and are implicated in regulating transcription, DNA replication, and DNA repair.[9,52] Histone lysine residues are subject to modification by one, two, or three methyl groups (Kme1, Kme2, Kme3). Similarly, arginine residues may be mono- (Rme1) or dimethylated. Given the dimethylated form is either in symmetric or asymmetric conformation (Rme2s, Rme2a) it can give rise to four states, one unmethylated and three methylated forms. Histone lysine methylation affects the transcriptional activity of DNA by functioning as a recognition template for effector proteins that modify the chromatin environment and determine the transcriptional outcome, expression, or activation. Conversely, arginine methylation affects transcription by preventing methylation at specific lysine residues and inhibits the binding of effector proteins/complexes.[34,48] The transcriptional outcome of histone methylation is a function of the degree of methylation and of the location of the modified residue. Hence, methylation at lysine residues H3K9, H3K27, and H4K20 is associated with transcriptional repression, whereas methylation at H3K4, H3K36, and H3K79 is linked to transcriptional activation.

The epigenome of a cell is not as static as its genome because the epigenome is determined by enzymes that are subject to modulation by environmental and exogenous factors such as small molecules. The plasticity of the epigenetic setup of a cell is exemplified by the finding that somatic cells, like fully differentiated fibroblasts, can be transformed into a pluripotent stem cell (so-called iPS cells) by expressing a combination of transcription factors.[99]

Proteins involved in epigenetic modulation are obviously vital for normal development and cellular function; their aberrant regulation or function may contribute or cause human disease. Histone methyltransferases that catalyze histone methylation and effector proteins that "read" histone methylations and trigger transcriptional outcomes have been known for two decades.

On the other hand, enzymes that catalyze removal of methylation from histones were discovered only in 2004.[93] Since then, several families of histone demethylases have been characterized and implicated in transcriptional regulation and in various other biological processes.[25,93,114]

Due to the essential regulatory role of histone demethylases, dysregulation or inappropriate positioning of these enzymes may contribute to or cause disease. This chapter reviews current knowledge of this class of enzymes and discusses their potential involvement in human diseases.

Table 5.1 provides an overview of histone demethylases that have been associated with disease. Some of these enzymes have not, as yet, been

P. Cloos
Biotech Research & Innovation Centre,
Copenhagen, Denmark
e-mail: paul.cloos@bric.dk

H.I. Roach et al. (eds.), *Epigenetic Aspects of Chronic Diseases*,
DOI: 10.1007/978-1-84882-644-1_5,

Table 5.1 Histone demethylases associated with disease

Name	Synonym	Activity	Involvement in disease	Evidence
LSD1	AOF2 KDM1 BHC110 KIAA0601	H3K4me2me1 H3K9me2/me1	Negative regulator of metastasis	Cellular assays Mouse models
FBXL10	JHDM1B KDM2B	H3K36me2/me1	Candidate tumor suppressor	Cellular assays
GASC1	JMJD2C KDM4C	H3K9me3/me2 H3K36me3/me2	Putative oncogene Candidate gene for squamous cell carcinoma, prostate cancer, breast cancer, and medulloblastoma	Studies on cells Copy number variations Expression arrays on human samples
JHDM2A	JMJD1A TSGA KDM3A	H3K9me2/me1	Candidate gene for azoospermia and globospermia	Knockout mice
JHDM2B	JMJD1B KDM3B TRIP8 5qCNA	H3K9me2/me1	Candidate tumor suppressor Candidate oncogene Candidate gene for acute myeloid leukemia and myelodysplasia	Studies on cells Mouse models Location at genomic hotspot (5q31)
JMJD1C		Unknown	Candidate gene for autism	Mutations in humans
HR		Unknown	AUC (OMIM203655) APL (OMIM209500) MUHH (OMIM146550)	Mutations in humans Knockout mice
JMJD2A	JHDM3A KDM4A	H3K9me3/me2	Putative oncogene Candidate gene for prostate cancer	Expression arrays on human samples
JMJD2B	KDM4B	H3K9me3/me2	Putative oncogene Candidate gene for prostate cancer and medulloblastoma	Copy number variations Expression arrays on human samples
JARID1B	PLU1 KDM5B	H3K4me3/me2	Putative oncogene	Studies on cells Mouse models
JARID1C	SMCX KDM5C	H3K4me3/me2	XLMR	Mutations in humans
JARID2	Jmj	Unknown	Candidate gene for CHD and Schizophrenia	Mutations in humans
PHF8	PHD8	H3K9me2/me1	XLMR Cleft lip/palate	Mutations in humans
PHF2	PHD2	H3K9me2/me1	Candidate gene for HSN1 and ESS1	Mutations in humans
UTX	KDM6A	H3K27me3/me2	Candidate tumor suppressor	Mutations in humans Expression arrays on human samples

Table 5.1 (continued)

Name	Synonym	Activity	Involvement in disease	Evidence
JMJD3	KDM6B	H3K27me3/me2	Candidate tumor suppressor	Location at genomic hotspot (17p13.1) Studies on cells Expression arrays on human samples
JMJD5		Unknown	Candidate tumor suppressor	Mouse models
HSPBAP1	PASS1	Unknown	Candidate gene for IE	Mutations in humans

The activity column provides the substrate specificity of the demethylase, i.e., JHDM2A catalyzes demethylation of lysine 9 on H3 from either dimethylated or monomethylated lysine

APL Arthricia with popular lesion, *AUC* Alopecia universalis congenital, *CHD* Congenital Heart Disease, *ESS1* multiple self-healing squamosus epitheloma, *HSN1* Hereditary neuropathy1, *IE* intractable epilepsy, *MUHH* Marie Unna Hereditary Hypotrichosis, *XLMR* X-linked mental retardation

characterized as histone demethylases, but feature motifs, as the JMJC protein domain, known to catalyze histone demethylation.

5.2 Histone Demethylases in Cancer

Many genes that encode histone demethylases are subject to aberration mutations, translocations, deletions, and abnormal expression in a variety of human cancers; this suggests that defective control of these enzymes contributes to the pathogenesis of cancers.

As reviewed below, some histone demethylases have been associated with specific cancer types, whereas others, e.g., GASC1, are linked to many different cancers, they may therefore be general tumor suppressors or oncoproteins.

5.2.1 The H3K9 Demethylase GASC1 Is a Putative Oncoprotein

The *gene-amplified in squamous cell carcinoma* (*GASC1*) encodes GASC1 (also denoted JMJD2C and KDM4C[4]), a member of the JMJD2 family of histone demethylases.[22,50,118]

The JMJD2 group of histone demethylases catalyzes the demethylation of trimethylated and dimethylated lysine 9 and of lysine 36 on histone H3 (H3K9me3/me2 and H3K36me3/me2). The biological significance of this dual specificity is not well understood. However, findings that connect GASC1 to oncogenesis point to its role in regulating H3K9 demethylation.

The GASC1 gene, as well as some other genes encoding members of the JMJD2 group, is subject to copy number variation and highly expressed in various malignancies, including carcinomas of the esophagus, prostate, breast, and medulloblastomas.[22,23,62,74,119,124] Moreover, the fact that GASC1 is necessary for the proliferation of cancer cells[22,62,119] points to a role of GASC1 and H3K9 demethylation in cancer pathogenesis.

Whether the amplification of GASC1 or other JMJD2 demethylases is causally involved in the pathogenesis of these cancer subtypes and what the molecular mechanism could be that induces carcinogenesis is not known. Some possibilities of how aberrant function arises are: (1) activation of oncogenes, (2) pushing cells toward a pluripotent state, and (3) repressing senescence.

5.2.1.1 Activation of Oncogenes

GASC1 and other JMJD2 histone demethylases catalyze the demethylation of the repressive H3K9me3 mark and are therefore likely to be transcriptional activators. It is therefore conceivable that their oncogenic function is related to their role as activators of an oncogene. In support of such a notion is the report[40] that GASC1 induces transcription of the oncogene MDM2, a p53-specific ubiquitin ligase. Ectopic expression of GASC1 was found to increase MDM2 and to reduce p53.[40] Genomic localization analysis (ChIP) indicated that GASC1 was recruited to the P2 promoter region of *MDM2*, causing this promoter area to become demethylated.[40] On the other hand, siRNA-mediated depletion of GASC1 decreased MDM2 expression. These findings link GASC1 to MDM2 expression and implicate it in the downregulation of the tumor suppressor p53.

5.2.1.2 Pushing Cells Toward a Pluripotent State

The so-called cancer stem cell hypothesis holds that a fraction of tumor cells with stem cell-like properties are primarily responsible for the establishment, progression, and recurrence of cancer. As stem or progenitor cells evolve to more specialized cell types in the course of differentiation, genome-wide chromatin changes take place including the removal of some and the setting up of other histone marks. This rearrangement of the epigenetic setup will modify the transcriptional program of the cell resulting in new phenotypes that characterize the resulting lineage.

Importantly, "reprogrammed" murine iPS cells can differentiate into any cell type and form viable chimeras; this indicates that biological potency and the epigenetic state of iPS cells are indistinguishable from those of ES cells.[66]

At the chromatin level, the cellular reprogramming of iPS cells is accompanied by a re-setting (erasure) of epigenetic marks (i.e., DNA and histone methylation). This epigenetic re-setting is probably mediated by epigenetic enzymes such as histone and DNA-demethylases,[23,98] which in turn are directed by transcriptional factors.

Alterations in histone methylation therefore represent important steps in normal cellular differentiation, but may also be at play in cancer progression where aberrant regulation of histone methylation/demethylation may alter the normal transcriptional programs. Demethylases thus regulate factors of importance to stem cells or in embryogenesis and early development, but in adult organisms, their importance may be lost or even become detrimental. Examples of such factors are *OCT4* and *NANOG* that are key for stem cell self-renewal.

GASC1 has been implicated in the transcriptional regulation of key factors essential for the self-renewal of stem cells. One of these is the pluripotency factor *Nanog* that acts on the *GASC1* gene, recruiting *GASC1* to the *Nanog* promoter in mouse ES cells.[64] The same authors found *Gasc1* to be a target of the reprogramming factor Oct4 in murine ES cells; this suggests that

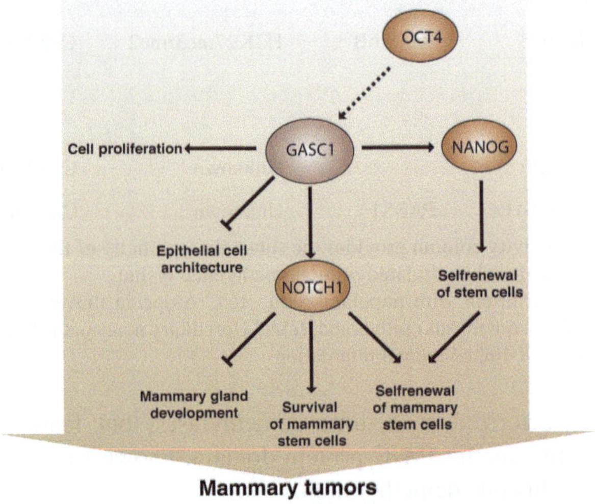

Fig. 5.1 Model for the role of histone GASC1 in self-renewal and breast cancer. GASC1 has been implicated in the regulation of factors important for self-renewal of stem cells as NANOG and NOTCH1 (see text). The induction of NOTCH1 prevents mammary gland development and leads to formation of breast tumors in animal models.[38] In addition, GASC1 may contribute to oncogenesis by stimulating cell proliferation and modulating epithelial cell architecture

Gasc1 may play a role in relation to stem cells.[64] Similarly, *NOTCH1* is a target gene of GASC1 in breast cancer cells, and NANOG is induced in mammosphere-forming mammary cells that overexpress GASC1[62] (Fig. 5.1).

These studies suggest that GASC1 functions in establishing a stem cell-like transcriptional signature that may contribute to oncogenic transformation.

5.2.1.3 Counteracting Senescence

Cellular senescence is a form of irreversible growth arrest and constitutes a barrier against excess cell proliferation induced by many stimuli, including activation of oncogenes and DNA damage. That senescence acts to repress carcinogenesis has been shown in a number of reports.[15,24,57,70]

The transcription factors pRB and p53 are important tumor suppressors and fulfill key roles in the induction of senescence. p53 induces the cyclin-dependent kinase inhibitor, p21 that in turn inhibits the cell cycle at G1. The active (hypophosphorylated) form of the pRb protein forms complex with E2F proteins, silencing gene transcription. These complexes also induce heterochromatin on E2F target promoters and so-called Senescence Associated Heterochromatin Foci (SAHFs) that are enriched in H3K9 trimethylation and constitute hallmarks of senescence.[71]

JMJD2 demethylases that erase H3K9me3 methylations may contribute to genomic instability and counteract the formation of SAHFs to prevent senescence induction, potentially paving the way for malignant transformation.[22] The essential role played by H3K9 trimethylation in SAHF formation and in senescene induction has been repeatedly reported.

Double knockout mice to the H3K9 methyltransferases *Suv39h1* and *Suv39h2* display loss of H3K9 trimethylation at pericentric chromatin and impairment of heterochromatin structures and genomic stability.[82] Furthermore, the loss of *Suv39H1* in transgenic mice expressing a constitutively active version of the oncogene K-RAS (RAS(G12D)) causes development of B cell lymphomas. This leads to a decrease in the ability of primary lymphocytes to enter senescence, a finding that links loss of H3K9 methylation to the inability to induce senescence.[15] Moreover, knockout mice for the gene encoding the H3K9 methyltransferases *RIZ1* develop a high incidence of diffuse, large B cell lymphoma and display a range of rare tumors.[95] This finding again links the loss of H3K9 methyltransferase activity and the consequent loss of H3K9 methylation to oncogenesis.

Similarly, the expression and activity of the H3K9 methyltransferase RIZ1 are reduced in many cancers by deletions, missense, and frameshift mutations, as well as by promoter methylation.[32,54,84]

Taken together, these findings indicate that loss of H3K9me3 as a consequence of increased GASC1 activity or expression may cause genomic instability and decrease the ability of cells to enter senescence, thereby contributing to oncogenesis.

5.2.2 The H3K27me3 Histone Demethylase JMJD3 Is a Tumor Suppressor Protein Involved in the Activation of the INK4/ARF Locus

Another demethylase implicated in the regulation of the senescence process is the H3K27 demethylase JMJD3.[2,3,6,55] Whereas heterochromatin foci and increased levels of repressive K9 histone marks are associated with senescence the induction of repressive K27 methylation at specific genetic loci and the resulting gene silencing has been linked with cell immortalization and cancer.[14,24,108]

Two studies[3,6] have shown that JMJD3 fulfills a crucial role in senescence induction as it catalyzes the active removal of H3K27me3 from the *INK4/ARF* tumor suppressor locus in response to stress- and oncogene-induced stimuli.

JMJD3 is rapidly induced in response to oncogenic BRAF or RAS.[3,6] Genomic localization analysis has shown that JMJD3 is recruited to the *INK4A/ARF* locus in cells stimulated with BRAF, along with the loss of polycomb group proteins, of H3K27 methylation, and an increase in the transcription of *INK4A* and *ARF*.[3,6] Similarly, shRNA-mediated depletion of *Jmjd3* in mouse embryonic fibroblasts causes suppression of *Ink4a* and *Arf* expression and immortalization of cells.[3,6] On the other hand, ectopic expression of *JMJD3* leads to increased levels of *INK4A* and of senescence.[3,6]

Together these findings indicate that the loss of H3K27me3 at the *INK4A/ARF* locus as cells undergo senescence is at least partially due to an increase in the association of the H3K27me3 demethylase with this genomic site (Fig. 5.2). Of note, the minimal promoter area necessary for JMJD3 induction in response to BRAF features a binding site for the AP-1 transcription factor. The latter is a heterodimer of c-JUN, and of c-FOS, both targets of the extracellular signal-regulated (ERK) kinase which is a component of the RAS-RAF-MEK-ERK kinase pathway.[3] These findings are consistent with and rationalize earlier reports that the AP-1 transcription factor fulfills key functions in regulating cell cycle control, transformation, and transcription of the *INK4A/ARF* locus.[5,44]

Together these studies show that reduced expression or loss-of-function of JMJD3 may play a role in some cancers, most probably by inhibiting H3K27 demethylation at the *INK4A/ARF* tumor suppressor locus and thus counteracting its activation. Indeed, JMJD3 expression is markedly reduced in cancers, including lung and liver carcinomas, as well as in a variety of hematopoietic malignancies.[3,6]

It is of interest that the *JMJD3* gene is sited on chromosome 17p, close to the *TP53* (*p53*) tumor suppressor gene, a genomic region that is among the most frequent for genetic cancer lesions.[73] Deletions at this site have been reported in a variety of human cancers,

Fig. 5.2 Model for the role of the histone H3K27 demethylase JMJD3 in the activation of the INK4A-ARF locus and senescence induction. (**a**) The RAS/RAF/MEK/ERK kinase pathway is induced in response to oncogene activation or other cellular stresses. Activated ERK translocates to the nucleus and catalyzes phosphorylation of transcription factors such as c-JUN and c-FOS. (**b**) In turn, these downstream transcription factors (probably AP-1) act to induce *JMJD3*. Finally, JMJD3 demethylates H3K27 at the *INK4A-ARF* locus to induce expression of *p16INK4A* and *p14ARF/p19ARF*, triggering senescence or apoptosis. Transcriptional induction of these genes is not only dependent on JMJD3 but will probably also require the action of methyltransferase(s) catalyzing H3K4 methylation (KMT2) and acetyltransferases catalyzing histone lysine acetylation. *Unlabeled shapes* represent other proteins that may interact with JMJD3

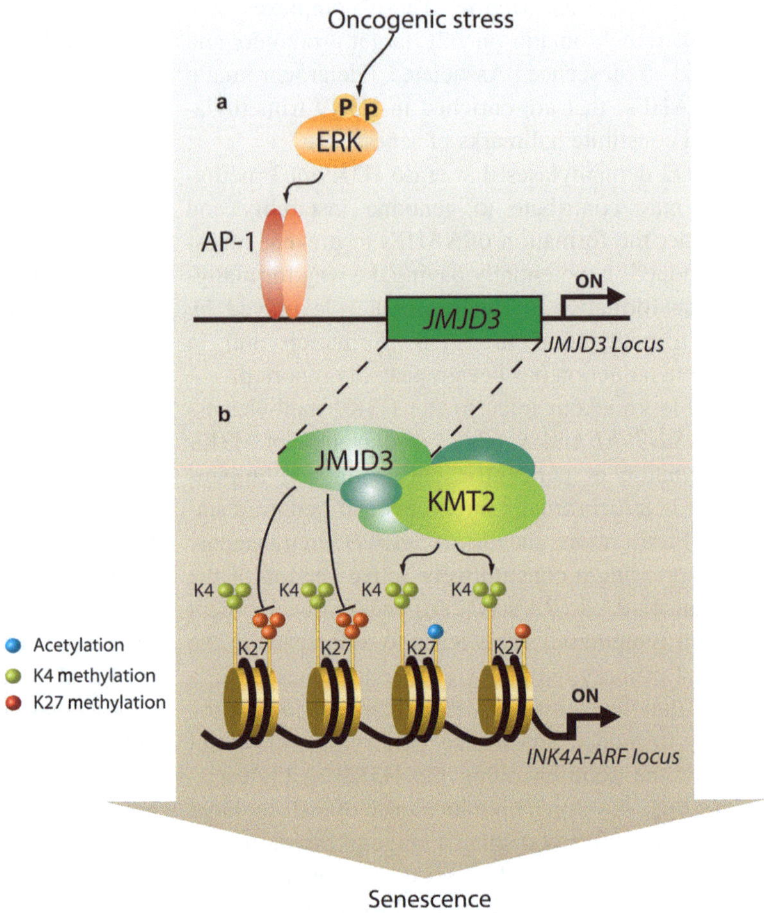

most of which include the *TP53* and *JMJD3* loci. Since JMJD3 is at the nexus of the INK4A/ARF tumor suppressor pathways, loss of both *TP53* and *JMJD3* will likely lead to inactivation of the pRB and p53 tumor suppressor pathways. This could explain why cancers with this deletion are highly aggressive.

5.2.3 The H3K27me3 Demethylase UTX Is a Putative Tumor Suppressor Protein

Several inactivating somatic mutations within the *UTX* gene encoding the histone H3K27me3 demethylase UTX have been reported in multiple cancer types.[109] These mutations included large deletions, nonsense mutations, frame-shifts insertion/deletions, and consensus splice site mutations causing aberrant splicing and premature termination codons.[108] These inactivating *UTX* mutations are found with the highest prevalence in multiple myelomas and squamous cell carcinomas, featuring mutation rates of 10% and 8%, respectively. However, *UTX* mutations have also been identified in other cancers, e.g., myeloid leukemias, breast and colorectal cancers as well as glioblastomas.[109]

In addition, as previously noted,[23] many cancers, including melanoma, prostate, brain and breast carcinoma, display a significant reduction in UTX expression.[103]

The reintroduction of *UTX* into cancer cell lines carrying inactivating *UTX* mutations leads to a significant decrease in cell proliferation, as well as to transcriptional changes (repression) of various H3K27me3 marked (polycomb target) genes. Clearly, *UTX* loss of function contributes to the transformation of phenotypes in these cancers. Of further note, chromatin localization (ChIP) analysis has revealed that H3K27me3 levels are significantly reduced because *SOX21* and *PCDH19* are transcriptionally induced when UTX is reintroduced. The expression changes would therefore appear due to the reconstituted H3K27 demethylase activity.[109] Moreover, these transcriptional changes have not been observed in cancer cell lines with wild type UTX. Whether and how UTX inactivation contributes to oncogenesis is still an open question.

One possible explanation is that the tumor suppressor activities of the H3K27 demethylases UTX are the result of their maintaining the cells in a non-pluripotent state. PcG proteins catalyzing K27 methylation are essential in maintaining cellular pluripotency by the repression of genes involved in cellular differentiation.[12,13,58,108] In theory, inappropriate function or localization of UTX or other K27 demethylases may push cells toward a pluripotent and malignant state, thereby contributing to carcinogenesis.

As UTX is part of mixed-lineage-leukemia (MLL2/3) complexes that catalyze histone H3K4 methylation,[20,41,79] another possible explanation is that loss of UTX leads to the instability of the complexes, causing loss of other complex partners such as MLL2/3. This can lead to further transcriptional deregulation and contribute to carcinogenesis.

5.2.4 The JHDM1B Histone Demethylase: A Tumor Suppressor or Oncoprotein?

The activity and biological function of JHDM1B (also denoted JMJD1b or KDM2B[4]) are subject to controversy.

Using retroviral insertional mutagenesis, Suzuki et al. identified *Jhdm1b* as candidate a tumor suppressor in BLM-deficient mice, a mouse model of leukemia.[96] Careful analysis of additional tumor samples revealed locus insertions that left the *Jhdm1b* coding region intact. It is unclear therefore whether tumorigenesis was triggered by activation or suppression of Jhdm1b function.

Frescas and co-workers reported that JHDM1B catalyzed demethylation of H3K4me3 in vivo and repressed rRNA genes. JHDM1B therefore acted as a tumor suppressor.[31] Another report implicated JHDM1B in the downregulation of c-Jun and tumor suppression.[53]

He and collaborators,[36] on the other hand, have demonstrated that JHDM1B functions as an H3K36me2/me1 demethylase both in vitro and in vivo and that depletion of Jhdm1b in mouse embryonic fibroblasts (MEFs) causes cell proliferation defects and induced senescence.[36] Likewise ectopic expression of Jhdm1b cooperates with Ras to transform primary MEFs.[36]

Jhdm1b also helps regulate cell proliferation and senescence by directly repressing expression of the p15Ink4b tumor suppressor. Jhdm1b therefore appears to act as a proto-oncogene. This inference derives support from a study which reported that Jmjc domain-containing oncoproteins immortalized embryonic fibroblasts via a Jmjc domain-dependent process.[86]

5.2.5 Prostate Cancer

GASC1 and the JMJD2 family of histone demethylases have been implicated in prostate cancer.[22] Their potential involvement in prostate carcinomas derives support from the finding that GASC1 interacts with the androgen receptor (AR) and functions as an essential co-activator of AR-induced transcription and cellular growth.[119] Studies by Metzger and co-workers[69] have implicated protein-kinase-C-related kinase 1 (PRK1) in GASC1-mediated activation of androgen-responsive genes, inasmuch as PRK1 catalyzes phosphorylation at threonine 11 on histone H3 in a ligand (androgen)-dependent manner (Fig. 5.3a). T11 phosphorylation in turn triggers demethylation of H3K9me3 by GASC1, probably by increasing its affinity for its histone substrate[67] (Fig. 5.3b). Other demethylases, LSD1 or JHDM1A with specificity for mono and dimethylated H3K9, may fully remove repressive H3K9 methylation and induce transcription (Fig. 5.3c)[68,69,123].

5.2.6 Breast Cancer

Liu et al.[62] have shown that *GASC1* is one of the amplified genes in the 9p23–24 region in breast cancer, particularly in basal-like subtypes. *GASC1* expression was significantly higher in aggressive, basal-like breast cancers, compared with nonbasal-like breast cancers. Moreover, when GASC1 is overexpressed in immortalized, non-transformed mammary epithelial MCF10A cells, a variety of transformed phenotypes are induced. These include growth factor-independent proliferation, anchorage-independent growth, altered morphogenesis in matrigel, and mammosphere-forming ability, a hallmark of breast cancer stem cells (Fig. 5.1).

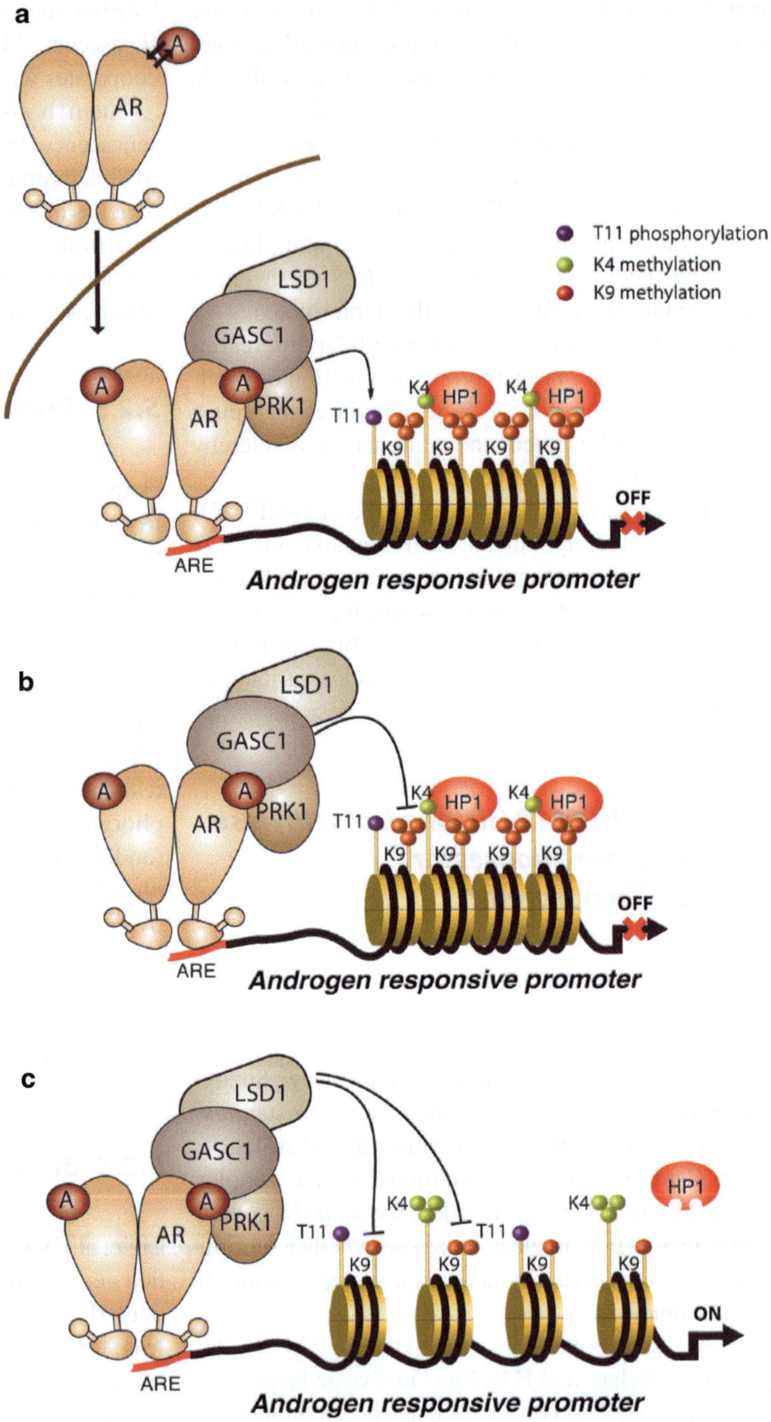

Fig. 5.3 Model for the role of GASC1 in the transcriptional activation of androgen-responsive genes. (**a**) When the androgen receptor (*AR*) binds to androgens (*A*), it translocates to the nucleus and is bound to androgen-responsive genes at the androgen-responsive elements (*ARE*). The androgen receptor interacts with the kinase PRK1 and the histone demethylase GASC1, and possibly other histone demethylases such as LSD1 or JHDM2A. The kinase PRK1 interacts with the AR in an androgen-dependent manner and catalyzes the phosphorylation of threonine 11 on histone H3 (T11). (**b**) PRK1-mediated phoshorylation of T11 will in turn trigger demethylation of trimethylated lysine 9 on histone 3 (K9) by GASC1, possibly by increasing its affinity to the substrate histone. (**c**) Dimethylated lysine 9 (K9) may be further demethylated to monomethylated or unmethylated states by LSD1 or JHDM2A. This causes dissociation of repressive factors as heterochromatin protein 1 (*HP1*) and leads to gene induction. Gene transcription will probably also require the action of a histone methyltransferase of the KMT2 family catalyzing trimethylation at H3K4. *Unlabeled shapes* represent other proteins that may interact with the GASC:AR complex

When GASC1 expression is forced, *NOTCH1* is induced, a factor that promotes self-renewal of human mammary stem cells.[26] This in turn is linked to the development of breast cancer and the induction of stem cell phenotypes[27,37,85,88,89] (Fig. 5.1).

Conversely, when GASC1 is depleted via shRNA, the proliferation of breast cancer cells is decreased and NOTCH1 strongly reduced. This finding further implicates GASC1 in *NOTCH1* regulation. Likewise, monomethylated H3K9, which results when GASC1 demethylates H3K9me3/me2, markedly accumulates on *NOTCH1* genomic loci in breast cancer cells when amplified by *GASC1*; this further validates *NOTCH1* as a bona fide target gene for GASC1.

It is of interest that mammosphere-forming breast cancer cells that express *GASC1* ectopically dramatically overexpress *NANOG,* compared with sphere-forming cells from control breast cancer cells.[62]

Nanog, as noted above, is a target of GASC1. GASC1 in murine ES cells is in turn a target of the reprogramming factor Oct4; linking GASC1 to a stem cell-like transcriptional signature.[64] In breast cancers, this ES-like signature is associated with high-grade ER-negative tumors, often of the basal-like subtype,[35] with *GASC1* amplifications frequently present.[62]

It is apparent that GASC1 is involved in the demethylation and induction of *NOTCH1* and linked to stem cell phenotypes in human breast cancer. This has led Liu and co-workers[62] to suggest that *GASC1* was a driving oncogene in the 9p23–24 amplicon in human breast cancer (Fig. 5.1).

LSD1 (also termed KDM1[4]) catalyzes the demethylation of H3K4me/2me1 through a FAD-dependent reaction.[93] LSD1 has also been assigned a role in gene activation of androgen receptor target genes[68] and may promote activation by demethylating the repressive H3K9me2/me1 marks. However, biochemical and structural studies of the recombinant enzyme have shown that it has no activity on H3 peptides that are methylated at Lys9.[28–30,93]

LSD1 may induce metastasis of breast cancer cells. Ectopic expression and the knockdown of *LSD1* in breast cancer cells in vitro have led to a threefold decrease and a fivefold increase of cell invasion, respectively.[115] Studies[115] on the gain of function and loss of function of *LSD1* in immunocompromised SCID mice have shown that spontaneous metastasis is suppressed in animals that carry an *LSD1* overexpressing tumor and that it is attenuated in mice with *LSD1*-depleted tumors. LSD1 suppresses the metastatic potential of breast cancer in vivo. Consistent with this inference is a statistically significant decrease of *LSD1* in human breast tumor samples; this was not found in adjacent normal tissue.

JARID1B, also known as PLU1 or KDM5B,[4] belongs to the JARID1 family of demethylases,[21,122] with specificity for activating H3K4me3/me2 histone methylations. The *JARID1B* gene was initially identified in a screen for genes regulated by the oncogene *c-ErbB2*. *JARID1B* is highly expressed in 90% of ductal breast carcinomas and has been associated with the malignant phenotype of breast cancer.[7,65]

Consistent with its role as an oncogene in breast cancer, *JARID1B* has been shown to be essential for the proliferation of the breast carcinoma cell line MCF-7 and for the proliferation of breast cancer cells in nude mice.[122]

JARID1B target genes have not yet been mapped in the whole genome, but several have been identified and implicated in breast cancer proliferation. These include *14-3-3*, *BRCA1*, *CAV1,* and *HOXA5,* which are induced when *JARID1B* is depleted via shRNA and H3K4 trimethylation genomic loci are increased.[122]

5.2.7 Hematopoietic Cancers

Using retroviral insertional mutagenesis, Suzuki et al.[96] identified *Jmjd5* as a candidate tumor suppressor in BLM-deficient mice, a mouse model of leukemia. *Jmjd5* encodes a member of the JMJC family, but it is not known whether it has histone demethylase activity. However, CO6H2.3, an ortholog of the human *Jmjd5* gene, has been shown to contribute to genomic stability in *C. elegans* somatic cells.[86]

The molecular mechanism that triggers oncogenesis as a consequence of loss-of-function mutations in *Jmjd5* is not yet known. However, the fact that *Jmjd5* loss decreased the survival of cells in response to treatment with the DNA-damaging agent MNNG implies that Jmjd5 may play a role in DNA repair.

van Zutven et al.[110] have reported on a patient with nucleoporin-98 (NUP98) translocations that induced a mutant fusion protein (NUP98-JARID1A) between NUP98 and the H3K4me3 demethylase JARID1A. The mutant NUP98-JARID1A protein contained the C-terminal PHD domain of JARID1A that binds to H3K4me3 and ensures the correct chromatin localization of the wild type JARID1A protein.

Wang et al.[112] fused the H3K4-trimethylation (H3K4me3)-binding PHD finger of JARID1A to NUP98. This generated a potent oncoprotein that prevented hematopoietic differentiation and induced acute myeloid leukemia in murine models. Interestingly, mutations in the PHD finger that prevented H3K4me3 from binding also abolished the leukemic transformation.

NUP98–PHD fusion also prevents the differentiation-associated removal of H3K4me3 at many loci that encode lineage-specific transcription factors, including *Hox(s)*, *Gata3*, *Meis1*, *Eya1,* and *Pbx1*. It also leads to increased transcription of the affected genes in murine hematopoietic stem/progenitor cells.[112] This is consistent with the proposal[23] that the mutant fusion protein contributes to the pathogenesis of AML-M7 leukemia by functioning as a dominant negative mutant that protects the H3K4me3 mark against JARID1-catalyzed demethylation.

5.2.8 Brain Cancer

GASC1 and *JMJD2B* are amplified in medulloblastoma.[74]

5.2.9 Renal Cancer

In a family with renal cell cancer, the *HSPBAP1* gene, which encodes one JMJC protein of the putative histone demethylases, is translocated, with a *DIRC3-HSPBAP1* fusion transcript as the result.[11] Whether HSPBAP1 functions as a histone demethylase and if or how its translocation contributes to the establishment or progression of renal cancers is not known.

5.3 Histone Demethylases in Drug Resistance

Subgroups of cancers acquire drug-resistance through various mechanisms including the activation of alternative survival pathways, permitting them to evade cancer therapy.

In a recent study, Sharma et al. identified RBP2/JARID1A encoding the H3K4me3 demethylase JARID1A as one of a few genes that showed increased expression in drug-resistant cell populations.[91]

A closer analysis revealed that drug-resistant cells display decreased H3K4me3 methylation levels consistent with an increased H3K4me3 demethylation activity in these cells. Moreover, shRNA-mediated depletion of JARID1A led to sensitization of drug-resistant cells to drugs, whereas transient expression of JARID1A led to a decreased sensitivity to treatment.[91]

In concert, these findings suggest that this histone demethylase may be involved in acquiring a drug-resistant chromatin state, and may constitute a potential drug-target for treatment of drug-resistant cancers.

5.4 Histone Demethylases in Neural Disorders

5.4.1 X-Linked Mental Retardation

JARID1C encodes JARID1C, also termed SMCX and KDM5C.[4] JARID1C belongs to the JARID1 family of histone demethylases that are specific for di- and trimethylated lysine 4 on histone 3 (H3K4me3/me2).[21,42,98]

Patients affected by X-linked mental retardation (XLMR) have been found to have mutations of the *JARID1C* gene.[17,43,90,107] One mutation (A388P) is in close proximity to the N-terminal PHD domain of the protein. Iwase and colleagues[42] have shown that the PHD domain of the wild type JARID1C binds to the repressive H3K9me3 mark, but not to the A388P mutant. The A388P mutation likely leads to an aberrant localization of JARID1C and to deregulation of target genes.

Likewise, the in vitro demethylation activity of the protein is markedly reduced upon introduction of the A388P or other XLMR-linked mutations into the wild type JARID1C.[42]

These results indicate that XLMR-associated mutations in JARID1C compromise both the catalytic activity and chromatin association of the protein, thereby contributing to XLMR pathogenesis. Moreover, *Jarid1C* depletion in rat cerebellar granule neurons brings about a significant reduction in the dendritic length of neurons. This was reversed when the wild type protein, but not the XLMR-linked mutants were reintroduced.[42] These studies provide further evidence on the functional links between *JARID1C* mutations and XLMR phenotypes.

Patients with X-linked mental retardation (XLMR) and cleft lip/palate have mutations in the *PHF8* gene.[51,56,94] This indicates an important role of *PHF8* in neural development and midline formation. Interestingly, these XLMR mutations cluster in exons that encode for the double-stranded beta-helix fold, believed to be important for the catalytic activity of the protein. These are likely to be compromised by mutations. Although a truncated E-coli expressed form of this protein demethylates various di- and monomethylated histone lysines in vitro,[63] the full-length protein appears to have exquisite specificity for the repressive H3K9me2/me1 histone methylations both in vitro and in vivo.[49] PHF8 is therefore likely to function as a transcriptional activator. Consistent with this notion, the PHD domain of PHF8 binds the activating H3K4me3 mark and co-localizes with H3K4me3 at transcription initiation sites.[49] Of further note, PHF8 binds another XMLR protein, ZNF711, and co-localizes with this at various target genes, including the XLMR gene JARID1C.[49] Of interest, F29B9.2 the *C. elegans* homolog of PHF8 is highly expressed in neurons, and mutant animals exhibit impaired locomotion.[49] In concert, these findings indicate that PHF8 is important for the regulation of XLMR-genes and neural development. Further studies are however needed to determine exactly how *PHF8* mutations contribute to XLMR.

5.4.2 Schizophrenia

A single nucleotide polymorphism located in the region that codes for the Arid/Bright domain of JARID2 has been associated with schizophrenia, implicating this gene in this disorder.[77,80] Whether this JMJC protein has catalytic activity is not known, but as mentioned below, JARID2 appears to be crucial for recruiting the H3K27 methyltransferase complex PRC2.[60,78,81,92]

5.4.3 Autism

Castermans et al.[19] have reported that a boy with autism had a balanced paracentric inversion, 46, XY, inv(10), with a breakpoint in chromosome 10q21.3. This breakpoint was located in the first intron of the *JMJD1C*. This chromosomal aberration caused a twofold decrease in expression of two of three *JMJD1C* transcripts. This member of the JmjC family has therefore been proposed as a candidate gene for autism.

Similarly, a missense mutation in exon 16 of the *JARID1C* gene has been identified in a patient with autism spectrum disorder (ASD).[1] This mutation results in an amino acid substitution at residue 766 (R766W), which is highly conserved among JARID1C orthologs from other species. JARID1C functions as a transcriptional repressor that contributes to REST-mediated neuronal gene regulation. JARID1C-regulated genes such as *SCN2A*, *CACNA1H*, *BDNF*, and *SLC18A1* are associated with autism and cognitive dysfunction.[1]

5.4.4 Epilepsy

The Heat-shock 27 (Hsp27)-associated protein 1 (HSPBAP1), also termed PASS1, is a member of the JmjC protein family, most members of which have been shown to be histone demethylases. HSPBAP1 has not, so far, been shown to have histone demethylase activity. In an effort to identify genes differentially expressed in intractable epilepsy (IE), Xi et al.[120] found that HSPBAP1 is expressed in the anterior temporal neocortex of patients with intractable epilepsy (IE) at a tenfold higher level than in normal controls.

HSPBAP1 reduces the neuroprotective function of Hsp27 in experimentally induced epileptic neuropathology. HSPBAP1 probably does so by inhibiting the ability of Hsp27 to protect cells against heat shock.[61] HSPBAP1 may therefore have a role in the regulation of hsp27 function and in IE development.

5.4.5 Neuropathy

The *PHF2* gene encodes PHF2, a member of the PHF family of JMJC histone demethylases with specificity for the H3K9me2/me1 repressive methylation.[117]

The *PHF2* gene is located on human chromosome 9q22, within the candidate region for hereditary neuropathy I (HSN1)[10,72] and for multiple self-healing squamous epitheloma (ESS1).[33] At present, however, it is not known what are the target genes of PHF2 and, indeed, whether PHF2 has a role in the pathogenesis of these two human diseases.

5.5 Male Infertility

The JMJC histone demethylase JHDM2A (JmjC-containing histone demethylase 2a), also known as

JMJD1A or KDM3A,[4] was cloned as a testis-specific gene transcript[37,123] and on the basis of its nuclear expression in round spermatids and co-localization with RNA polymerase II, has been identified as a transcriptional activator. This is consistent with its catalyzing the removal of the repressive H3K9 me2/me1 methylations.

Jhdm2a genetrap mice are viable, but males exhibit smaller testes and are functionally infertile.[76] Although Jhdm2a interacts with the androgen receptor, functioning as a transcriptional co-activator of the androgen receptor, neither androstenedione and testosterone, nor leutenising hormone (LH), follicle-stimulating hormone (FSH), or estradiol were affected in *Jhdm2a* genetrap mice. Rather, spermatids from genetrap mice exhibit chromatin condensation defects, as a result of which the number of mature spermatids is reduced.

Most sperm cells from *Jhdm2a* genetrap mice are immotile and have abnormally shaped heads.[76] Global H3K9 methylation levels are unaltered. Jhdm2a may therefore induce the transcriptional activation of some genes that are needed for chromatin condensation. In support of this notion is the finding that in sperm cells from *Jhdm2a −/−* mice two testis-specific basic proteins, transition protein 1 (Tnp1) and protamine 1 (Prm1), are necessary for histones to be replaced in the final stages of chromatin condensation.[76]

Jhdm2a is recruited to the promoters of both Tnp1 and Prm1 in round spermatids, but this is not true in sperm cells from *Jhdm2a* null mice. Moreover, H3K9 methylation is markedly increased at these promoters in spermatids from *Jhdm2a* knockout mice as compared to controls. This indicates that Jhdm2a binds to and activates these promoters by catalyzing H3K9 demethylation.[76]

In summary, *Jhdm2a* is indispensable for spermatogenesis. Disruption in mice leads to male infertility similar to azoospermia and globospermia in humans. Studies are needed to determine whether disruption of *Jhdm2a* is involved in human infertility.

5.6 Demethylases in Obesity and Related Diseases

Obesity and type 2 diabetes (T2D) are caused, on one hand, by precipitating factors, such as high caloric intake and inadequate physical activity, and, on the other, by predisposing factors including genetics and inappropriate fetal development. In addition, epigenetic mechanisms probably are key to predisposing or protecting against obesity and T2D.

Recently, the H3K9me2/me1 demethylase has been implicated in energy expenditure and obesity in mice, inasmuch as deletion of *Jhdm2a* led to obesity and a type II diabetes-like syndrome[39,101] (see Fig. 5.4). Homozygous *Jhdm2a* null mice gain weight rapidly and may reach three times the weight of wild type controls. *Jhdm2a* null mice also display an increase in both adipocyte size and number. Excess weight and increased fat deposits occur even on a restricted diet that is adequate for weight maintenance in wild type mice.

Jhdm2a −/− mice also display hyperphagia, a diabetes-like syndrome of hyperglycemia, glucose intolerance, and hyperinsulinemia. The latter only develops after obesity is established and is probably a secondary result.

Jhdm2a −/− mice display a higher respiratory quotient, fasted induced hyperthermia, and inability to respond to cold-induced thermogenesis; this indicates impaired ability to burn fat for energy production.[39,101] In fact, expression of several genes involved in mitochondrial functions, including *Ppara,* are decreased in the mutant mice.

Ucp1, a key gene involved in β-adrenergic signaling-mediated thermogenesis in brown adipose tissue, is markedly upregulated when wild type mice are exposed to low temperatures. But cold-induced *Ucp1* upregulation is completely abolished in *Jhdm2a* knockout mice. ChIP analysis has shown that *Ucp1* is a target gene of Jhdm2a and that Jhdm2a induces transcriptional activation by removing H3K9 methylation from the promoter.[101]

The exact molecular mechanisms leading to obesity, hyperglycemia, and a T2D-like syndrome in adult *Jhdm2a −/−* mice are not fully understood. However, because *Jhdm2a* catalyzes the removal of the repressive H3K9me2/1 histone methylation, obesity and other phenotypic expressions are likely to involve repression of the genes involved in metabolism, energy expenditure, and fat storage. Some discrepancies in the findings of the two studies[39,101] notwithstanding, both clearly attribute a significant role for Jhdm2a in regulating systemic metabolic control, including the *Ppara* and β-adrenergic and β signaling pathways (Fig. 5.4).

More studies are needed to determine whether Jhdm2a is involved in thermogenesis and energy

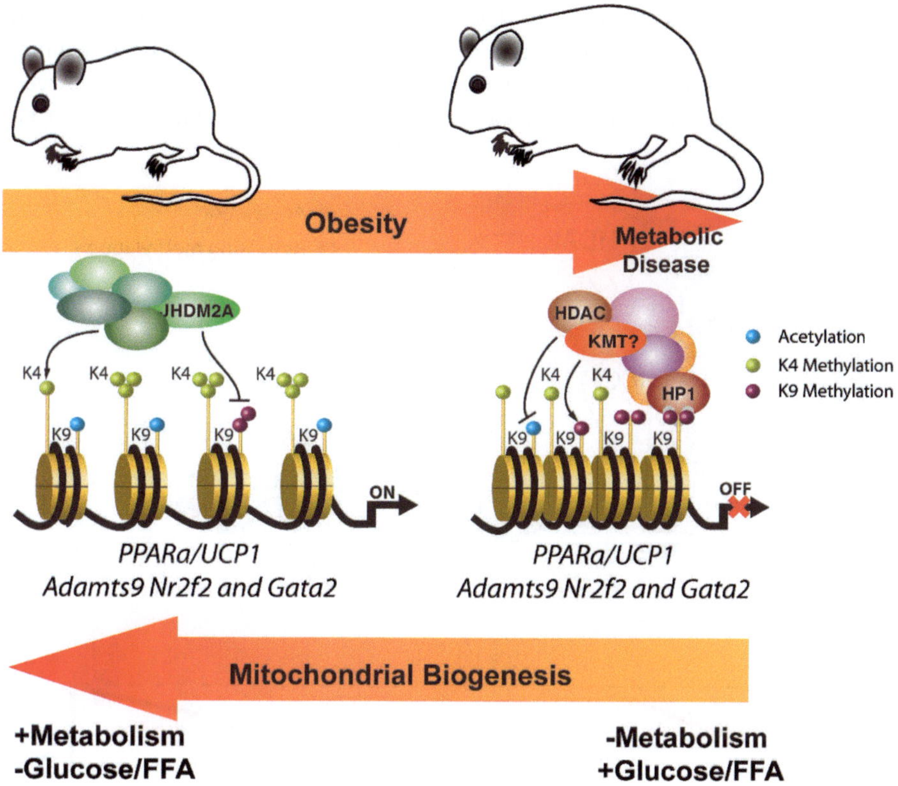

Fig. 5.4 Model for the role of JHDM2a in the regulation of mitochondrial biogenesis and obesity. Removal of repressive H3K9 dimethylation and monomethylation by JHDM2A lead to activation of genes involved in metabolism, energy expenditure, or fat storage, such as *PPARα*, *UCP1 Adamtsp*, *Nrf2*, and *Gata2*.[75,105,121] Gene induction acts to increase mitochondrial biogenesis and energy consumption. Conversely, loss of JHDM2A-activity or H3K9 methylation by methyltransferase(s) may lead to the recruitment of repressive effector proteins like HP1 and, along with other repressive activities as by the histone deacetylases (*HDAC*), cause gene silencing. As a result, metabolism decreases and glucose and free fatty acids (*FFA*) levels increase. This leads to obesity and in turn triggers type II diabetes. *Unlabeled shapes* represent other proteins within the JHDM2A or KMT complexes

expenditure in man and whether Jhdm2a loss-of-function plays a role in human obesity and T2D.

5.7 Congenital Heart Disease

Individuals who carry single nucleotide polymorphisms in exon 6 of the *JARID2* gene have a threefold increase in the risk for congenital heart diseases (CHD).[111] *JARID2* encodes the JMJC protein JARID2, which acts as transcriptional repressor in cardiomyocyte proliferation and mouse embryonic development.[45–47,106] So far JARID2 has not been assigned any histone demethylase activity; however, several recent studies have shown that JARID2 binds the polycomb repressive complex 2 (PRC2) and that the protein is essential for the recruitment of this complex to its target genes and for modulating the H3K27 methyltransferase activity of the PRC2 complex.[60,78,81,92] JARID2 knockout mice display heart abnormalities similar to human CHD, including ventricular septal defects, noncompaction of the ventricular wall, double outlet right ventricle, and dilated atria.[59,100] These findings point to a role of JARID2 in the pathophysiology of CHD in mouse, but whether this protein is involved in human CHD is not known.

5.8 Allopecia/Arthricia

The gene *Hairless* (*HR*) encodes the protein Hairless (HR) which belongs to the JMJC group of which 15 are known to catalyze histone demethylation. To date HR has not been shown to have histone demethylase activity, but it may be implicated in repression of gene transcription and therefore may have a role in epigenetic regulation.

The requirement of *HR* for hair maintenance in humans and animals has been demonstrated in many studies, some of which are reviewed below. A wild hairless mouse mutant was described as early as 1926.[16,102]

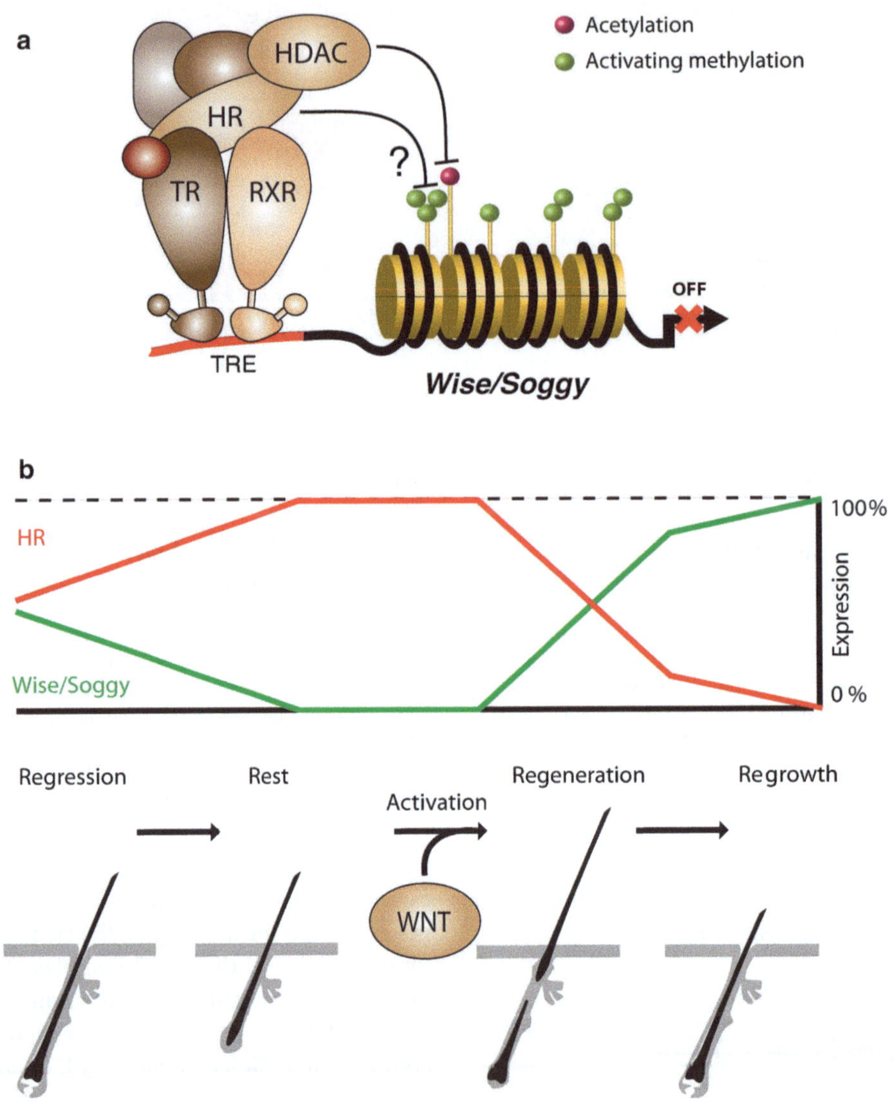

Fig. 5.5 Model for the role of Hairless (*HR*) in the regulation of transcription in hair re-growth/ hair cycling. (**a**) Model of transcriptional regulation by HR. HR interacts with the thyroid hormone receptor (*TR*): RXR heterodimer that binds DNA at genes featuring a thyroid hormone response element (*TRE*). HR also binds other histone-modifying enzymes, such as histone deacetylases (*HDAC*), that repress the target genes. *Unlabeled shapes* represent other proteins that may interact with the HR:TR complex. (**b**) Model for HR action in hair re-growth. The *upper panel* summarizes HR protein expression through the hair cycle in relation to *Wise* and *Soggy* mRNA expression. Hair growth occurs in non-synchronized cycles that consist of three phases: a growth phase (anagen), a shortening phase (catagen), and a resting phase, after which hair is shedded and a new cycle is initiated. HR protein is induced during hair regression (catagen), concomitant with a decrease of *Wise/Soggy* mRNA. HR protein expression during rest (telogen) will repress *Wise/Soggy* mRNA expression, allowing activation of Wnt signaling. Subsequent to the reactivation of hair growth, HR protein is downregulated leading to the induction of Wise/Soggy. In turn, the increase in *Wise/Soggy* expression inhibits Wnt signaling. In *HR* mutants, uncontrolled expression of *Wise/Soggy* and of other Wnt inhibitors prevents hair re-growth because Wnt signaling is suppressed (Modified from Thompson et al.[104])

Half a century later, Stoye and co-workers established that the hairless phenotype seen in these mice arose from a stable proviral integration of the *HR* gene into intron 6, causing aberrant splicing and severely reduced *HR* mRNA transcripts.[18] Later, *HR* knockout mice were described and reported to have a phenotype closely resembling the HRS/J mice.[8] Hair follicles develop normally in these mice forming normal visible hair. However, following the initial hair production cycle, follicles degenerate and never regenerate, with complete hair loss occurring at around 3 weeks age.

The molecular mechanisms that trigger re-initiation of hair growth would require repression of HR in several genes. Candidate HR target genes include *Soggy* and *Wnt modulator in surface ectoderm* (*Wise*), believed to downregulate the Wnt signaling pathway and to be involved in hair follicle regeneration. Both genes are significantly increased in skin from *HR* knockout mice and restored to wild type levels by the reintroduction of *HR* in *HR* null mice.[8]

On the basis of these findings, Thompson and co-workers put forward a model in which HR acts as a transcriptional repressor of the *Soggy* and *Wise* genes, thus upregulating Wnt signaling and hair follicle regeneration[8] (Fig. 5.5).

In humans, HR deregulation closely recapitulates the mouse phenotypes. Many nonsense, missense, insertion, and deletion mutations of the human *HR* gene have been described, all of which cause a massive loss of hair follicles and a concomitant accumulation of cyst-like follicles that lead to the hair loss disorders alopecia universalis congenita (AUC; OMIM 203655) arthricia with papular lesions (APL; OMIM 209500) and Marie Unna hereditary hypotrichosis (MUHH; OMIM 146550).[77,116] The molecular mechanism by which HR downregulates transcription is not known.

Conceivably HR is involved in the recruitment of corepressors as it binds to repressors such as histone deacetylases.[87,113]

Alternatively, HR may demethylate an activating histone methylation. In support of this possibility is the fact that five *HR*-mutations are known to cause APL in humans. These mutations reside within the JmjC domain that may catalyze histone demethylation.

It is therefore possible that HR is a histone demethylase that acts as a transcriptional corepressor essential for normal regulation of hair growth.

5.9 Summary

Although histone demethylases have only been known for less than a decade, a raft of studies have already linked the aberrant expression or function of these enzymes to a variety of human diseases. Coming years will certainly provide more detailed insights into the role of histone demethylases in health and disease and determine the potential of these interesting proteins as targets for drug treatments.

References

1. Adegbola A, Gao H, Sommer S, Browning M. A novel mutation in JARID1C/SMCX in a patient with autism spectrum disorder (ASD). *Am J Med Genet.* 2008;146A: 505-511.
2. Agger K, Cloos PA, Christensen J, et al. UTX and JMJD3 are histone H3K27 demethylases involved in HOX gene regulation and development. *Nature.* 2007;449:731-734.
3. Agger K, Cloos PA, Rudkjaer L, et al. The H3K27me3 demethylase JMJD3 contributes to the activation of the INK4A-ARF locus in response to oncogene- and stress-induced senescence. *Genes Dev.* 2009;23:1171-1176.
4. Allis CD, Berger SL, Cote J, et al. New nomenclature for chromatin-modifying enzymes. *Cell.* 2007;131:633-636.
5. Ameyar-Zazoua M, Wisniewska MB, Bakiri L, Wagner EF, Yaniv M, Weitzman JB. AP-1 dimers regulate transcription of the p14/p19ARF tumor suppressor gene. *Oncogene.* 2005;24:2298-2306.
6. Barradas M, Anderton E, Acosta JC, et al. Histone demethylase JMJD3 contributes to epigenetic control of INK4a/ARF by oncogenic RAS. *Genes Dev.* 2009;23:1177-1182.
7. Barrett A, Madsen B, Copier J, et al. PLU-1 nuclear protein, which is upregulated in breast cancer, shows restricted expression in normal human adult tissues: a new cancer/testis antigen? *Int J Cancer.* 2002;101:581-588.
8. Beaudoin GM 3rd, Sisk JM, Coulombe PA, Thompson CC. Hairless triggers reactivation of hair growth by promoting Wnt signaling. *Proc Natl Acad Sci USA.* 2005;102: 14653-14658.
9. Berger SL. The complex language of chromatin regulation during transcription. *Nature.* 2007;447:407-412.
10. Blair IP, Dawkins JL, Nicholson GA. Fine mapping of the hereditary sensory neuropathy type I locus on chromosome 9q22.1–>q22.3: exclusion of GAS1 and XPA. *Cytogenet Cell Genet.* 1997;78:140-144.
11. Bodmer D, Schepens M, Eleveld MJ, Schoenmakers EF, Geurts van Kessel A. Disruption of a novel gene, DIRC3, and expression of DIRC3-HSPBAP1 fusion transcripts in a case of familial renal cell cancer and t(2;3)(q35;q21). *Genes Chromosomes Cancer.* 2003;38:107-116.
12. Boyer LA, Plath K, Zeitlinger J, et al. Polycomb complexes repress developmental regulators in murine embryonic stem cells. *Nature.* 2006;441:349-353.

13. Bracken AP, Dietrich N, Pasini D, Hansen KH, Helin K. Genome-wide mapping of Polycomb target genes unravels their roles in cell fate transitions. *Genes Dev.* 2006;20: 1123-1136.
14. Bracken AP, Kleine-Kohlbrecher D, Dietrich N, et al. The Polycomb group proteins bind throughout the INK4A-ARF locus and are disassociated in senescent cells. *Genes Dev.* 2007;21:525-530.
15. Braig M, Lee S, Loddenkemper C, et al. Oncogene-induced senescence as an initial barrier in lymphoma development. *Nature.* 2005;436:660-665.
16. Brooke HC. Hairless mice. *J Hered.* 1926;17:173-174.
17. Brown CJ, Miller AP, Carrel L, Rupert JL, Davies KE, Willard HF. The DXS423E gene in Xp11.21 escapes X chromosome inactivation. *Hum Mol Genet.* 1995;4:251-255.
18. Cachon-Gonzalez MB, Fenner S, Coffin JM, Moran C, Best S, Stoye JP. Structure and expression of the hairless gene of mice. *Proc Natl Acad Sci USA.* 1994;91:7717-7721.
19. Castermans D, Vermeesch JR, Fryns JP, et al. Identification and characterization of the TRIP8 and REEP3 genes on chromosome 10q21.3 as novel candidate genes for autism. *Eur J Hum Genet.* 2007;15:422-431.
20. Cho YW, Hong T, Hong S, et al. PTIP associates with MLL3- and MLL4-containing histone H3 lysine 4 methyltransferase complex. *J Biol Chem.* 2007;282:20395-20406.
21. Christensen J, Agger K, Cloos PA, et al. RBP2 belongs to a family of demethylases, specific for tri-and dimethylated lysine 4 on histone 3. *Cell.* 2007;128:1063-1076.
22. Cloos PA, Christensen J, Agger K, et al. The putative oncogene GASC1 demethylates tri- and dimethylated lysine 9 on histone H3. *Nature.* 2006;442:307-311.
23. Cloos PA, Christensen J, Agger K, Helin K. Erasing the methyl mark: histone demethylases at the center of cellular differentiation and disease. *Genes Dev.* 2008;22: 1115-1140.
24. Collado M, Gil J, Efeyan A, et al. Tumour biology: senescence in premalignant tumours. *Nature.* 2005;436:642.
25. Cuthbert GL, Daujat S, Snowden AW, et al. Histone deimination antagonizes arginine methylation. *Cell.* 2004;118: 545-553.
26. Dontu G, Jackson KW, McNicholas E, Kawamura MJ, Abdallah WM, Wicha MS. Role of Notch signaling in cell-fate determination of human mammary stem/progenitor cells. *Breast Cancer Res.* 2004;6:R605-R615.
27. Farnie G, Clarke RB. Mammary stem cells and breast cancer – role of Notch signalling. *Stem Cell Rev.* 2007;3:169-175.
28. Forneris F, Binda C, Dall'Aglio A, Fraaije MW, Battaglioli E, Mattevi A. A highly specific mechanism of histone H3-K4 recognition by histone demethylase LSD1. *J Biol Chem.* 2006;281:35289-35295.
29. Forneris F, Binda C, Vanoni MA, Battaglioli E, Mattevi A. Human histone demethylase LSD1 reads the histone code. *J Biol Chem.* 2005;280:41360-41365.
30. Forneris F, Binda C, Vanoni MA, Mattevi A, Battaglioli E. Histone demethylation catalysed by LSD1 is a flavin-dependent oxidative process. *FEBS Lett.* 2005;579:2203-2207.
31. Frescas D, Guardavaccaro D, Bassermann F, Koyama-Nasu R, Pagano M. JHDM1B/FBXL10 is a nucleolar protein that represses transcription of ribosomal RNA genes. *Nature.* 2007;450:309-313.
32. Geli J, Nord B, Frisk T, et al. Deletions and altered expression of the RIZ1 tumour suppressor gene in 1p36 in pheochromocytomas and abdominal paragangliomas. *Int J Oncol.* 2005;26:1385-1391.
33. Goudie DR, Yuille MA, Leversha MA, et al. Multiple self-healing squamous epitheliomata (ESS1) mapped to chromosome 9q22-q31 in families with common ancestry. *Nat Genet.* 1993;3:165-169.
34. Guccione E, Bassi C, Casadio F, et al. Methylation of histone H3R2 by PRMT6 and H3K4 by an MLL complex are mutually exclusive. *Nature.* 2007;449:933-937.
35. Han W, Jung EM, Cho J, et al. DNA copy number alterations and expression of relevant genes in triple-negative breast cancer. *Genes Chromosomes Cancer.* 2008;47:490-499.
36. He J, Kallin EM, Tsukada Y-I, Zhang Y. The H3K36 demethylase Jhdm1b/Kdm2b regulates cell proliferation and senescence through p15Ink4b. *Nat Struct Mol Biol.* 2008;15(11):1169-1175.
37. Hoog C, Schalling M, Grunder-Brundell E, Daneholt B. Analysis of a murine male germ cell-specific transcript that encodes a putative zinc finger protein. *Mol Reprod Dev.* 1991;30:173-181.
38. Hu C, Dievart A, Lupien M, Calvo E, Tremblay G, Jolicoeur P. Overexpression of activated murine Notch1 and Notch3 in transgenic mice blocks mammary gland development and induces mammary tumors. *Am J Pathol.* 2006;168:973-990.
39. Inagaki T, Tachibana M, Magoori K, et al. Obesity and metabolic syndrome in histone demethylase JHDM2a-deficient mice. *Genes Cells.* 2009;14:991-1001.
40. Ishimura A, Terashima M, Kimura H, et al. Jmjd2c histone demethylase enhances the expression of Mdm2 oncogene. *Biochem Biophys Res Commun.* 2009;389:366-371.
41. Issaeva I, Zonis Y, Rozovskaia T, et al. Knockdown of ALR (MLL2) reveals ALR target genes and leads to alterations in cell adhesion and growth. *Mol Cell Biol.* 2007;27:1889-1903.
42. Iwase S, Lan F, Bayliss P, et al. The X-linked mental retardation gene SMCX/JARID1C defines a family of histone H3 lysine 4 demethylases. *Cell.* 2007;128:1077-1088.
43. Jensen LR, Amende M, Gurok U, et al. Mutations in the JARID1C gene, which is involved in transcriptional regulation and chromatin remodeling, cause X-linked mental retardation. *Am J Hum Genet.* 2005;76:227-236.
44. Johnson R, Spiegelman B, Hanahan D, Wisdom R. Cellular transformation and malignancy induced by ras require c-jun. *Mol Cell Biol.* 1996;16:4504-4511.
45. Jung J, Kim TG, Lyons GE, Kim HR, Lee Y. Jumonji regulates cardiomyocyte proliferation via interaction with retinoblastoma protein. *J Biol Chem.* 2005;280:30916-30923.
46. Jung J, Mysliwiec MR, Lee Y. Roles of JUMONJI in mouse embryonic development. *Dev Dyn.* 2005;232:21-32.
47. Kim TG, Chen J, Sadoshima J, Lee Y. Jumonji represses atrial natriuretic factor gene expression by inhibiting transcriptional activities of cardiac transcription factors. *Mol Cell Biol.* 2004;24:10151-10160.
48. Kirmizis A, Santos-Rosa H, Penkett CJ, et al. Arginine methylation at histone H3R2 controls deposition of H3K4 trimethylation. *Nature.* 2007;449:928-932.
49. Kleine-Kohlbrecher D, Christensen J, Vandamme J, et al. A functional link between the histone demethylase PHF8 and the transcription factor ZNF711 in X-linked mental retardation. *Mol Cell.* 2010;38(2):165-178.

50. Klose RJ, Yamane K, Bae Y, et al. The transcriptional repressor JHDM3A demethylates trimethyl histone H3 lysine 9 and lysine 36. *Nature*. 2006;442:312-316.
51. Koivisto AM, Ala-Mello S, Lemmela S, Komu HA, Rautio J, Jarvela I. Screening of mutations in the PHF8 gene and identification of a novel mutation in a Finnish family with XLMR and cleft lip/cleft palate. *Clin Genet*. 2007;72: 145-149.
52. Kouzarides T. Chromatin modifications and their function. *Cell*. 2007;128:693-705.
53. Koyama-Nasu R, David G, Tanese N. The F-box protein Fbl10 is a novel transcriptional repressor of c-Jun. *Nat Cell Biol*. 2007;9:1074-1080.
54. Lal G, Padmanabha L, Smith BJ, et al. RIZ1 is epigenetically inactivated by promoter hypermethylation in thyroid carcinoma. *Cancer*. 2006;107:2752-2759.
55. Lan F, Bayliss PE, Rinn JL, et al. A histone H3 lysine 27 demethylase regulates animal posterior development. *Nature*. 2007;449:689-694.
56. Laumonnier F, Holbert S, Ronce N, et al. Mutations in PHF8 are associated with X linked mental retardation and cleft lip/cleft palate. *J Med Genet*. 2005;42:780-786.
57. Lazzerini Denchi E, Attwooll C, Pasini D, Helin K. Deregulated E2F activity induces hyperplasia and senescence-like features in the mouse pituitary gland. *Mol Cell Biol*. 2005;25:2660-2672.
58. Lee TI, Jenner RG, Boyer LA, et al. Control of developmental regulators by polycomb in human embryonic stem cells. *Cell*. 2006;125:301-313.
59. Lee Y, Song AJ, Baker R, Micales B, Conway SJ, Lyons GE. Jumonji, a nuclear protein that is necessary for normal heart development. *Circ Res*. 2000;86:932-938.
60. Li G, Margueron R, Ku M, Chambon P, Bernstein BE, Reinberg D. Jarid2 and PRC2, partners in regulating gene expression. *Genes Dev*. 2010;24:368-380.
61. Liu C, Gilmont RR, Benndorf R, Welsh MJ. Identification and characterization of a novel protein from Sertoli cells, PASS1, that associates with mammalian small stress protein hsp27. *J Biol Chem*. 2000;275:18724-18731.
62. Liu G, Bollig-Fischer A, Kreike B, et al. Genomic amplification and oncogenic properties of the GASC1 histone demethylase gene in breast cancer. *Oncogene*. 2009;28:4491-4500.
63. Loenarz C, Ge W, Coleman ML, et al. PHF8 a gene associated with cleft lip/palate and mental retardation encodes for an N(epsilon)dimethyllysine demethylase. *Hum Mol Genet*. 2010;19:217-222.
64. Loh YH, Zhang W, Chen X, George J, Ng HH. Jmjd1a and Jmjd2c histone H3 Lys 9 demethylases regulate self-renewal in embryonic stem cells. *Genes Dev*. 2007;21: 2545-2557.
65. Lu PJ, Sundquist K, Baeckstrom D, et al. A novel gene (PLU-1) containing highly conserved putative DNA/chromatin binding motifs is specifically up-regulated in breast cancer. *J Biol Chem*. 1999;274:15633-15645.
66. Maherali N, Hochedlinger K. Guidelines and techniques for the generation of induced pluripotent stem cells. *Cell Stem Cell*. 2008;3:595-605.
67. Malloy PJ, Wang J, Jensen K, Feldman D. Modulation of vitamin d receptor activity by the corepressor hairless: differential effects of hairless isoforms. *Endocrinology*. 2009;150:4950-4957.
68. Metzger E, Wissmann M, Yin N, et al. LSD1 demethylates repressive histone marks to promote androgen-receptor-dependent transcription. *Nature*. 2005;437:436-439.
69. Metzger E, Yin N, Wissmann M, et al. Phosphorylation of histone H3 at threonine 11 establishes a novel chromatin mark for transcriptional regulation. *Nat Cell Biol*. 2008;10: 53-60.
70. Michaloglou C, Vredeveld LC, Soengas MS, et al. BRAFE600-associated senescence-like cell cycle arrest of human naevi. *Nature*. 2005;436:720-724.
71. Narita M, Nunez S, Heard E, et al. Rb-mediated heterochromatin formation and silencing of E2F target genes during cellular senescence. *Cell*. 2003;113:703-716.
72. Nicholson GA, Dawkins JL, Blair IP, et al. The gene for hereditary sensory neuropathy type I (HSN-I) maps to chromosome 9q22.1-q22.3. *Nat Genet*. 1996;13:101-104.
73. Nigro JM, Baker SJ, Preisinger AC, et al. Mutations in the p53 gene occur in diverse human tumour types. *Nature*. 1989;342:705-708.
74. Northcott PA, Nakahara Y, Wu X, et al. Multiple recurrent genetic events converge on control of histone lysine methylation in medulloblastoma. *Nat Genet*. 2009;41:465-472.
75. Ohguchi H, Tanaka T, Uchida A, et al. Hepatocyte nuclear factor 4alpha contributes to thyroid hormone homeostasis by cooperatively regulating the type 1 iodothyronine deiodinase gene with GATA4 and Kruppel-like transcription factor 9. *Mol Cell Biol*. 2008;28:3917-3931.
76. Okada Y, Scott G, Ray MK, Mishina Y, Zhang Y. Histone demethylase JHDM2A is critical for Tnp1 and Prm1 transcription and spermatogenesis. *Nature*. 2007;450:119-123.
77. Panteleyev AA, Botchkareva NV, Sundberg JP, Christiano AM, Paus R. The role of the hairless (hr) gene in the regulation of hair follicle catagen transformation. *Am J Pathol*. 1999;155:159-171.
78. Pasini D, Cloos PA, Walfridsson J, et al. JARID2 regulates binding of the Polycomb repressive complex 2 to target genes in ES cells. *Nature*. 2010;464:306-310.
79. Patel SR, Kim D, Levitan I, Dressler GR. The BRCT-domain containing protein PTIP links PAX2 to a histone H3, lysine 4 methyltransferase complex. *Dev Cell*. 2007;13:580-592.
80. Pedrosa E, Ye K, Nolan KA, et al. Positive association of schizophrenia to JARID2 gene. *Am J Med Genet B Neuropsychiatr Genet*. 2007;144:45-51.
81. Peng JC, Valouev A, Swigut T, et al. Jarid2/ Jumonji coordinates control of PRC2 enzymatic activity and target gene occupancy in pluripotent cells. *Cell*. 2009;139:1290-1302.
82. Peters AH, O'Carroll D, Scherthan H, et al. Loss of the Suv39h histone methyltransferases impairs mammalian heterochromatin and genome stability. *Cell*. 2001;107: 323-337.
83. Pfau R et al. Members of a family of JmjC domain-containing oncoproteins immortalize embryonic fibroblasts via a JmjC domain-dependent process. *Proc Natl Acad Sci U S A*. 2008;105:1907-1912.
84. Poetsch M, Dittberner T, Woenckhaus C. Frameshift mutations of RIZ, but no point mutations in RIZ1 exons in malignant melanomas with deletions in 1p36. *Oncogene*. 2002;21:3038-3042.
85. Politi K, Feirt N, Kitajewski J. Notch in mammary gland development and breast cancer. *Semin Cancer Biol*. 2004;14:341-347.

86. Pothof J, van Haaften G, Thijssen K, et al. Identification of genes that protect the C. elegans genome against mutations by genome-wide RNAi. *Genes Dev*. 2003;17:443-448.
87. Potter GB, Beaudoin GM 3rd, DeRenzo CL, Zarach JM, Chen SH, Thompson CC. The hairless gene mutated in congenital hair loss disorders encodes a novel nuclear receptor corepressor. *Genes Dev*. 2001;15:2687-2701.
88. Rizzo P, Osipo C, Foreman K, Golde T, Osborne B, Miele L. Rational targeting of Notch signaling in cancer. *Oncogene*. 2008;27:5124-5131.
89. Sansone P, Storci G, Giovannini C, et al. p66Shc/Notch-3 interplay controls self-renewal and hypoxia survival in human stem/progenitor cells of the mammary gland expanded in vitro as mammospheres. *Stem Cells*. 2007;25: 807-815.
90. Santos C, Rodriguez-Revenga L, Madrigal I, Badenas C, Pineda M, Mila M. A novel mutation in JARID1C gene associated with mental retardation. *Eur J Hum Genet*. 2006;14:583-586.
91. Sharma SV, Lee DY, Li B, et al. A chromatin-mediated reversible drug-tolerant state in cancer cell subpopulations. *Cell*. 2010;141:69-80.
92. Shen X, Kim W, Fujiwara Y, et al. Jumonji modulates polycomb activity and self-renewal versus differentiation of stem cells. *Cell*. 2009;139:1303-1314.
93. Shi Y, Lan F, Matson C, et al. Histone demethylation mediated by the nuclear amine oxidase homolog LSD1. *Cell*. 2004;119:941-953.
94. Siderius LE, Hamel BC, van Bokhoven H, et al. X-linked mental retardation associated with cleft lip/palate maps to Xp11.3-q21.3. *Am J Med Genet*. 1999;85:216-220.
95. Steele-Perkins G, Fang W, Yang XH, et al. Tumor formation and inactivation of RIZ1, an Rb-binding member of a nuclear protein methyltransferase superfamily. *Genes Dev*. 2001;15:2250-2262.
96. Suzuki T, Minehata K, Akagi K, Jenkins NA, Copeland NG. Tumor suppressor gene identification using retroviral insertional mutagenesis in Blm-deficient mice. *Embo J*. 2006;25: 3422-3431.
97. Tahiliani M, Koh KP, Shen Y, et al. Conversion of 5-methylcytosine to 5-hydroxymethylcytosine by the MLL partner TET1. *Science*. 2009;324:930-935.
98. Tahiliani M, Mei P, Fang R, et al. The histone H3K4 demethylase SMCX links REST target genes to X-linked mental retardation. *Nature*. 2007;447:601-605.
99. Takahashi K, Yamanaka S. Induction of pluripotent stem cells from mouse embryonic and adult fibroblast cultures by defined factors. *Cell*. 2006;126:663-676.
100. Takeuchi T, Kojima M, Nakajima K, Kondo S. jumonji gene is essential for the neurulation and cardiac development of mouse embryos with a C3H/He background. *Mech Dev*. 1999;86:29-38.
101. Tateishi K, Okada Y, Kallin EM, Zhang Y. Role of Jhdm2a in regulating metabolic gene expression and obesity resistance. *Nature*. 2009;458:757-761.
102. The Jackson Laboratory homepage (http://www.jaxmice.jax.org/strain/000673.html)
103. The Oncomine homepage (https://www.oncomine.org)
104. Thompson CC, Sisk JM, Beaudoin GM 3rd. Hairless and Wnt signaling: allies in epithelial stem cell differentiation. *Cell Cycle*. 2006;5:1913-1917.
105. Tong Q, Dalgin G, Xu H, Ting CN, Leiden JM, Hotamisligil GS. Function of GATA transcription factors in preadipocyte-adipocyte transition. *Science*. 2000;290: 134-138.
106. Toyoda M, Shirato H, Nakajima K, et al. Jumonji down-regulates cardiac cell proliferation by repressing cyclin D1 expression. *Dev Cell*. 2003;5:85-97.
107. Tzschach A, Lenzner S, Moser B, et al. Novel JARID1C/SMCX mutations in patients with X-linked mental retardation. *Hum Mutat*. 2006;27:389.
108. Valk-Lingbeek ME, Bruggeman SW, van Lohuizen M. Stem cells and cancer; the polycomb connection. *Cell*. 2004;118:409-418.
109. van Haaften G, Dalgliesh GL, Davies H, et al. Somatic mutations of the histone H3K27 demethylase gene UTX in human cancer. *Nat Genet*. 2009;41:521-523.
110. van Zutven LJ, Onen E, Velthuizen SC, et al. Identification of NUP98 abnormalities in acute leukemia: JARID1A (12p13) as a new partner gene. *Genes Chromosomes Cancer*. 2006;45:437-446.
111. Volcik KA, Zhu H, Finnell RH, Shaw GM, Canfield M, Lammer EJ. Evaluation of the jumonji gene and risk for spina bifida and congenital heart defects. *Am J Med Genet A*. 2004;126:215-217.
112. Wang GG, Song J, Wang Z, et al. Haematopoietic malignancies caused by dysregulation of a chromatin-binding PHD finger. *Nature*. 2009;59:847-851.
113. Wang J, Malloy PJ, Feldman D. Interactions of the vitamin D receptor with the corepressor hairless: analysis of hairless mutants in atrichia with papular lesions. *J Biol Chem*. 2007;282:25231-25239.
114. Wang Y, Wysocka J, Sayegh J, et al. Human PAD4 regulates histone arginine methylation levels via demethylimination. *Science*. 2004;306:279-283.
115. Wang Y, Zhang H, Chen Y, et al. LSD1 is subunit of the NURD complex and targets the metastasis programs in breast cancer. *Cell*. 2009;138:660-672.
116. Wen Y, Liu Y, Xu Y, et al. Loss-of-function mutations of an inhibitory upstream ORF in the human hairless transcript cause Marie Unna hereditary hypotrichosis. *Nat Genet*. 2009;41:228-233.
117. Wen H, Li J, Song T, et al. Recognition of histone H3K4 trimethylation by the plant homeodomain of PHF2 modulates histone demethylation. *J Biol Chem*. 2010;285: 9322-9326.
118. Whetstine JR, Nottke A, Lan F, et al. Reversal of histone lysine trimethylation by the JMJD2 family of histone demethylases. *Cell*. 2006;125:467-481.
119. Wissmann M, Yin N, Muller JM, et al. Cooperative demethylation by JMJD2C and LSD1 promotes androgen receptor-dependent gene expression. *Nat Cell Biol*. 2007;9: 347-353.
120. Xi ZQ, Sun JJ, Wang XF, et al. HSPBAP1 is found extensively in the anterior temporal neocortex of patients with intractable epilepsy. *Synapse*. 2007;61:741-747.

121. Xu Z, Yu S, Hsu CH, Eguchi J, Rosen ED. The orphan nuclear receptor chicken ovalbumin upstream promoter-transcription factor II is a critical regulator of adipogenesis. *Proc Natl Acad Sci USA*. 2008;105:2421-2426.
122. Yamane K, Tateishi K, Klose RJ, et al. PLU-1 is an H3K4 demethylase involved in transcriptional repression and breast cancer cell proliferation. *Mol Cell*. 2007; 25:801-812.
123. Yamane K, Toumazou C, Tsukada Y, et al. JHDM2A, a JmjC-containing H3K9 demethylase, facilitates transcription activation by androgen receptor. *Cell*. 2006; 125:483-495.
124. Yang ZQ, Imoto I, Pimkhaokham A, et al. A novel amplicon at 9p23–24 in squamous cell carcinoma of the esophagus that lies proximal to GASC1 and harbors NFIB. *Jpn J Cancer Res*. 2001;92:423-428.

6 Autoimmune Diseases

Travis Hughes and Amr H. Sawalha

6.1 Introduction

Epigenetic mechanisms play a central regulatory role in the immune system, ensuring that the immune response can clear pathogens or foreign material from the system, yet maintain self-tolerance. Autoimmunity is defined as the production of antibodies against a host of self-antigens and arises when the immune system can no longer distinguish between foreign and self-antigens. Autoimmunity results from the aberrant activation of the innate or adaptive immune systems, with excessive proliferation of antibody-producing cells as the result.

6.2 Epigenetic Regulation of T Helper Cell Subsets

T cell populations are dynamic and diverse. Epigenetic plasticity of T helper lineages contributes to the diversity and malleability of CD4+ T cell populations. T helper responses enhance both cell-mediated and antibody-mediated immune responses. Naïve T cells differentiate into several effector subsets characterized by distinct cytokine expression patterns. T helper cells have traditionally been divided into two main classes, Th1 and Th2, but more recently, Th17 and T_{FH} cells have been added to the lineage. These subsets serve a specific role in modulating the immune response.

A.H. Sawalha (✉)
Department of Medicine, Division of Rheumatology, and Department of Pathology, University of Oklahoma Health Sciences Center, and
Department of Veterans Affairs Medical Center, and Arthritis and Clinical Immunology Program, Oklahoma Medical Research Foundation, Oklahoma City, OK 73104, USA
e-mai: amr-sawalha@omrf.ouhsc.edu

Effector T cell populations vary widely in characteristics, largely due to epigenetic modifications at key regulatory loci. Bivalent epigenetic modifications contribute to the plasticity of these effector T cell populations by inducing a primed chromatin state at key regulatory loci. H3K4me3, a permissive epigenetic mark, and H3K27me3, a repressive mark, act concurrently at the T-bet and GATA3 loci in naïve CD4+ T cells.[96] However, H3K4me3 comes to predominate with lineage differentiation. This indicates that histone modification patterns contribute to lineage commitment among CD4+ T cells.

Changes in chromatin structure correspond to T helper lineage differentiation. BRG1/ SWI-SNF chromatin remodeling takes place at the IFN-gamma locus in Th1 subsets in a process mediated by STAT4 signaling.[106] BRG1 chromatin remodeling regulates activation of the Th2 locus control region (LCR) via a transcription factor–dependent manner.[95] Additionally, as a Th2 locus control region becomes demethylated, polarization of lineage increases (Fig. 6.1).[33] Interestingly, this demethylation is independent of the cell cycle.

6.2.1 Regulatory T Cells

Regulatory T cells (Treg's) are made up of CD4+ CD25+ T cells that express the transcription factor FOXP3. They regulate potentially deleterious effects of the immune system and maintain self-tolerance. Regulatory T cells arise both from thymic development and transient acquisition of suppressor function. Natural (thymic) regulatory T cells are critical in maintaining self-tolerance. TGF-beta signaling is essential for the development of thymic regulatory T cells.[41]

H.I. Roach et al. (eds.), *Epigenetic Aspects of Chronic Diseases*,
DOI: 10.1007/978-1-84882-644-1_6, © Springer-Verlag London Limited 2011

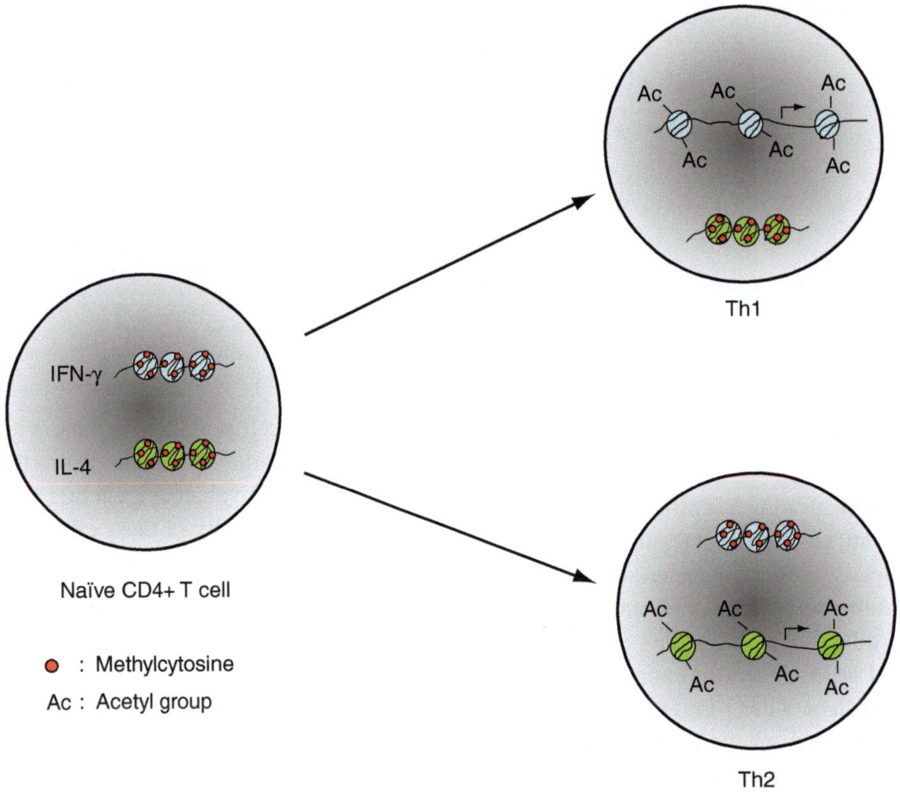

Fig. 6.1 Epigenetic modifications regulate T helper cell differentiation. Above, DNA methylation maintains transcriptional repression of lineage-specific cytokines in naïve CD4+ T cells. Upon activation and lineage differentiation, a reduction in DNA methylation and an increase in histone acetylation are observed at specific loci. Under Th1 polarizing conditions, IFN-gamma becomes expressed due to a loss of repressive DNA methylation and a gain of acetylated histone tails. While IFN-gamma is expressed in Th1 cells, IL-4 remains transcriptionally inactive due to epigenetic repression. Conversely, under Th2 polarizing conditions, IL-4 is acetylated and DNA methylation is reduced leading to its expression, while IFN-gamma is not expressed. These epigenetic modifications occur in a largely transcription factor–dependent manner mediated by T-bet in Th1cells and GATA3 in the Th2 lineage (From Sawalha[74]; Reprinted by permission of Taylor & Francis)

Both transient and thymic acquisition of suppressor function are dependent on the induction of FOXP3 expression. The longevity of the transiently acquired suppressor function affects non-thymic regulatory T cells in human autoimmunity, because diminution of suppressor function contributes to autoimmunity.

In mice, hypermethylation of an enhancer region located upstream of the FOXP3 gene correlates directly with a reduction in the expression and proliferation of regulatory T cells in vivo.[38] Specifically, the region found to be differentially methylated between murine naive CD4+ T cells and nTregs is located within a CpG island 5.5 kb upstream of the FOXP3 transcription start site. In humans, a specific pattern of demethylation of the FOXP3 promoter characterizes stable Treg cell populations.[29] The methylation of a specific site, the TSDR (Treg-specific demethylated region), largely determines the stability and suppressive function of FOXP3 expression. The TSDR is demethylated in stable regulatory T cells that have undergone thymic development and selection, but is only partially demethylated in cells with transient FOXP3 expression.[16] The degree of TSDR hypomethylation along with prolonged TGF-beta exposure determines the durability of the induced regulatory T cell lineage.[64] The degree of methylation at this site may help to explain differences between natural (thymic) and induced (peripheral) Treg cell populations.[27]

Given histone deacetylase inhibitors (HDACi) ameliorate autoimmunity and inflammation,[93] they may constitute a potential therapeutic agent. Treatment with HDAC inhibitors supports proliferation of FOXP3+

regulatory T cells, with the activated T cells providing the suppressor function.[48]

6.2.2 T Helper 17 Cells

Th17 cells, also referred to as inflammatory helper T cells, are characterized by the elevated expression of IL-17 and IL-17F. Normally, Th17 cells, in addition to recruiting neutrophils, clear pathogens in the course of infection.[62] Th17 proliferation may induce lupus flairs, at the same time antagonizing Treg cell function. Th17 levels increase the mononuclear cells of peripheral blood of lupus patients who have experienced flares, especially in patients with vasculitis.[104]

Interestingly, the relationship between regulatory T cells and Th17 cells is not only antagonistic, but also reciprocal. Stimulation of a highly purified subset of regulatory T cells with exogenous rIL-2/rIL-15 led to a subset of IL-17-producing cells.[35] As IL-17 expression increased, FOXP3 expression decreased. Interestingly, these changes were counteracted by addition of HDACs.[35]

IL-6 and TGF-beta act in concert to induce the proliferation and expansion of Th-17 cells.[7] In response to IL-6 and TGF-beta,[1] chromatin is remodeled. This in turn leads to the upregulation of IL-17 and IL-17F in inflammatory T cells. Together, TGF-beta and IL-6 also work to reduce FOXP3 promoter occupancy, a process counteracted by the addition of HDAC inhibitors.[73] IL-6/IL-6R/STAT3 signaling inhibits FOXP3 expression.[100] In the presence of IL-6, the proliferation of FOXP3+ Tregs is markedly reduced. In the absence of IL-6, lupus-prone mice retain an active regulatory T cell population and resist the development of autoimmunity.[36]

Epigenetically induced variations in IL-6 affect autoimmunity. The demethylation of a single CpG dinucleotide has been linked to an increase in IL-6 mRNA in the peripheral blood mononuclear cells of a rheumatoid arthritis patient.[58] IL-6, by modulating DNA methylation, is also involved in an increase in B cells in autoimmune arthritis. IL-6 modulates CD5 expression through an alteration of its promoter methylation.[18] On the basis of these findings, IL-6 seems a promising therapeutic candidate to modulate autoimmunity.

Interestingly, activation of the Aryl Hydrocarbon Receptor (AHR), closely associated with dioxin exposure, leads to increased proliferation of Th17 cells and contributes to the development of Th17-mediated autoimmunity.[91] This constitutes evidence for a tangible link between environmental exposure and the development of autoimmunity.

6.2.3 Follicular Helper T Cells (T_{FH})

Follicular helper T cells function to support B cell–mediated immunity and antibody class switching.[65] The follicular helper T cell lineage, characterized by the expression of CXCR5, PD1, ICOS, and IL-21, arises from either uncommitted CD4+ T cells or already polarized T helper cells.[105] BCL6 seems to be the master transcription factor that initiates the T_{FH} lineage.[59,101] BCL6 expression also reduce the expression of other lineage associated transcription factors,[37] and BCL6 as well as follicular cell proliferation is downregulated by BLIMP1.[30]

In autoimmunity, the T_{FH} lineage contributes to excess germinal center (GC) formation. This leads to an increase in the selection of antibody-producing B cells in *sanroque* mice,[39] an autoimmune strain of mice with an M199R mutation in the Roquin gene. Deletion of one allele of BCL6 ameliorates autoimmunity in these mice. Roquin, an E3 ubiquitin ligase, represses the development of autoimmunity by limiting inducible T cell co-stimulator (ICOS) translation,[102] a factor that promotes T cell activation and expansion and thus autoimmunity.[32] ICOS regulates the expression of c-Maf and IL-21, which contribute to the development of Th17 and T_{FH} cells.[6]

Therapeutic intervention aimed at the limitation of the T_{FH} lineage and ICOS+ T cells attenuates autoimmune disease manifestations in NZB/NZW mice.[26]

6.3 Epigenetic Dysregulation in Systemic Lupus Erythematosus

Systemic lupus erythematosus is an autoimmune disorder that affects multiple organs, including the skin, kidneys, heart, lungs, and the nervous system, and that is characterized by the formation of autoantibodies against multiple nuclear antigens. The etiology of lupus is not entirely understood, but genetic and epigenetic factors probably play a role in the development of lupus. Genome-wide association studies are revealing an increasing number of polymorphisms that are positively associated with disease susceptibility in lupus,[21] but lupus also has a large epigenetic component.[22,83] This means that environmental exposure may initiate lupus in genetically predisposed individuals.

6.3.1 Genes Overexpressed in Lupus Due to Promoter Demethylation

DNA methylation, the covalent addition of a methyl group to CpG cytosine residues, is most often associated with a repressive role in the regulation of gene expression. Hypermethylation of promoter regions typically inhibits gene expression, whereas hypomethylation permits gene expression. Generalized DNA hypomethylation due to epigenetic dysregulation is seen in lupus CD4+ T cells and results in overexpression of methylation-sensitive genes and T cell autoreactivity.[74] Reduced promoter methylation is often accompanied by overexpression of genes that encode for cell surface–expressed molecules and effector cytokines. This contributes to autoimmunity.

Lymphocyte function–associated antigen 1(LFA-1) is overexpressed in lupus CD4+ T cells; this leads to autoreactivity.[67] LFA-1 consists of two subunits, ITGAL (CD11a) and CD18. Overexpression of either subunit is sufficient to induce T cell autoreactivity. Cells transfected with a vector to overexpress LFA-1 display autoreactivity similar to that of procainamide-treated cells.[103] The gene that encodes ITGAL is hypomethylated in lupus T cells. Promoter methylation for ITGAL is significantly reduced in individuals with active lupus.[43] Treatment of T cells with DNA methylation inhibitors 5-azaC and procainamide induces similar hypomethylation of promoter and ~threefold overexpression of CD11a.[42,43]

Perforin is a cytotoxic effector molecule that is expressed by natural killer cells (NK's) and select T cell subsets. The expression of perforin is regulated by the methylation of its promoter sequences.[44] Perforin is overexpressed in lupus CD4+ T cells; this corresponds to promoter demethylation.[31]

CD70 is a protein that plays a role in T cell co-stimulation, B cell CD27 stimulation, regulation of cytotoxic activity of NK's, and the regulation of immunoglobulin production.[34,56,98] It is overexpressed in response to promoter hypomethylation in both idiopathic and drug-induced lupus.[45] CD70 is overexpressed in normal and lupus CD4+ T cells treated with 5-azaC.[61] CD4+ T cells from a patient with subacute cutaneous lupus erythematosus also exhibit CD70 demethylation and overexpression.[49]

DNA methylation is defective and CD70 is overexpressed in MRL/lpr lupus-prone mice, with DNMT1 reduced and CD70 expression increased.[76] These findings provide evidence that CD70 expression is regulated by DNA methylation and that reduced methyltransferase activity leads to overexpression of CD70, events that contribute to autoimmunity.

Demethylation and overexpression of the killer Ig-like receptor molecule (KIR2DL4) have recently been reported in lupus CD4+ and CD8+ T cells.[5] This corresponds to disease activity in lupus patients, and may contribute to pathogenesis due to an increase in autologous macrophage killing and in IFN-gamma expression.[5]

6.3.2 Demethylation of the Inactive X Chromosome and Lupus

Sex chromosome complement strongly affects lupus susceptibility. Thus, XX mice are more prone to develop lupus than phenotypically female XY sry-negative mice.[80] In females, high levels of DNA methylation inactivate one of the two X chromosomes.[53] Some genes are demethylated on the inactive X chromosome in lupus patients. Loss of methylation-dependent imprinting leads to partial reactivation of the quiescent X chromosome. This may explain the 9:1 female bias observed in lupus.

CD40LG, a B cell co-stimulatory protein located on the X chromosome, is overexpressed in female but not in male CD4+ T cells of lupus patients. CD40LG is not methylated in males, but in women, it is methylated on the inactive X chromosome and not methylated on the active X chromosome. When cultured T cells from normal healthy men and women were treated with 5-azaC, a DNA methyltransferase inhibitor, CD40LG expression was increased only in CD4+ T cells from the women, as true also in CD4+ T cells from female lupus patients.[46] The overexpression of CD40LG in B cells induces autoantibodies and causes nephritis in mice.[24] Along the same line, human T cells that overexpress CD40LG induced autologous B cell activation and plasma cell differentiation in vitro.[108]

Additional evidence for the role of a gene-dose effect on the X chromosome in lupus comes from Klinefelter's Syndrome, a condition in which phenotypically male individuals exhibit sex chromosome trisomy (47, XXY). A study investigating the prevalence of Klinefelter's Syndrome among male lupus patients found that individuals with chromosome trisomy have a ten-fold higher risk of developing lupus than their 46 (XY) counterparts.[79]

6.3.3 Drug-Induced Lupus Caused by DNA Methylation Inhibitors

In 1986, treatment of CD4+ T cells with 5-azacytidine (5-azaC), a DNA methylation inhibitor, was first shown to cause hypomethylation and induce T cell autoreactivity.[66] Lupus is largely an idiopathic condition, but a number of drugs induce a lupus-like disease. An example is hydralazine, a blood pressure medication, prolonged treatment with which induces lupus in some patients. Procainamide is another drug that causes lupus. Both procainamide and hydralazine inhibit DNA methylation in T cells in vivo. Procainamide induces autoreactivity in CD4+ T cells by inhibiting DNA methyltransferase enzymes. This leads to hypomethylation and increased gene expression.[11,78] This pattern of hypomethylation and autoreactivity is similar to that seen in idiopathic lupus CD4+ T cells.[68] Treatment of Th2 cells with procainamide and 5-azaC induces autoreactivity in vitro. When transferred, these cells induce lupus-like phenotypes in vivo.[104]

A closer understanding of the mechanisms underlying drug-induced lupus has helped shed light on the etiology of idiopathic lupus. Hydralazine induces autoimmunity by inhibiting the extracellular signal–regulated kinase (ERK) pathway. The expression of DNMT1 is, at least in part, regulated by the ERK pathway. Treatment of T cells with ERK inhibitors or hydralazine reduces phosphorylation of ERK, lowers DNMT1 expression levels, induces a generalized hypomethylation of DNA in CD4+ T cells, increases LFA-1 expression, and subsequently induces T cell autoreactivity.[12] The role played by a defective T cell ERK signaling pathway in the pathogenesis of idiopathic lupus is described below.

6.3.4 ERK Pathway Signaling, DNA Methylation, and Lupus

Mitogen-activated protein kinases (MAPKs) help maintain normal immune system homeostasis.[14] These signaling pathways help drive the changes in gene expression that take place as immune cells differentiate. For example, the ERK signaling pathway is essential for CD4+ T cells to give rise to Th2.[14,76] MAPK also has a regulatory function through its recruitment of the AP-1 transcription factor. In addition, MAPK modulates methyltransferase activity.[13]

Differential expression of Interferon-regulated genes in the spleen is due to reduced T cell ERK pathway signaling in transgenic mice which express a dominant-negative form of MEK. These mice also exhibit reduced DNMT1 expression, overexpression of other methylation-sensitive genes associated with human lupus, and the production of anti-dsDNA antibodies (Fig. 6.2).[76]

The defect in ERK signaling seen in idiopathic and hydralazine-induced lupus is linked to aberrant PKC-delta phosphorylation. Reduced phosphorylation of PKC-delta and reduced ERK signaling cause increased CD70 expression in response to CD70 promoter demethylation.[19] PKC-delta knockout mice also have increased B cell proliferation and develop autoimmunity.[52] Overall levels of active ERK and JNK are positively correlated to the disease activity index of systemic lupus erythmatosus (SLEDAI).[54]

6.3.5 Histone Protein Modifications and Lupus

Modification of histone proteins modulates gene expression and overall chromatin architecture. Histone modifications are related to DNA methylation and other epigenetic modifications. The association of the histone methyltransferases G9a and SUV39H1 with DNMT1 during maintenance methylation provides evidence for significant interaction between DNA methylation and H3K9 methylation.[15] G9a-deficient

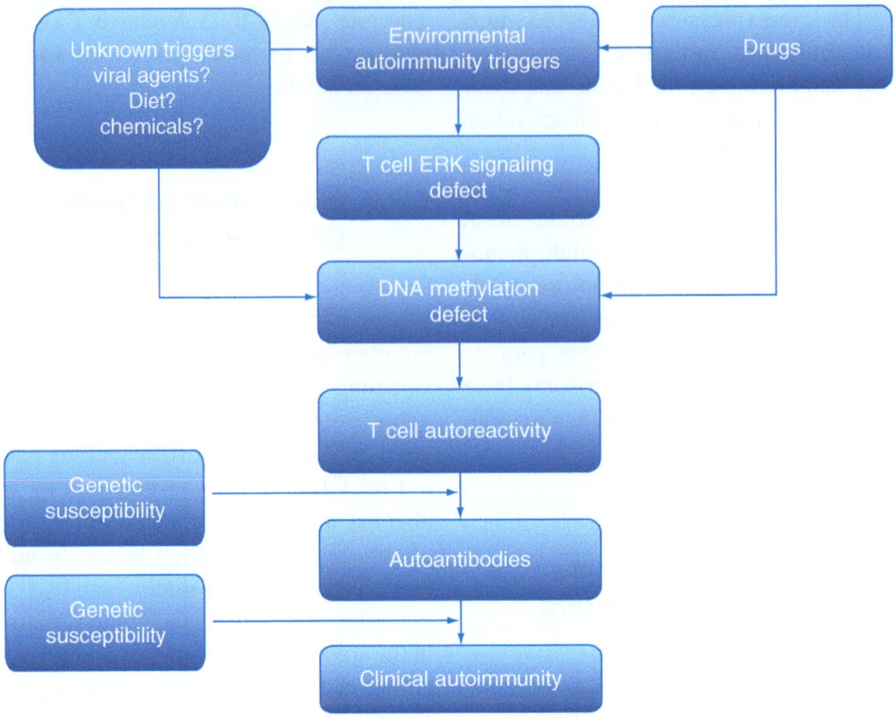

Fig. 6.2 Hypothetical scheme that demonstrates a central role for a defect in DNA methylation in the pathogenesis of autoimmunity. Environmental triggers, such as the lupus-inducing drug hydralazine, and possibly other unidentified triggers, lead to defects in the ERK signaling pathway in T cells. This induces a DNA methylation defect, as ERK signaling regulates DNA methyltransferase I (*DNMT1*) expression. As a result of reduced DNMT1 expression, T cell methylation-sensitive genes are overexpressed. This results in T cell autoreactivity, B cell help, and autoantibody production. Other environmental triggers can induce a T cell methylation defect without affecting ERK signaling. An example is procainamide which inhibits DNMT1 directly. The production of autoantibodies that predate disease is influenced by the genetic susceptibility of the host. After autoantibodies are produced, and if genetic susceptibility loci are present, patients will develop the clinical features of autoimmune disease

lymphocytes display unperturbed development in T cells, whereas B lymphocyte maturation and plasma cell differentiation are marginally attenuated.[89]

Polycomb group proteins appear to form another link between DNA methylation and histone modification. DNMT1 associates directly with EZH2, a histone methyltransferase that mediates H3K27 trimethylation, a repressive epigenetic mark.[92] EZH2 is a member of the polycomb repressor complex (PRC2) which mediates repression through the induction of large (>10 kb) chromatin loop structures.[87] EED, another member of PRC2 that can recognize H3K27me3, is essential for the persistence of histone modifications through cell division.[51]

Abnormal patterns of histone modification have been observed in lupus CD4+ T cells. These include reductions in histone acetylation, H3K9 methylation, p300 expression, and EZH2 expression.[25] Interestingly, the loss of epigenetic marks was correlated with disease activity. Moreover, the loss of p300 acetyltransferase activity in B cells results in a lupus-like autoimmune disease in mice.[17]

6.3.6 Chromatin Remodeling Complexes and Lupus

The deletion of Mta2, a component of the NuRD repressor complex, leads to a lupus-like autoimmune disease characterized by skin lesions, autoantibody formation, and renal infiltration of lymphocytes.[47] Furthermore, the recruitment of the NuRD repressor complex is initiated by the localization of Methyl-Binding Domain Protein 2 (MBD2). These findings highlight the importance of maintaining methylation-dependent repression at specific loci so as to prevent autoimmunity.

6.4 MECP2: The Role of Methyl-Binding Domain Proteins in Epigenetics and Autoimmunity

Methyl-CpG-Binding Protein 2 (MECP2) most commonly known for its causative role in Rett syndrome, is a methyl-CpG-binding protein[20] that binds to methylated DNA, where it acts either as an inducer or repressor of transcription. Through the recruitment of the Sin3a/HDAC co-repressor complex, MECP2 represses transcription by causing DNA to become more tightly coiled and less accessible to transcriptional machinery (Fig. 6.3).[57] Interestingly, MECP2 is also an activator of transcription, through recruitment of CREB1.[9] In fact, MECP2 more often functions as an activator than as a repressor.

The role of MECP2 in the regulation of methylated genes makes it an interesting target for lupus research. Genotyping assays within Xq28, which contains the gene that encodes MECP2, have revealed a haplotype that is positively associated with the susceptibility to developing lupus.[77] Risk MECP2 haplotype variants are strongly correlated with aberrant gene expression in EBV-transformed human B cell lines from lupus patients.[94] The same lupus-associated MECP2 genetic polymorphism has also been found to increase susceptibility to Sjogren's syndrome (unpublished observation). It therefore seems variants within MECP2 affect human autoimmunity.

Other methyl-binding domain proteins are also overexpressed in lupus patients. Increases in MBD2 and MBD4 transcripts in lupus CD4+ T cells are inversely correlated to global methylation levels in lupus CD4+ T cells, as reported from a cohort of 29 lupus patients.[3]

6.5 Interference RNA Signatures of Autoimmunity

The role of micro-RNA in the etiology of autoimmunity is poorly understood, yet it is apparent that micro-RNA play an important role in the regulation of normal immune function and in the suppression of autoimmunity.[69] Micro-RNA plays a critical role in maintaining regulatory T cell suppressor function in lineage-committed cells. Dicer-deficient CD4+ T cells

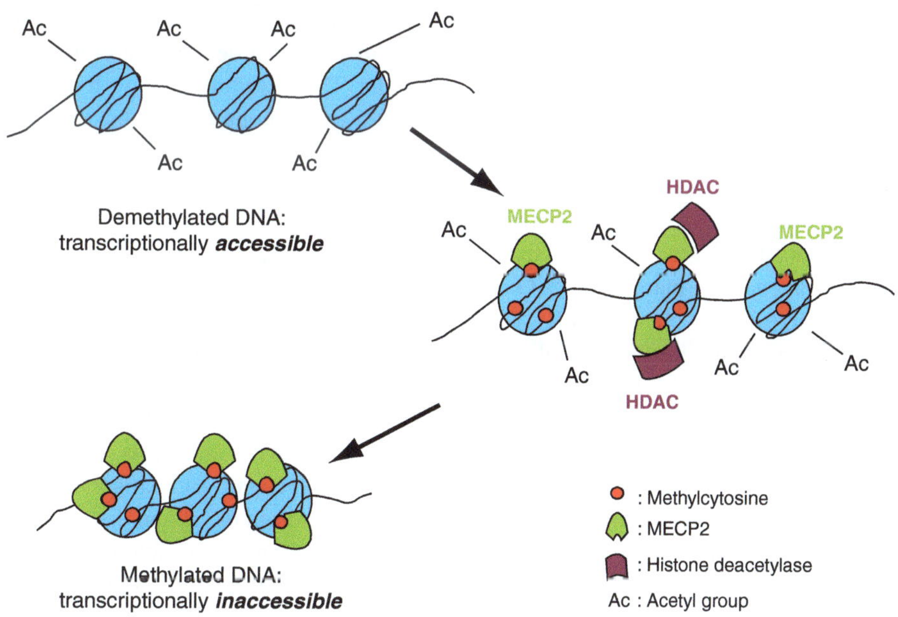

Fig. 6.3 DNA methylation is a negative regulator of transcription. In the absence of DNA methylation, chromatin maintains an open conformation that is transcriptionally accessible. However, upon the addition of a methyl group to CpG cytosine residues, chromatin becomes transcriptionally inaccessible due to nucleosome remodeling mediated by methyl-binding domain proteins. One such protein, Methyl-CpG-Binding Protein 2 MECP2, localizes to methylated DNA where it can recruit histone deacetylases (*HDAC*) which remove acetyl groups from histone tails. This leads to close chromatin conformation and transcriptionally inaccessible nucleosome remodeling (From Sawalha[74]; Reprinted by permission of Taylor & Francis)

show reduced differentiation and proliferation upon stimulation[84] and lose lineage commitment and suppressor function; this leads to systemic autoimmunity in mice.[40,107] It is particularly interesting that a single micro-RNA can affect the translation of multiple genes.[2] In autoimmunity, Su-antibodies target the interference RNA machinery.[28] These antibodies recognize Dicer and Argonaut group proteins in vitro. This finding suggests that in autoimmunity, there is a loss of proper interference RNA function.

When the innate immune system is activated, miRNA-146 expression, induced by NF-kappaB, downregulates certain downstream cytokines.[83] Micro-RNA-146 also downregulates type-I interferon expression.[10] Underexpression of micro-RNA-146 may account for some of the loss of regulation of type-I interferon expression seen in lupus.[84] An excess of type-I interferon expression, along with induction of interferon-regulated genes, may play a role in lupus and the progression of this disease.[4] Micro-RNA-31 has a binding site in the 3′ UTR of FOXP3 mRNA. When binding occurs, FOXP3 expression is downregulated.[71]

The disruption of interference RNA signatures during viral latency may contribute to the development of autoimmunity. Epstein–Barr virus infection has been implicated as an environmental factor that contributes to the development of lupus.[22] Interestingly, a membrane protein of the latent virus inhibits the expression of miRNA-146a.[8] EBV latency has also been found to lead to epigenetic repression of Bim in B cells through H3K27me3 and subsequent DNA methylation.[63]

The deletion of micro-RNA-17-92 in a mouse model has revealed that this cluster has an essential role in B cell development and maturation.[90] Overexpression of the micro-RNA cluster miR-17-92 causes autoimmunity in mice and has been implicated in lymphoproliferative disease.[97] Moreover, overexpression of miR-17-92 leads to a reduction in PTEN and BIM, two proteins that are critical for the maintenance of stable lymphocyte populations.

Micro-RNA-155 appears to play an integral role in immune system homeostasis and function.[60] This was shown in a study with bic/micro-RNA-155 knockout mice, whose pathogen response was significantly attenuated in response to a lack of bic, the precursor of micro-RNA-155.[70] Micro-RNA-155 regulates germinal center activity in mice[85] and lack of micro-RNA-155 causes a number of genes to be differentially expressed. Half of these genes must have micro-RNA-155 binding sites in their 5′ UTR's.[91] Whereas the lack of micro-RNA-155 inhibits plasma cell differentiation and expansion, overexpression in stimulated CD4+ T cells is refractory to nTreg-mediated suppression. Inhibition of micro-RNA-155 in CD4+ T cells makes them more responsive to nTreg-suppression.[81]

6.6 TLR9, CPG DNA, and Autoimmunity

Another potential factor in the development of autoimmunity is TLR9 which activates the immune system through signal transduction.[23] Aberrant recognition of CpG motifs may lead to increased B cell activation, stimulating the production of anti-dsDNA antibodies that are commonly seen in autoimmune disorders. Interferon-alpha sensitizes B cells to recognize self-CpG DNA by TLR9.[88] Lack of TLR9 attenuates the severity of the disease manifestations as when brought about by the transfer of alloreactive T cells in the chronic-graft vs. host model of Lupus.[50] ERK pathway signaling is essential limiting BCR induction through TLR9. This also prevents antibody-mediated autoimmunity by limiting the proliferation of antibody-secreting cells.[72] Inasmuch as lupus is characterized by a T cell DNA methylation defect, Rui et al. raise the question whether TLR9 stimulation by hypomethylated DNA released from apoptotic T cells contributes to autoimmunity in lupus patients.

6.7 Conclusion

The dysregulation of epigenetic mechanisms that govern gene expression and immune cell proliferation contributes to the diversity of clinical manifestations in autoimmune disorders. It is the interplay between genetic background and epigenetic modulation that shapes the mosaic of human disease. It is highly probable that environmental factors predispose to disease in genetically susceptible individuals by altering epigenetic mechanisms. How major alterations in fundamental and global basic biological processes such as epigenetic regulation lead to specific disease manifestations without at the same time altering the existence of the living cell remains a mystery. Our understanding of epigenetic mechanisms and their alterations in disease is, as yet, therefore still in its infancy. But when the etiology, the extent, and the consequences of epigenetic defects in autoimmune diseases are understood, targets for novel therapeutic options are likely to become available.

References

1. Akimzhanov AM, Yang XO, Dong C. Chromatin remodeling of interleukin-17 (IL-17)-IL-17F cytokine gene locus during inflammatory helper T cell differentiation. *J Biol Chem*. 2007;282(9):5969-5972.
2. Baek D, Villen J, Shin C, Camargo FD, Gygi SP, Bartel DP. The impact of microRNAs on protein output. *Nature*. 2008;455(7209):64-71.
3. Balada E, Ordi-Ros J, Serrano-Acedo S, Martinez-Lostao L, Vilardell-Tarres M. Transcript overexpression of the MBD2 and MBD4 genes in CD4+ T cells from systemic lupus erythematosus patients. *J Leukoc Biol*. 2007;81(6):1609-1616.
4. Banchereau J, Pascual V. Type I interferon in systemic lupus erythematosus and other autoimmune diseases. *Immunity*. 2006;25(3):383-392.
5. Basu D, Liu Y, Wu A, et al. Stimulatory and inhibitory killer Ig-like receptor molecules are expressed and functional on lupus T cells. *J Immunol*. 2009;183(5):3481-3487.
6. Bauquet AT, Jin H, Paterson AM, et al. The costimulatory molecule ICOS regulates the expression of c-Maf and IL-21 in the development of follicular T helper cells and TH-17 cells. *Nat Immunol*. 2009;10(2):167-175.
7. Bettelli E, Carrier Y, Gao W, et al. Reciprocal developmental pathways for the generation of pathogenic effector TH17 and regulatory T cells. *Nature*. 2006;441(7090):235-238.
8. Cameron JE, Yin Q, Fewell C, et al. Epstein-Barr virus latent membrane protein 1 induces cellular MicroRNA miR-146a, a modulator of lymphocyte signaling pathways. *J Virol*. 2008;82(4):1946-1958.
9. Chahrour M, Jung SY, Shaw C, et al. MECP2, a key contributor to neurological disease, activates and represses transcription. *Science*. 2008;320(5880):1224-1229.
10. Chan EK, Satoh M, Pauley KM. Contrast in aberrant microRNA expression in systemic lupus erythematosus and rheumatoid arthritis: is microRNA-146 all we need? *Arthritis Rheum*. 2009;60(4):912-915.
11. Cornacchia E, Golbus J, Maybaum J, Strahler J, Hanash S, Richardson B. Hydralazine and procainamide inhibit T cell DNA methylation and induce autoreactivity. *J Immunol*. 1988;140(7):2197-2200.
12. Deng C, Lu Q, Zhang Z, et al. Hydralazine may induce autoimmunity by inhibiting extracellular signal-regulated kinase pathway signaling. *Arthritis Rheum*. 2003;48(3):746-756.
13. Deng C, Yang J, Scott J, Hanash S, Richardson BC. Role of the ras-MAPK signaling pathway in the DNA methyltransferase response to DNA hypomethylation. *Biol Chem*. 1998;379(8-9):1113-1120.
14. Dong C, Davis RJ, Flavell RA. MAP kinases in the immune response. *Annu Rev Immunol*. 2002;20:55-72.
15. Esteve PO, Chin HG, Smallwood A, et al. Direct interaction between DNMT1 and G9a coordinates DNA and histone methylation during replication. *Genes Dev*. 2006;20(22):3089-3103.
16. Floess S, Freyer J, Siewert C, et al. Epigenetic control of the foxp3 locus in regulatory T cells. *PLoS Biol*. 2007;5(2):e38.
17. Forster N, Gallinat S, Jablonska J, Weiss S, Elsasser HP, Lutz W. p300 protein acetyltransferase activity suppresses systemic lupus erythematosus-like autoimmune disease in mice. *J Immunol*. 2007;178(11):6941-6948.
18. Garaud S, Le Dantec C, Jousse-Joulin S, et al. IL-6 modulates CD5 expression in B cells from patients with lupus by regulating DNA methylation. *J Immunol*. 2009;182(9):5623-5632.
19. Gorelik G, Fang JY, Wu A, Sawalha AH, Richardson B. Impaired T cell protein kinase C delta activation decreases ERK pathway signaling in idiopathic and hydralazine-induced lupus. *J Immunol*. 2007;179(8):5553-5563.
20. Gura T. Gene defect linked to Rett syndrome. *Science*. 1999;286(5437):27.
21. Harley JB, Alarcon-Riquelme ME, Criswell LA, et al. Genome-wide association scan in women with systemic lupus erythematosus identifies susceptibility variants in ITGAM, PXK, KIAA1542 and other loci. *Nat Genet*. 2008;40(2):204-210.
22. Harley JB, James JA. Epstein-Barr virus infection induces lupus autoimmunity. *Bull NYU Hosp Jt Dis*. 2006;64(1-2):45-50.
23. He B, Qiao X, Cerutti A. CpG DNA induces IgG class switch DNA recombination by activating human B cells through an innate pathway that requires TLR9 and cooperates with IL-10. *J Immunol*. 2004;173(7):4479-4491.
24. Higuchi T, Aiba Y, Nomura T, et al. Cutting Edge: ectopic expression of CD40 ligand on B cells induces lupus-like autoimmune disease. *J Immunol*. 2002;168(1):9-12.
25. Hu N, Qiu X, Luo Y, et al. Abnormal histone modification patterns in lupus CD4+ T cells. *J Rheumatol*. 2008;35(5):804-810.
26. Hu YL, Metz DP, Chung J, Siu G, Zhang M. B7RP-1 blockade ameliorates autoimmunity through regulation of follicular helper T cells. *J Immunol*. 2009;182(3):1421-1428.
27. Huehn J, Polansky JK, Hamann A. Epigenetic control of FOXP3 expression: the key to a stable regulatory T-cell lineage? *Nat Rev Immunol*. 2009;9(2):83-89.
28. Jakymiw A, Ikeda K, Fritzler MJ, Reeves WH, Satoh M, Chan EK. Autoimmune targeting of key components of RNA interference. *Arthritis Res Ther*. 2006;8(4):R87.
29. Janson PC, Winerdal ME, Marits P, Thorn M, Ohlsson R, Winqvist O. FOXP3 promoter demethylation reveals the committed Treg population in humans. *PLoS One*. 2008;3(2):e1612.
30. Johnston RJ, Poholek AC, DiToro D, et al. Bcl6 and Blimp-1 are reciprocal and antagonistic regulators of T follicular helper cell differentiation. *Science*. 2009;325(5943):1006-1010.
31. Kaplan MJ, Lu Q, Wu A, Attwood J, Richardson B. Demethylation of promoter regulatory elements contributes to perforin overexpression in CD4+ lupus T cells. *J Immunol*. 2004;172(6):3652-3661.

32. Kawamoto M, Harigai M, Hara M, et al. Expression and function of inducible co-stimulator in patients with systemic lupus erythematosus: possible involvement in excessive interferon-gamma and anti-double-stranded DNA antibody production. *Arthritis Res Ther*. 2006;8(3):R62.
33. Kim ST, Fields PE, Flavell RA. Demethylation of a specific hypersensitive site in the Th2 locus control region. *Proc Natl Acad Sci USA*. 2007;104(43):17052-17057.
34. Kobata T, Jacquot S, Kozlowski S, Agematsu K, Schlossman SF, Morimoto C. CD27–CD70 interactions regulate B-cell activation by T cells. *Proc Natl Acad Sci USA*. 1995;92(24): 11249-11253.
35. Koenen HJ, Smeets RL, Vink PM, van Rijssen E, Boots AM, Joosten I. Human CD25highFoxp3pos regulatory T cells differentiate into IL-17-producing cells. *Blood*. 2008;112(6): 2340-2352.
36. Korn T, Mitsdoerffer M, Croxford AL, et al. IL-6 controls Th17 immunity in vivo by inhibiting the conversion of conventional T cells into Foxp3+ regulatory T cells. *Proc Natl Acad Sci USA*. 2008;105(47):18460-18465.
37. Kusam S, Toney LM, Sato H, Dent AL. Inhibition of Th2 differentiation and GATA-3 expression by BCL-6. *J Immunol*. 2003;170(5):2435-2441.
38. Lal G, Zhang N, van der Touw W, et al. Epigenetic regulation of Foxp3 expression in regulatory T cells by DNA methylation. *J Immunol*. 2009;182(1):259-273.
39. Linterman MA, Rigby RJ, Wong RK, et al. Follicular helper T cells are required for systemic autoimmunity. *J Exp Med*. 2009;206(3):561-576.
40. Liston A, Lu LF, O'Carroll D, Tarakhovsky A, Rudensky AY. Dicer-dependent microRNA pathway safeguards regulatory T cell function. *J Exp Med*. 2008;205(9):1993-2004.
41. Liu Y, Zhang P, Li J, Kulkarni AB, Perruche S, Chen W. A critical function for TGF-beta signaling in the development of natural CD4+CD25+Foxp3+ regulatory T cells. *Nat Immunol*. 2008;9(6):632-640.
42. Lu Q, Kaplan M, Ray D, Zacharek S, Gutsch D, Richardson B. Demethylation of ITGAL (CD11a) regulatory sequences in systemic lupus erythematosus. *Arthritis Rheum*. 2002;46(5): 1282-1291.
43. Lu Q, Ray D, Gutsch D, Richardson B. Effect of DNA methylation and chromatin structure on ITGAL expression. *Blood*. 2002;99(12):4503-4508.
44. Lu Q, Wu A, Ray D, et al. DNA methylation and chromatin structure regulate T cell perforin gene expression. *J Immunol*. 2003;170(10):5124-5132.
45. Lu Q, Wu A, Richardson BC. Demethylation of the same promoter sequence increases CD70 expression in lupus T cells and T cells treated with lupus-inducing drugs. *J Immunol*. 2005;174(10):6212-6219.
46. Lu Q, Wu A, Tesmer L, Ray D, Yousif N, Richardson B. Demethylation of CD40LG on the inactive X in T cells from women with lupus. *J Immunol*. 2007;179(9):6352-6358.
47. Lu X, Kovalev GI, Chang H, et al. Inactivation of NuRD component Mta2 causes abnormal T cell activation and lupus-like autoimmune disease in mice. *J Biol Chem*. 2008;283(20):13825-13833.
48. Lucas JL, Mirshahpanah P, Haas-Stapleton E, Asadullah K, Zollner TM, Numerof RP. Induction of Foxp3+ regulatory T cells with histone deacetylase inhibitors. *Cell Immunol*. 2009;257(1–2):97-104.
49. Luo Y, Zhao M, Lu Q. Demethylation of promoter regulatory elements contributes to CD70 overexpression in CD4+ T cells from patients with subacute cutaneous lupus erythematosus. *Clin Exp Dermatol*. 2010;35(4):425-430.
50. Ma Z, Chen F, Madaio MP, Cohen PL, Eisenberg RA. Modulation of autoimmunity by TLR9 in the chronic graft-vs-host model of systemic lupus erythematosus. *J Immunol*. 2006;177(10):7444-7450.
51. Margueron R, Justin N, Ohno K, et al. Role of the polycomb protein EED in the propagation of repressive histone marks. *Nature*. 2009;461(7265):762-767.
52. Miyamoto A, Nakayama K, Imaki H, et al. Increased proliferation of B cells and auto-immunity in mice lacking protein kinase Cdelta. *Nature*. 2002;416(6883):865-869.
53. Mohandas T, Sparkes RS, Shapiro LJ. Reactivation of an inactive human X chromosome: evidence for X inactivation by DNA methylation. *Science*. 1981;211(4480): 393-396.
54. Molad Y, Amit-Vasina M, Bloch O, Yona E, Rapoport MJ. Increased ERK and JNK activities correlate with disease activity in patients with systemic lupus erythematosus. *Ann Rheum Dis*. 2010;69(1):175-180.
55. Muljo SA, Ansel KM, Kanellopoulou C, Livingston DM, Rao A, Rajewsky K. Aberrant T cell differentiation in the absence of Dicer. *J Exp Med*. 2005;202(2):261-269.
56. Nagumo H, Agematsu K, Shinozaki K, et al. CD27/CD70 interaction augments IgE secretion by promoting the differentiation of memory B cells into plasma cells. *J Immunol*. 1998;161(12):6496-6502.
57. Nan X, Ng HH, Johnson CA, et al. Transcriptional repression by the methyl-CpG-binding protein MECP2 involves a histone deacetylase complex. *Nature*. 1998;393(6683): 386-389.
58. Nile CJ, Read RC, Akil M, Duff GW, Wilson AG. Methylation status of a single CpG site in the IL6 promoter is related to IL6 messenger RNA levels and rheumatoid arthritis. *Arthritis Rheum*. 2008;58(9):2686-2693.
59. Nurieva RI, Chung Y, Martinez GJ, et al. Bcl6 mediates the development of T follicular helper cells. *Science*. 2009;325(5943):1001-1005.
60. O'Connell RM, Taganov KD, Boldin MP, Cheng G, Baltimore D. MicroRNA-155 is induced during the macrophage inflammatory response. *Proc Natl Acad Sci USA*. 2007;104(5):1604-1609.
61. Oelke K, Lu Q, Richardson D, et al. Overexpression of CD70 and overstimulation of IgG synthesis by lupus T cells and T cells treated with DNA methylation inhibitors. *Arthritis Rheum*. 2004;50(6):1850-1860.
62. Oukka M. Th17 cells in immunity and autoimmunity. *Ann Rheum Dis*. 2008;67(Suppl 3):iii26-iii29.
63. Paschos K, Smith P, Anderton E, Middeldorp JM, White RE, Allday MJ. Epstein-barr virus latency in B cells leads to epigenetic repression and CpG methylation of the tumour suppressor gene Bim. *PLoS Pathog*. 2009;5(6): e1000492.
64. Polansky JK, Kretschmer K, Freyer J, et al. DNA methylation controls Foxp3 gene expression. *Eur J Immunol*. 2008;38(6):1654-1663.
65. Reinhardt RL, Liang HE, Locksley RM. Cytokine-secreting follicular T cells shape the antibody repertoire. *Nat Immunol*. 2009;10(4):385-393.

66. Richardson B. Effect of an inhibitor of DNA methylation on T cells. II. 5-Azacytidine induces self-reactivity in antigen-specific T4+ cells. *Hum Immunol.* 1986;17(4):456-470.
67. Richardson B, Powers D, Hooper F, Yung RL, O'Rourke K. Lymphocyte function-associated antigen 1 overexpression and T cell autoreactivity. *Arthritis Rheum.* 1994;37(9): 1363-1372.
68. Richardson BC, Strahler JR, Pivirotto TS, et al. Phenotypic and functional similarities between 5-azacytidine-treated T cells and a T cell subset in patients with active systemic lupus erythematosus. *Arthritis Rheum.* 1992;35(6):647-662.
69. Rigby RJ, Vinuesa CG. SiLEncing SLE: the power and promise of small noncoding RNAs. *Curr Opin Rheumatol.* 2008;20(5):526-531.
70. Rodriguez A, Vigorito E, Clare S, et al. Requirement of bic/microRNA-155 for normal immune function. *Science.* 2007;316(5824):608-611.
71. Rouas R, Fayyad-Kazan H, El Zein N, et al. Human natural Treg microRNA signature: role of microRNA-31 and microRNA-21 in FOXP3 expression. *Eur J Immunol.* 2009; 39(6):1608-1618.
72. Rui L, Vinuesa CG, Blasioli J, Goodnow CC. Resistance to CpG DNA-induced autoimmunity through tolerogenic B cell antigen receptor ERK signaling. *Nat Immunol.* 2003;4(6): 594-600.
73. Samanta A, Li B, Song X, et al. TGF-beta and IL-6 signals modulate chromatin binding and promoter occupancy by acetylated FOXP3. *Proc Natl Acad Sci USA.* 2008;105(37): 14023-14027.
74. Sawalha AH. Epigenetics and T-cell immunity. *Autoimmunity.* 2008;41(4):245-252.
75. Sawalha AH, Jeffries M. Defective DNA methylation and CD70 overexpression in CD4+ T cells in MRL/lpr lupus-prone mice. *Eur J Immunol.* 2007;37(5):1407-1413.
76. Sawalha AH, Jeffries M, Webb R, et al. Defective T-cell ERK signaling induces interferon-regulated gene expression and overexpression of methylation-sensitive genes similar to lupus patients. *Genes Immun.* 2008;9(4):368-378.
77. Sawalha AH, Webb R, Han S, et al. Common variants within MECP2 confer risk of systemic lupus erythematosus. *PLoS One.* 2008;3(3):e1727.
78. Scheinbart LS, Johnson MA, Gross LA, Edelstein SR, Richardson BC. Procainamide inhibits DNA methyltransferase in a human T cell line. *J Rheumatol.* 1991;18(4): 530-534.
79. Scofield RH, Bruner GR, Namjou B, et al. Klinefelter's syndrome (47, XXY) in male systemic lupus erythematosus patients: support for the notion of a gene-dose effect from the X chromosome. *Arthritis Rheum.* 2008;58(8):2511-2517.
80. Smith-Bouvier DL, Divekar AA, Sasidhar M, et al. A role for sex chromosome complement in the female bias in autoimmune disease. *J Exp Med.* 2008;205(5):1099-1108.
81. Stahl HF, Fauti T, Ullrich N, et al. miR-155 inhibition sensitizes CD4+ Th cells for TREG mediated suppression. *PLoS One.* 2009;4(9):e7158.
82. Strickland FM, Richardson BC. Epigenetics in human autoimmunity. Epigenetics in autoimmunity – DNA methylation in systemic lupus erythematosus and beyond. *Autoimmunity.* 2008;41(4):278-286.
83. Taganov KD, Boldin MP, Chang KJ, Baltimore D. NF-kappaB-dependent induction of microRNA miR-146, an inhibitor targeted to signaling proteins of innate immune responses. *Proc Natl Acad Sci USA.* 2006;103(33):12481-12486.
84. Tang Y, Luo X, Cui H, et al. MicroRNA-146A contributes to abnormal activation of the type I interferon pathway in human lupus by targeting the key signaling proteins. *Arthritis Rheum.* 2009;60(4):1065-1075.
85. Thai TH, Calado DP, Casola S, et al. Regulation of the germinal center response by microRNA-155. *Science.* 2007;316(5824):604-608.
86. Thomas LR, Miyashita H, Cobb RM, et al. Functional analysis of histone methyltransferase g9a in B and T lymphocytes. *J Immunol.* 2008;181(1):485-493.
87. Tiwari VK, McGarvey KM, Licchesi JD, et al. PcG proteins, DNA methylation, and gene repression by chromatin looping. *PLoS Biol.* 2008;6(12):2911-2927.
88. Uccellini MB, Busconi L, Green NM, et al. Autoreactive B cells discriminate CpG-rich and CpG-poor DNA and this response is modulated by IFN-alpha. *J Immunol.* 2008; 181(9):5875-5884.
89. Veldhoen M, Hirota K, Westendorf AM, et al. The aryl hydrocarbon receptor links TH17-cell-mediated autoimmunity to environmental toxins. *Nature.* 2008;453(7191):106-109.
90. Ventura A, Young AG, Winslow MM, et al. Targeted deletion reveals essential and overlapping functions of the miR-17 through 92 family of miRNA clusters. *Cell.* 2008;132(5):875-886.
91. Vigorito E, Perks KL, Abreu-Goodger C, et al. microRNA-155 regulates the generation of immunoglobulin class-switched plasma cells. *Immunity.* 2007;27(6):847-859.
92. Vire E, Brenner C, Deplus R, et al. The Polycomb group protein EZH2 directly controls DNA methylation. *Nature.* 2006;439(7078):871-874.
93. Wang L, Tao R, Hancock WW. Using histone deacetylase inhibitors to enhance Foxp3(+) regulatory T-cell function and induce allograft tolerance. *Immunol Cell Biol.* 2009;87(3):195-202.
94. Webb R, Wren JD, Jeffries M, et al. Variants within MECP2, a key transcription regulator, are associated with increased susceptibility to lupus and differential gene expression in patients with systemic lupus erythematosus. *Arthritis Rheum.* 2009;60(4):1076-1084.
95. Wei G, Wei L, Zhu J, et al. Global mapping of H3K4me3 and H3K27me3 reveals specificity and plasticity in lineage fate determination of differentiating CD4+ T cells. *Immunity.* 2009;30(1):155-167.
96. Wurster AL, Pazin MJ. BRG1-mediated chromatin remodeling regulates differentiation and gene expression of T helper cells. *Mol Cell Biol.* 2008;28(24):7274-7285.
97. Xiao C, Srinivasan L, Calado DP, et al. Lymphoproliferative disease and autoimmunity in mice with increased miR-17-92 expression in lymphocytes. *Nat Immunol.* 2008;9(4): 405-414.
98. Yang FC, Agematsu K, Nakazawa T, et al. CD27/CD70 interaction directly induces natural killer cell killing activity. *Immunology.* 1996;88(2):289-293.
99. Yang J, Chu Y, Yang X, et al. Th17 and natural Treg cell population dynamics in systemic lupus erythematosus. *Arthritis Rheum.* 2009;60(5):1472-1483.

100. Yang XO, Nurieva R, Martinez GJ, et al. Molecular antagonism and plasticity of regulatory and inflammatory T cell programs. *Immunity*. 2008;29(1):44-56.
101. Yu D, Rao S, Tsai LM, et al. The transcriptional repressor Bcl-6 directs T follicular helper cell lineage commitment. *Immunity*. 2009;31(3):457-468.
102. Yu D, Tan AH, Hu X, et al. Roquin represses autoimmunity by limiting inducible T-cell co-stimulator messenger RNA. *Nature*. 2007;450(7167):299-303.
103. Yung R, Powers D, Johnson K, et al. Mechanisms of drug-induced lupus. II. T cells overexpressing lymphocyte function-associated antigen 1 become autoreactive and cause a lupuslike disease in syngeneic mice. *J Clin Invest*. 1996;97(12):2866-2871.
104. Yung RL, Quddus J, Chrisp CE, Johnson KJ, Richardson BC. Mechanism of drug-induced lupus. I. Cloned Th2 cells modified with DNA methylation inhibitors in vitro cause autoimmunity in vivo. *J Immunol*. 1995;154(6):3025-3035.
105. Zaretsky AG, Taylor JJ, King IL, Marshall FA, Mohrs M, Pearce EJ. T follicular helper cells differentiate from Th2 cells in response to helminth antigens. *J Exp Med*. 2009;206(5):991-999.
106. Zhang F, Boothby M. T helper type 1-specific Brg1 recruitment and remodeling of nucleosomes positioned at the IFN-gamma promoter are Stat4 dependent. *J Exp Med*. 2006;203(6):1493-1505.
107. Zhou X, Jeker LT, Fife BT, et al. Selective miRNA disruption in T reg cells leads to uncontrolled autoimmunity. *J Exp Med*. 2008;205(9):1983-1991.
108. Zhou Y, Yuan J, Pan Y, et al. T cell CD40LG gene expression and the production of IgG by autologous B cells in systemic lupus erythematosus. *Clin Immunol*. 2009;132(3):362-370.

Epigenetics of Rheumatoid Arthritis 7

Aleksander M. Grabiec, Paul P. Tak, and Kris A. Reedquist

7.1 Pathobiology of Rheumatoid Arthritis

Rheumatoid arthritis (RA) is an autoimmune disorder characterized by chronic inflammation of the joints that leads to progressive and irreversible destruction of bone and cartilage, with joint pain, stiffness, and swelling as predominant symptoms. It is one of the most common autoimmune diseases affecting approximately 1% of the world's population. Although the exact mechanisms that lead to the pathological changes in RA patients are still poorly understood, there is general consensus that RA is the outcome of a combination of genetic susceptibility factors, of aberrant regulation of the immune system leading to autoantibody production, and of environmental factors, such as smoking or inappropriate nutrition.[11,19]

The principal site of pathological changes characteristic of RA is the synovium.[50] The disease involves the recruitment and accumulation of activated immune cells in the synovial sublining, along with hyperplastic growth of the intimal lining layer that harbors macrophages and fibroblast-like synoviocytes (FLS). The inflammatory infiltrate in the synovial sublining consists of macrophages, T cells, and plasma cells, with a few other cell types, such as B cells, dendritic cells, mast cells, natural killer cells, and neutrophils. Excessive infiltration of activated immune cells is accompanied by deregulated expression of multiple adhesion molecules, proinflammatory cytokines, and chemokines, such as tumor necrosis factor-alpha (TNF-α), interleukin-1beta (IL-1β), IL-6, IL-8, and receptor activator of nuclear factor-kappaB (NF-κB) ligand (RANKL).[6] The inflammatory cells that accumulate in the joint are characterized by enhanced retention and impaired apoptosis.[20,40] Pathological processes in the rheumatoid joint are also associated with enhanced angiogenesis and the transformation of blood vessels into high endothelial venules that facilitate inflammatory cell infiltration of the synovial sublining. The most typical histological symptom of RA is the formation of pannus – an invasive region of the synovial lining at the junction between bone and synovium that contains differentiated macrophages, osteoclasts, and FLS, all of which can readily invade surrounding tissues and secrete high quantities of matrix-degrading enzymes, such as matrix metalloproteinases (MMPs), thereby further contributing to joint destruction (see Fig. 7.1).[22,46]

As our understanding of the specific role of the components of this complex inflammatory network has improved, it has become possible to develop therapeutic strategies to target cytokines (infliximab, adalimumab, etanercept), co-stimulatory molecules (abatacept), and B cells (rituximab).[27,60] Moreover, a growing body of experimental evidence has demonstrated that activation of conserved intracellular signaling pathways, including NF-κB, the mitogen-activated protein kinases (MAPKs), phosphatidylinositol 3-kinases (PI3-K), the Janus tyrosine kinase (JAK)/signal transducers, and activators of transcription (STAT) pathway, is altered in RA. Improvements in our understanding of contributions of these pathways to inflammation and joint erosion in RA have initiated intensive research directed toward identification of potential new

A.M. Grabiec (✉)
Division of Clinical Immunology and Rheumatology, Academic Medical Center, University of Amsterdam, Amsterdam, The Netherlands
e-mail: a.m.grabiec@amc.uva.nl

H.I. Roach et al. (eds.), *Epigenetic Aspects of Chronic Diseases*,
DOI: 10.1007/978-1-84882-644-1_7,

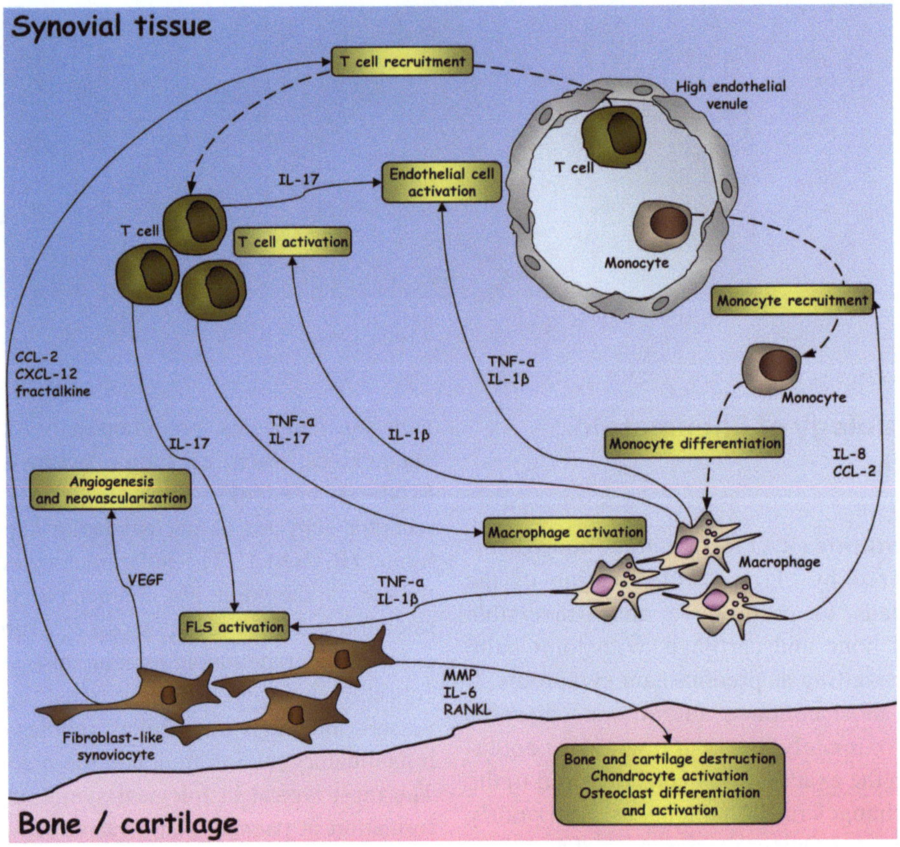

Fig. 7.1 Overview of the pathogenesis of RA. Monocytes and T cells are attracted to the rheumatoid joint and migrate via endothelial venules to the synovial tissue (*blue area*). Monocytes differentiate into macrophages, become activated, and secrete chemokines that attract more inflammatory cells to the joint. Cytokines in turn activate T cells, FLS, and stimulate expression of adhesion molecules in endothelial cells. Activated FLS secrete chemokines, angiogenic factors, cytokines, and matrix-degrading enzymes, all of which act to perpetuate inflammation, cause cartilage degradation and bone erosion (*pink area*). Activated T cells further contribute to inflammatory stimulation of macrophages, FLS and the endothelium, particularly through secretion of IL-17, *IL-1β* interleukin-1beta, *CCL-2* CC chemokine ligand-2, *CXCL-12* CXC chemokine ligand-12, *MMP* matrix metalloproteinase, *RANKL* receptor activator of nuclear factor-κB ligand, *TNF-α* tumor necrosis factor-alpha, *VEGF* vascular endothelial growth factor

targets for therapeutic intervention.[41,51,55] However, as a substantial fraction of RA patients remains non-responsive to currently available therapeutic strategies, and given that genetic predisposition contributes relatively little to the development of RA, there is growing interest in alternative mechanisms of gene regulation that may influence susceptibility to RA, disease severity, and response to treatment. Recognition of aberrant epigenetic modifications, defined as heritable changes that modulate gene expression without altering the DNA sequence, has led to studies that have identified altered epigenetic mechanisms involved in the pathobiology of RA.[49]

7.2 DNA Methylation: Global Versus Promoter-Specific Modifications

Methylation of CpG dinucleotide clusters within the 5′ regulatory regions of genes, catalyzed by DNA methyltransferases (Dnmts), is the best-characterized mechanism of epigenetic gene silencing. DNA methylation influences gene transcription either by direct disruption of interactions between transcription factors and DNA or by recruitment of transcriptional co-repressors. Alterations in DNA methylation patterns, i.e., hypomethylation leading to chromosomal instability and the expression of proto-oncogenes, or hypermethylation-mediated silencing of tumor suppressor genes, have been firmly associated with cancer. Moreover, chronic inflammatory disorders, including RA, are now known to involve similar changes in the methylation status of promoter regions of genes that regulate activation and survival of cells that contribute to the pathogenesis of these diseases.[4]

Several studies have shown that DNA methylation level is decreased in RA cells. First, the direct analysis of global DNA methylation status has revealed that the number of methylated CpG regions in RA T cells is lower than in cells from controls.[42] Second, the expression of the endogenous retrotransposable element long interspersed nuclear element-1 (LINE-1) reported in RA synovial fluid[35] and in RA synovial tissue at the site of inflammation[2] provides an indication of decreased methylation level in RA. Retrotransposons are genetic elements of retroviral origin that integrate into the host genome and alter its gene expression. In normal cells, retrotransposon function is silenced by DNA methylation, but in pathological conditions where DNA methylation is deregulated, such as cancer, retrotransposons can be reactivated and cause insertional mutagenesis. Even though the exact LINE-1 promoter methylation status in RA synovial tissue has not yet been determined, the expression of LINE-1 in stromal cells derived from RA synovial tissue constitutes clear evidence that DNA hypomethylation regulates gene expression in RA. Moreover, the observation that enforced expression of LINE-1 sequences in primarily LINE-1-negative FLS results in elevated expression of the p38δ MAP kinase may provide a link between LINE-1 expression and the highly activated phenotype of RA FLS.[21]

Hypomethylation-mediated aberrant gene induction also characterizes "senescent" T cells in RA. High numbers of proinflammatory CD28-negative "senescent" T cells, expressing enhanced levels of cytotoxic perforin, the CD70 costimulatory protein, and inflammatory interferon-γ, which display high cytolytic activity against endothelium, are readily detected in chronic inflammatory diseases. Dnmt1 and Dnmt3a expressions are decreased in this T cell subset of RA patients.[24] An important contribution of this epigenetic phenomenon to the pathologic function of "senescent" T cells has been confirmed in vitro: Dnmt1 and Dnmt3a gene silencing in T cells from healthy individuals leads to hypomethylation of the regulatory regions and, consequently, elevated expression of genes responsible for inflammatory and cytolytic activities of these cells.[24]

Decreased methylation status has also been observed in the promoter region of the *IL-6* gene in peripheral blood mononuclear cells (PBMCs) from RA patients.[36] Given the strong inverse correlation between *IL-6* promoter methylation and *IL-6* mRNA expression in LPS-stimulated macrophages, hypomethylation of the *IL-6* regulatory region in RA patients partly explains excessive production of this cytokine in the inflamed joint. A similar epigenetic phenomenon is responsible for elevated expression of ephrinB1 in both synovial and peripheral blood-derived lymphocytes in RA patients, compared to its expression in patients with osteoarthritis (OA) or in healthy controls. Ephrins constitute a family of membrane-bound ligands of the erythropoietin-producing human hepatocellular carcinoma (Eph) receptors that trigger signaling pathways essential for T cell–T cell interactions and T cell migration, and excessive Eph receptor signaling contributes to inflammatory processes.[16] Analysis of the methylation status of the *ephrinB1* promoter in T cells has revealed significant differences between RA patients and healthy individuals: cells isolated from RA patients contain a lower percentage of methylated CpG islands within the *ephrinB1* regulatory region.[18] These methylation events have a functional role in the modulation of *ephrinB1* transcription and increased ephrinB1 expression plays a role in RA pathogenesis, at least in the murine collagen antibody-induced arthritis model.[18] Therefore, deregulation of the ephrin/Eph receptor system via hypomethylation of the *ephrinB1* regulatory region represents yet another example of how aberrant DNA methylation may contribute significantly to the persistence of inflammation in RA.

The findings described above suggest a general proinflammatory role for DNA hypomethylation in RA, but counter-examples are also emerging. The hypermethylation of specific gene promoter regions, leading to repressed transcription of the target gene, also plays a role in the pathogenesis of numerous diseases. Gene silencing via promoter-specific hypermethylation has been characterized in great detail for several tumors, with effects on proliferation, cell cycle, DNA repair, and drug resistance. A similar mechanism may enhance relative resistance to apoptosis of cells present at the site of inflammation in RA. The expression of death receptor 3 (DR3), a member of the apoptosis-inducing TNF receptor superfamily, is reduced in synovial cells obtained from RA patients, compared to normal controls or patients with non-inflammatory OA.[52] Detailed analyses of the *DR3* promoter region revealed specific hypermethylation of a single CpG island in RA synovial cells and demonstrated that this particular CpG region is essential for transactivation and eventual expression of the *DR3* gene.[52] These findings provide not only clear evidence that DR3 expression is suppressed in RA via a specific epigenetic event, but also identify a molecular mechanism of how apoptosis is impaired in the RA synovium.

Thus, there is evidence that DNA methylation status is disturbed in RA patients, at the site of inflammation and in the periphery. Evidence is accumulating to suggest that increases or decreases in the methylation status of specific gene promoters can contribute to pathogenic gene expression (see Fig. 7.2). However, it is still unclear which changes in DNA methylation patterns represent transient modifications, made in response to inflammatory stimuli or repair processes, and which represent persistent, stable inheritable gene promoter modifications. Both transient and stable changes affecting important methylation patterns have been observed in kinetic studies of the differentiation of Th1 and Th2 T lymphocytes,[10] and may also occur in other cells that infiltrate the inflamed RA synovium.

7.3 Aberrant MicroRNA Expression in RA

The discovery of microRNAs (miRNAs), a new class of small, endogenous, single-strand non-coding RNA molecules, has improved our understanding of the complex post-transcriptional control of gene expression. Currently, more than 300 miRNAs have been

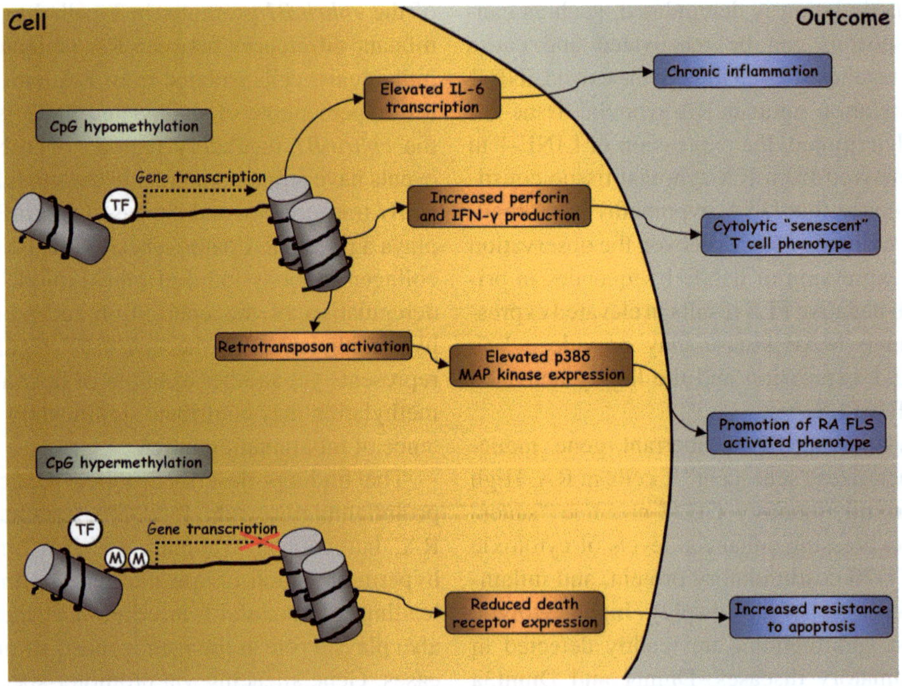

Fig. 7.2 Potential roles of altered DNA methylation in the pathogenesis of RA. Both aberrant hypomethylation of gene promoter regions, leading to elevated gene transcription, and promoter hypermethylation, causing transcriptional gene suppression, occur in RA patients. Reduced methylation of regulatory gene regions results in increased production of cytokines which contribute to chronic activation of immune cells. Hypomethylation-induced expression of signaling proteins causes pathological activation of RA stromal cells. At the same time, hypermethylation-mediated suppression of genes that regulate apoptosis can lead to increased relative resistance of RA FLS to apoptosis. *IL-6* interleukin-6, *M* methylation, *MAP kinase* mitogen-activated protein kinase, *IFN-γ* interferon-gamma, *TF* transcription factor

identified in humans. While deregulated expression of several miRNAs in cancer has demonstrated important roles for these molecules in cell differentiation, proliferation, and survival, a growing body of experimental evidence also indicates involvement of specific miRNAs in the regulation of immune responses and in the pathogenesis of chronic inflammatory diseases.[38,44]

Two miRNAs, miRNA-146, and miRNA-155, previously described as induced by inflammatory mediators and regulated by the NF-κB pathway,[47] have been associated with the pathobiology of RA. Expression of both miRNA-146 and miRNA-155 is elevated in RA synovial tissue compared to tissue obtained from OA patients, differences that are maintained in RA FLS in vitro. Expression of both miRNAs is induced by different but overlapping profiles of inflammatory stimuli, including TNF-α, IL-1β, and lipopolysaccharide (LPS).[48] Moreover, miRNA-146 expression strongly correlates with TNF-α mRNA levels in the synovial tissue of RA patients.[33] In addition to miRNA-146 and miRNA-155, other miRNAs are also increased in PBMCs from RA patients as compared to controls,[39] implying that these changes are systemic and not restricted to the site of inflammation. Even though expression of certain miRNAs may represent an interesting marker of RA disease, the pathogenic roles of miRNA-146 and miRNA-155 expression in RA have been challenged by two recent findings. First, transfection of RA FLS with miRNA-155 unexpectedly repressed both the basal and the inducible production of MMP-1 and MMP-3. This suggests that enhanced expression of miRNA-155 observed in RA serves to dampen destructive cellular responses, rather than to promote tissue destruction.[48] Second, analysis of TNF receptor–associated factor-6 (TRAF-6) and IL-1 receptor-associated kinase-1 (IRAK-1) expression, two known targets of miRNA-146, revealed no differences in mRNA and protein levels between PBMCs from RA patients and controls.[39] This is surprising, given that both TRAF-6 and IRAK-1 are needed by human monocytes to produce TNF-α. Therefore, the enhanced TNF-α production observed in RA patients may not be due to miRNA-146 upregulation, but rather may indicate instead an impaired ability of miRNA-146 to suppress expression of a subset of gene targets. Evidence consistent with this latter proposal has emerged from micro-array studies of gene expression in RA and in systemic lupus erythematosus (SLE), another chronic inflammatory disease. In SLE, where miRNA-146 is downregulated,[53] the potential contribution of decreased miR-146 expression to disease development might be relatively straightforward. miRNA-146 also negatively regulates type I interferon pathways, and an increase in type I interferon is a major contributor to the inflammatory processes characteristic for SLE, correlating with disease activity.[8] Curiously, gene array analyses of PBMCs from RA patients have revealed a subset of patients who also display a type I interferon profile.[57] It will be of interest to determine whether defective expression or function of miRNA-146 segregates in this group of patients.

Based on their anti-inflammatory properties, other miRNAs may prove therapeutically useful in RA. miRNA-346 is expressed in RA FLS following LPS stimulation, and decreases IL-18 mRNA stability by suppressing expression of Bruton's tyrosine kinase.[3] Local introduction of this miRNA at sites of inflammation in RA may have therapeutic value. Another miRNA that may be important in RA is miRNA-124a, identified during a screening of miRNAs that are expressed in FLS of RA and OA patients. Whereas previous studies dealt with the enhanced expression of miRNAs, Nakamachi et al.[31] have reported a decrease in miRNA-124a expression in RA. Transfection with miRNA-124a significantly suppresses cell proliferation and induces cell cycle arrest in RA FLS, but not in OA FLS. This suggests that depressed expression of miRNA124a might promote RA FLS proliferation during invasive pannus formation. This study also revealed cyclin-dependent kinase 2 (CDK-2), one of the key G1/S phase transition regulators, as a direct target for transcriptional silencing by miRNA-124a.[31] Overexpression of miRNA-124a also affects cellular activation status: production of CC chemokine ligand-2 (CCL-2, also known as monocyte chemoattractant protein-1), a chemokine present in high amounts in the synovial fluid of RA patients, is strongly suppressed in miRNA-124a-transfected RA FLS via direct interaction with the 3′ untranslated region of the *CCL-2* mRNA.[31] These findings indicate that miRNA-124a may not only constitute an interesting disease marker in RA, but also a potential therapeutic target. Curiously, miRNA-124a expression is silenced in some cancer cell lines via hypermethylation of the *miRNA-124a* gene.[25] If this is also true in RA, it would suggest a complex interplay between different epigenetic regulatory mechanisms.

It still remains unclear whether aberrant expression of these miRNAs in RA enhances production of inflammatory mediators or is the result of persistent exposure of cells to disease-specific inflammatory stimuli. Moreover, there is early evidence that changes in miRNA expression may represent an ineffective attempt at repairing damaged tissue (see Fig. 7.3). Initial proof of principle experiments have also shown that miRNAs do modulate chronic inflammatory disease: intra-articular administration of miRNA-15a in the murine autoantibody-induced model of arthritis has led to apoptosis of synovial cells.[30] Administration of double-stranded miRNA-15a, a potent regulator of the anti-apoptotic Bcl-2, significantly increased miRNA-15a expression in the joints of these mice, along with a reduction in Bcl-2 and increased activation of caspase-3 in the synovium.[30] The increasing clinical application of gene therapy techniques in RA implies that similar strategies for targeting miRNA may become available in the foreseeable future.[56]

7.4 Perturbation and Manipulation of Histone Acetyl Transferase and Histone Decacetylase Activities in RA

Analyses of the expression and activity of histone acetyl transferases (HATs) and histone decacetylases (HDACs) in cells and tissue from patients suffering from inflammatory lung diseases have demonstrated a shift in balance toward enzymatic activities that favor protein acetylation and contribute to disease pathology (Chap. 10). These seminal findings have sparked interest in the potential contributions of HAT and HDAC to other chronic inflammatory diseases, including RA. HAT activity does not differ in RA patients, OA patients, and healthy individuals, but HDAC activity and the HDAC/HAT activity ratio are markedly decreased in tissue obtained from RA patients, as compared to control groups.[15] Reduced HDAC activity in RA patient synovial tissue parallels a decrease in HDAC1 and

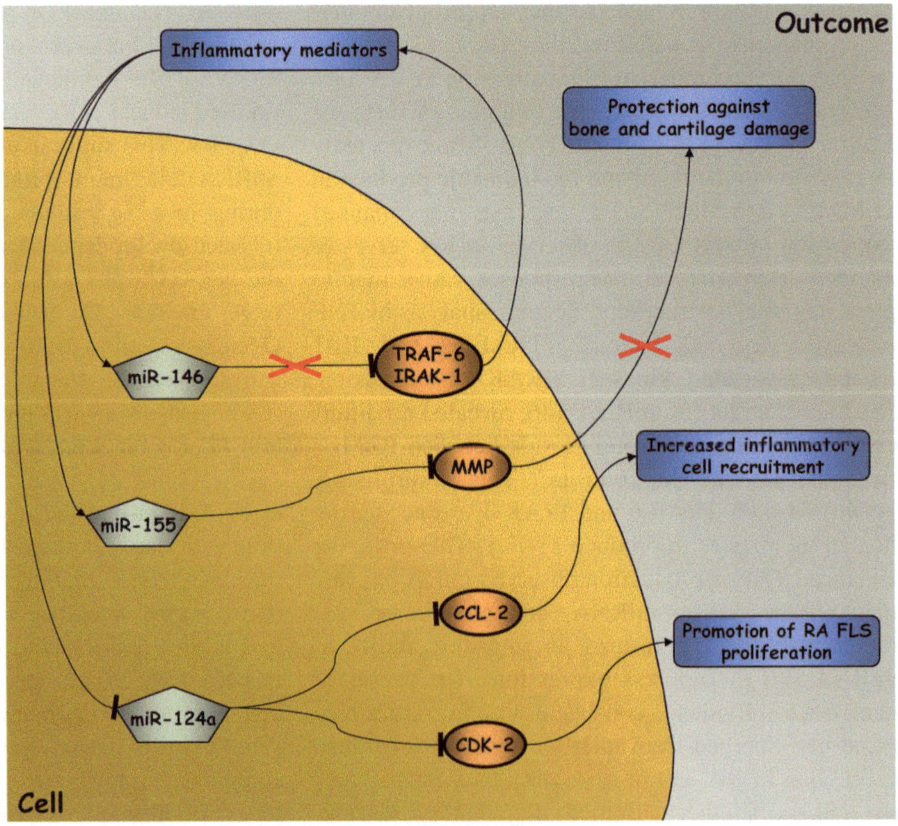

Fig. 7.3 Alterations in microRNA expression in RA synovial tissue. Elevated levels of miRNA-146 and miRNA-155 may represent an ineffective attempt to reduce inflammation and tissue damage. miRNA-146 fails to downregulate expression of proinflammatory TRAF-6 and IRAK-1, but miRNA-155-mediated MMP suppression does not protect against bone and cartilage degradation in RA patients. The suppressed expression of miRNA-124a observed in RA may contribute to both pathogenic cell proliferation and enhanced recruitment of immune cells to the rheumatoid joint. *CDK-2* cyclin-dependent kinase-2, *IRAK-1* interleukin-1 receptor–associated kinase-1, *CCL-2* CC chemokine ligand-2, *MMP* matrix metalloproteinase, *TRAF-6* TNF receptor–associated factor-6

HDAC2 protein expression in RA, which is not observed in OA. This difference was confirmed by immunohistochemical analysis, with the observed decrease most prominent in synovial macrophages.[15] Conceivably a decrease in HDAC activity in RA synovial macrophages could lead to increased histone acetylation and greater access of transcription factors to inflammatory gene promoters. This inference derives support from our study that macrophages and, to a lesser extent, FLS display high protein acetylation status, whereas synovial T cells and B cells have much lower levels of acetylated proteins. RA synovial cells, predominantly localized at the intimal lining layer, contain high levels of acetylated histone 3 and acetylated histone 4[13] (see Fig. 7.4). The finding that HDAC expression and activity are depressed in RA synovial tissue, similar to what is found in patients with COPD,[1] suggests the possibility that a localized shift in the balance of HDAC and HAT activity favors protein acetylation and may constitute a general feature of chronic inflammatory disease. However, larger independent studies are needed to confirm that synovial acetylation homeostasis is disturbed in RA and that these disturbances are related to disease severity and inflammation parameters. Interestingly, a later report[14] has shown that mRNA and protein expressions of HDAC1 are elevated in RA FLS, as compared to OA FLS, but that the levels of other HDAC isoforms in FLS are similar in the two patient groups. It is generally accepted that the in vivo semi-transformed pathogenic phenotype of FLS, and accompanying gene expression patterns found in vivo, are maintained in vitro.[29] While this observation obviously needs to be confirmed by in situ analysis of HDAC1 expression in FLS in RA synovial tissue, the contrast between the available studies stresses the need for more extensive studies as this field progresses.[14,15] It is also unclear whether altered HAT or HDAC activities are sufficient to contribute to inflammatory processes or are a consequence of the inflammatory process. In this regard, it is noteworthy that TNF-α can stimulate proteosome-dependent degradation of HDAC1.[59]

7.4.1 Effects of Histone Deacetylase Inhibitors on Cells of RA Synovium

Further verification that HDAC activity is depressed in RA synovial tissue would be important for developing

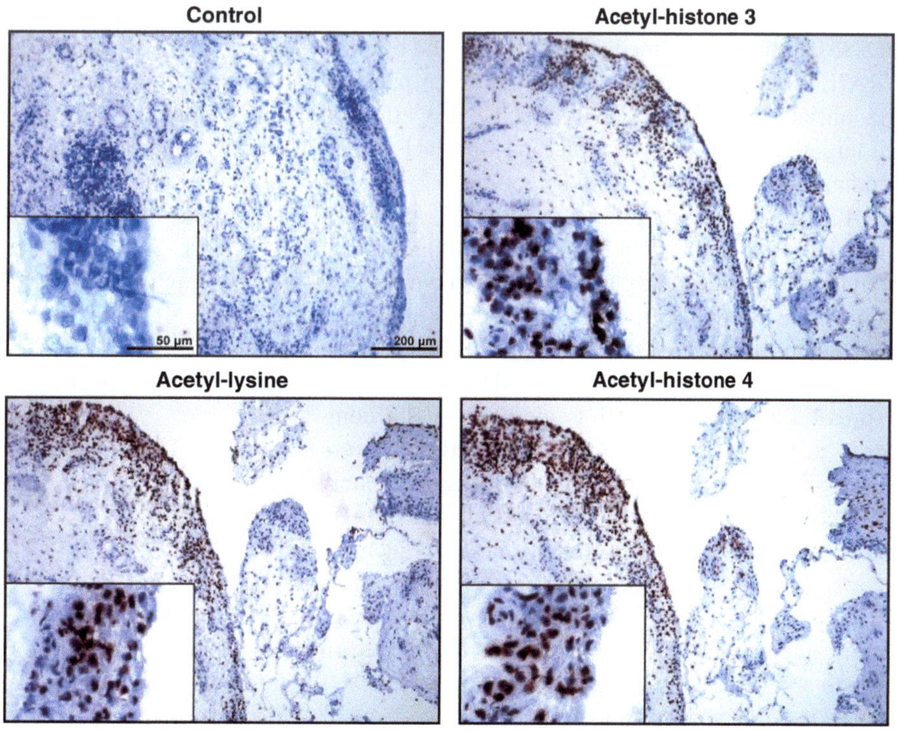

Fig. 7.4 Hyperacetylation of cellular proteins, histone 3 and histone 4 in RA synovial tissue. Immunohistochemical staining of RA synovial tissue with control rabbit antibody and antibodies against acetyl-lysine, acetyl-histone 3, and acetyl-histone 4. Stainings were developed using biotin tyramide enhancement and horseradish peroxidase, followed by amino-ethylcarbazole (*red*). They were subsequently counterstained with Mayer's hematoxylin (*blue*).[45] ×100 and ×400 (*insets*) magnifications are presented

therapeutic strategies that target protein acetylation processes. Because the decrease in HDAC activity in RA tissue might render patients refractory to anti-inflammatory effects of HDAC inhibitors (HDACis), experimental efforts should be directed toward restoring HDAC expression and the normal balance between HAT and HDAC activity in RA patient synovial tissue.[15] On the other hand, even if HDAC expression decreases, it is still possible that the remaining HDAC activity is needed for cellular processes such as activation, proliferation, or survival.[13] If that is the case, it would mean that RA synovial cells may be more sensitive than cells from clinically uninvolved tissues and depressed HDAC activity might then represent the weak point in the pathobiology of RA. As discussed below, HDACi have demonstrated potent anti-arthritic effects in animal models, and the effects of these compounds are now being studied extensively in RA synovial cells.

FK228 and trichostatin A (TSA), pan-specific HDACis, induce cell cycle arrest in RA FLS, associated with elevated expression of cell cycle inhibitors $p16^{Ink4a}$ and $p21^{Waf1/Cip1}$.[17,28,37] Incubation of RA FLS with HDACis fails to induce apoptosis, but TSA potently sensitizes FLS to apoptosis induced by TNF-related apoptosis-inducing ligand (TRAIL).[17] TSA can also sensitize FLS to Fas-mediated apoptosis, possibly via transcriptional downregulation of Flip protein.[28] In the context of the well-documented resistance of RA FLS to death receptor–mediated apoptosis and because TRAIL is elevated in the synovial fluid of RA patients, the inhibition of HDAC activity may be very effective in targeting FLS proliferation and survival at the inflammation site in RA. It is as yet unclear whether HDACis mediate their effects in FLS (or other synovial cells) via bulk action on total cellular HDAC activity, or via signaling pathways and gene transcription mediated by specific HDACs. Analysis of mRNA expression of all 11 HDAC isoforms in RA FLS cultures has revealed that HDAC1 and HDAC2 are highly expressed and that HDAC1 is elevated in RA compared to OA FLS. When HDAC1 and HDAC2 were genetically silenced, cell proliferation was moderately reduced and some apoptosis was induced.[14] If HDAC1 expression is also elevated in vivo and if this increase is associated with inflammatory gene transcription, HDAC1 could become a promising target for therapeutic intervention. However, because the degree of growth arrest observed following silencing of HDAC1 and HDAC2 is modest compared to HDACi effects,[9,17] other HDAC isoforms may play a major role in the regulation of FLS proliferation and viability. In either event, it is interesting that HDACis inhibit cellular activation at the site of inflammation, but have milder and less persistent action in non-involved tissues. The HDACi-induced cell cycle arrest is reversible in normal FLS after the drug has been removed, but persists in FLS isolated from joints of arthritic rats.[9] HDACis can affect not only FLS proliferation, but also their pro-angiogenic properties via suppression of cytokine-induced expression of hypoxia-inducible factor-1α (HIF-1α) and vascular endothelial growth factor (VEGF).[26] Together, these studies underscore the important role epigenetic mechanisms play in regulating FLS proliferation, survival, and activation. They also contradict the prediction that synovial cells may be resistant to HDACi treatment.

Data that HDACis impact on the activation and survival of other RA synovial cells is emerging. Potentially important for efforts to slow joint destruction in RA, HDACis modulate the catabolism of cartilage and collagen by chondrocytes. HDACis suppress the IL-1β- and oncostatin M–induced production of MMPs and aggrecan-degrading enzymes in human primary chondrocytes.[61] HDACis also block production of nitric oxide and prostaglandin E2 in OA chondrocytes, as well as cartilage explants that have been stimulated with inflammatory cytokines.[7] Additionally, HDACis exert anti-inflammatory effects on RA synovial macrophages. Both class I/II HDACis and, intriguingly, the class III sirtuin HDACi nicotinamide potently suppress production of IL-6 and TNF-α by RA macrophages. HDACis also block secretion of inflammatory mediators, including IL-6, chemokines IL-8, CCL-2, CCL-5 and CXC chemokine ligand-12, and pro-angiogenic VEGF by intact RA synovial biopsy explants.[12] It is thus apparent that cells involved in RA persistence and joint destruction are sensitive to the anti-inflammatory effects of HDACis (see Fig. 7.5).

7.4.2 Targeting HDAC Activity in Animal Models of Arthritis

Early studies in cancer cell lines and animal tumor models have shown that HDACis not only induce differentiation, cell cycle arrest, and/or apoptosis, but can also potently suppress production of inflammatory

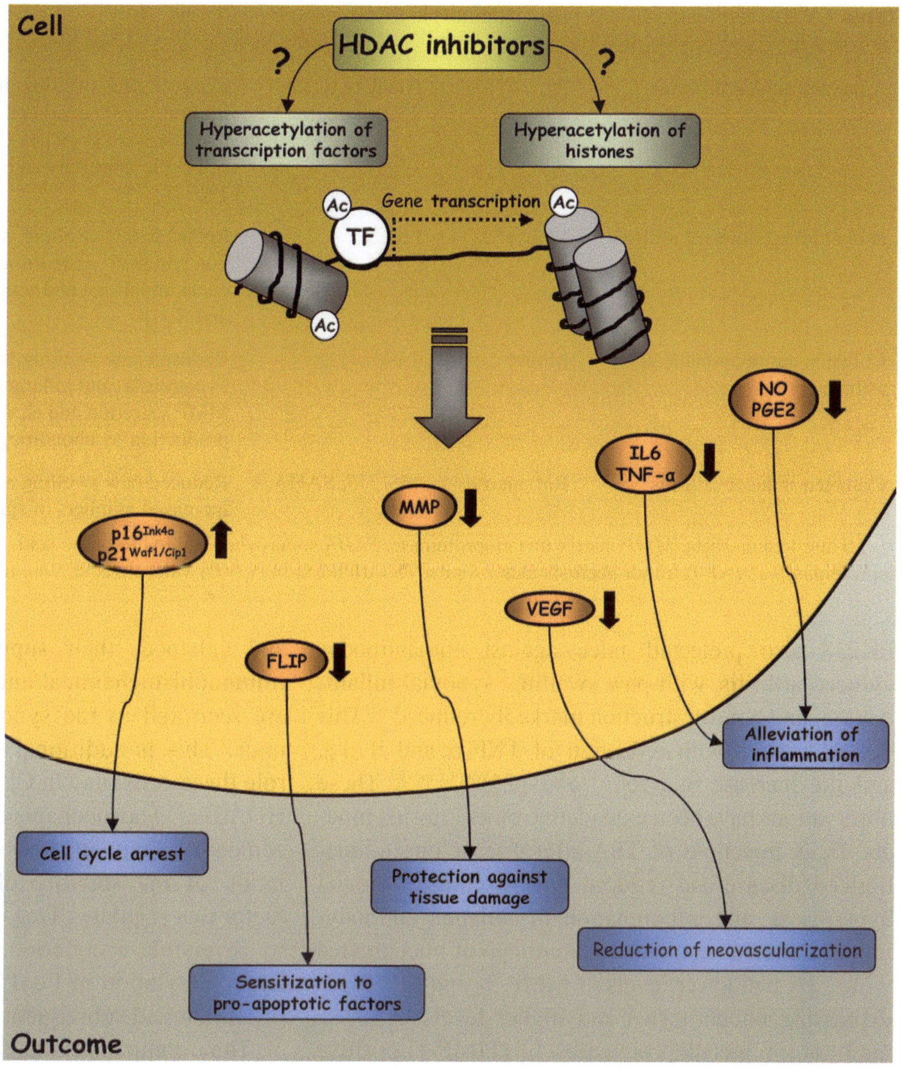

Fig. 7.5 Overview of the effects of HDACis on synovial cells of RA patients and their potential therapeutic consequences. Treatment with HDACis induces hyperacetylation (*Ac*) of histones and non-histone proteins, mainly transcription factors (*TF*). Hyperacetylation of histones generally leads to activation of gene expression, but hyperacetylation of TF can either induce or suppress target genes. In RA synovial cells, HDACis upregulate cell cycle inhibitors and downregulate anti-apoptotic proteins, matrix-degrading enzymes, and inflammatory and angiogenic mediators. *IL-6* interleukin-6, *MMP* matrix metalloproteinase, *NO* nitric oxide, *PGE2* prostaglandin E2, *TNF-α* tumor necrosis factor-alpha, *VEGF* vascular endothelial growth factor

mediators, thus providing a rationale for experiments to study the anti-inflammatory potential of these compounds in animal models of RA. In a pioneering study, Chung and co-workers[9] examined the effects of topical administration of TSA and phenylbutyrate ointments on the development of adjuvant-induced arthritis in rats. Although neither HDACi prevented disease onset following prophylactic application, paw swelling was markedly reduced. This was associated with suppression of pannus formation, subintimal inflammatory cell infiltration, synovial cell hyperplasia, and bone damage. At the molecular level, treatment with HDACis induced expression of the cell cycle inhibitors $p16^{INK4}$ and $p21^{WAF1/Cip1}$ and inhibited production of TNF-α. Induction of $p16^{INK4}$ and $p21^{WAF1/Cip1}$ expression by TSA and phenylbutyrate occurred only in the arthritic rats and not in controls.[9] The observation that HDACi treatment had no effect in the control animals is consistent with the possibility that these compounds target only inflamed cells.

HDACis have not only prophylactic, but also therapeutic potential in animal models of arthritis. This is illustrated in a study[32] that showed that intravenous administration of the depsipeptide FK228 prior to arthritis induction reduced paw swelling and minimized the reduction of animal body weight. This was not the case when FK228 was applied after disease onset. However, both prophylactic and therapeutic administration of FK228 significantly blocked disease-related bone destruction.[32] A single therapeutic dose of

Table 7.1 Effects of histone deacetylase inhibitors in animal models of arthritis

Disease model	Species	Inhibitor	Effect	Ref.
Adjuvant-induced arthritis	Rat	Phenylbutyrate, TSA, FK228	Reduced paw swelling, inflammatory cell infiltration, synovial cell hyperplasia, and bone damage associated with induction of cell cycle inhibitors and suppression of TNF-α	9, 32
Autoantibody-induced arthritis	Mouse	FK228	Reduced paw swelling, synovial inflammation and bone destruction, induction of cell cycle inhibitors, and suppression of TNF-α and IL-1β	37
Collagen antibody-induced arthritis	Mouse	TSA	Reduced paw swelling, synovial hyperplasia, and inflammation; suppressed MMP secretion and elevated TIMP-1 production by chondrocytes	34
Collagen-induced arthritis	Rat, mouse	MS-275, SAHA, VPA	Reduced paw swelling and bone erosions, increased numbers of splenic Tregs	23, 43

IL-1β interleukin-1beta, *MMP* matrix metalloproteinase, *SAHA* suberoylanilide hydroxamic acid, *TIMP-1* tissue inhibitor of metalloproteinases-1, *TNF-α* tumor necrosis factor-alpha, *TSA* trichostatin A, *VPA* valproic acid, *Treg* regulatory T cell

FK228 also protected mice against autoantibody-induced arthritis, with paw swelling, synovial inflammation, and bone destruction markedly reduced.[37] This was associated with reduction of TNF-α and IL-1β, and the increase in $p16^{INK4}$ and $p21^{WAF1/Cip1}$.[37] These observations have been extended to other arthritis models. Daily injections of TSA after disease onset, ameliorated disease and reduced paw swelling, synovial hyperplasia, and inflammation in collagen antibody-induced arthritis in mice. The cartilage of mice treated with TSA had lower levels of MMP-3- and MMP-13-expressing chondrocytes and higher levels of tissue inhibitor of metalloproteinases-1- (TIMP-1)-positive chondrocytes. This indicates that HDACis prevent cartilage destruction by modulating gene expression in chondrocytes.[34] Moreover, Lin and co-workers[23] have shown that MS-275 is therapeutically effective in collagen-induced arthritis (CIA) in rats, preventing further increase in paw swelling and reducing bone erosion in a dose-dependent manner. In the same study, MS-275 and another HDACi, suberoylanilide hydroxamic acid (SAHA), ameliorated CIA in both mice and rats in a prophylactic protocol, with MS-275 providing almost total protection against arthritis.

HDACis may also exert protective effects in animal models of arthritis by directly modulating T cell function.[54] Therapeutic administration of valproic acid (VPA), an HDACi clinically approved for cancer treatment, increases the numbers of murine splenic anti-inflammatory regulatory T cells (Tregs) in vivo and enhances their suppressive function in vitro. Immunohistochemical analysis has shown that Tregs are recruited to the synovial tissue in VPA-treated mice. This is additional evidence for the important role these cells play in CIA amelioration mediated by HDACis.[43] The mechanism of HDACi-induced enhancement of Treg suppressor phenotype involves regulation of the stability of FoxP3, the transcription factor that regulates Treg differentiation and function, by reversible acetylation. HDACi treatment induces hyperacetylation of FoxP3, preventing its poly-ubiquitination and subsequent proteosomal degradation.[58]

Thus, compounds that represent each chemical class of HDACis strongly reduce inflammation and prevent joint destruction in all currently available animal arthritis models (see Table 7.1). Since HDACis have been successfully used in animal models for other inflammatory diseases, including asthma, colitis, SLE, multiple sclerosis, and systemic endotoxic shock, it seems evident that inhibition of HDAC activity leads to suppression of inflammatory processes.[5] Given that HDACis are effective not only prophylactically, but also therapeutically, HDACis may find application in arthritis treatment.

7.5 Conclusions

Epigenetic mechanisms, including DNA methylation, histone modifications, and miRNA expression, are essential for the regulation of gene expression in health and disease as well as crucial for an effective, self-limiting immune response. Many experiments have shown that epigenetic mechanisms are associated with pathogenic alterations in gene expression that lead to chronic inflammation in RA and its persistence. However, the mechanisms involved are quite complex. Hypomethylation of gene promoter regions leads to elevated expression of genes responsible for inflammatory cell activation in RA, while hypermethylation-induced silencing of genes regulating cell survival appears to contribute to synovial cell resistance to apoptosis. Furthermore, aberrantly expressed miRNAs that are involved in inflammatory processes either do not regulate their target genes in RA patients or, when alterations in their expression are mimicked in vitro, do not lead to an increase in inflammation. Finally, even though HDAC expression is depressed in RA synovial tissue, thereby contributing perhaps to enhanced inflammatory gene expression, HDACis have a strong therapeutic effect in animal models of RA. It is therefore difficult to predict how to exploit epigenetic processes for RA treatment. As far as DNA methylation is concerned, no animal experiments have been reported to provide information on the therapeutic efficacy of compounds that modulate DNA methylation. Small pilot studies analyzing RA synovial tissue need to be followed up with large patient cohorts, with the additional aim of determining whether changes in epigenetic regulatory pathways contribute to the disease process or are secondary to inflammatory events. Studies of patients with early synovitis may provide an answer. In conclusion, studies on the role of epigenetics in RA have produced some compelling observations, but much more work is still needed to decipher the mechanisms involved and translate these findings to therapeutic strategies.

References

1. Adcock IM, Tsaprouni L, Bhavsar P, Ito K. Epigenetic regulation of airway inflammation. *Curr Opin Immunol*. 2007; 19:694-700.
2. Ali M, Veale DJ, Reece RJ, et al. Overexpression of transcripts containing LINE-1 in the synovia of patients with rheumatoid arthritis. *Ann Rheum Dis*. 2003;62:663-666.
3. Alsaleh G, Suffert G, Semaan N, et al. Bruton's tyrosine kinase is involved in miR-346-related regulation of IL-18 release by lipopolysaccharide-activated rheumatoid fibroblast-like synoviocytes. *J Immunol*. 2009;182:5088-5097.
4. Backdahl L, Bushell A, Beck S. Inflammatory signalling as mediator of epigenetic modulation in tissue-specific chronic inflammation. *Int J Biochem Cell Biol*. 2009;41:176-184.
5. Blanchard F, Chipoy C. Histone deacetylase inhibitors: new drugs for the treatment of inflammatory diseases? *Drug Discov Today*. 2005;10:197-204.
6. Buckley CD. Michael Mason prize essay 2003. Why do leucocytes accumulate within chronically inflamed joints? *Rheumatology (Oxford)*. 2003;42:1433-1444.
7. Chabane N, Zayed N, Afif H, et al. Histone deacetylase inhibitors suppress interleukin-1beta-induced nitric oxide and prostaglandin E2 production in human chondrocytes. *Osteoarthritis Cartilage*. 2008;16:1267-1274.
8. Chan EK, Satoh M, Pauley KM. Contrast in aberrant microRNA expression in systemic lupus erythematosus and rheumatoid arthritis: is microRNA-146 all we need? *Arthritis Rheum*. 2009;60:912-915.
9. Chung YL, Lee MY, Wang AJ, Yao LF. A therapeutic strategy uses histone deacetylase inhibitors to modulate the expression of genes involved in the pathogenesis of rheumatoid arthritis. *Mol Ther*. 2003;8:707-717.
10. Fields PE, Lee GR, Kim ST, Bartsevich VV, Flavell RA. Th2-specific chromatin remodeling and enhancer activity in the Th2 cytokine locus control region. *Immunity*. 2004;21:865-876.
11. Firestein GS. Evolving concepts of rheumatoid arthritis. *Nature*. 2003;423:356-361.
12. Grabiec AM, Krausz S, de Jager W, et al. Histone deacetylase inhibitors suppress inflammatory activation of rheumatoid arthritis patient synovial macrophages and tissue. *J Immunol*. 2010;184:2718-2728.
13. Grabiec AM, Tak PP, Reedquist KA. Targeting histone deacetylase activity in rheumatoid arthritis and asthma as prototypes of inflammatory disease: should we keep our HATs on? *Arthritis Res Ther*. 2008;10:226.
14. Horiuchi M, Morinobu A, Chin T, Sakai Y, Kurosaka M, Kumagai S. Expression and function of histone deacetylases in rheumatoid arthritis synovial fibroblasts. *J Rheumatol*. 2009;36:1580-1589.
15. Huber LC, Brock M, Hemmatazad H, et al. Histone deacetylase/acetylase activity in total synovial tissue derived from rheumatoid arthritis and osteoarthritis patients. *Arthritis Rheum*. 2007;56:1087-1093.
16. Ivanov AI, Romanovsky AA. Putative dual role of ephrin-Eph receptor interactions in inflammation. *IUBMB Life*. 2006;58:389-394.
17. Jungel A, Baresova V, Ospelt C, et al. Trichostatin A sensitises rheumatoid arthritis synovial fibroblasts for TRAIL-induced apoptosis. *Ann Rheum Dis*. 2006;65:910-912.
18. Kitamura T, Kabuyama Y, Kamataki A, et al. Enhancement of lymphocyte migration and cytokine production by ephrinB1 system in rheumatoid arthritis. *Am J Physiol Cell Physiol*. 2008;294:C189-C196.
19. Klareskog L, Catrina AI, Paget S. Rheumatoid arthritis. *Lancet*. 2009;373:659-672.
20. Korb A, Pavenstadt H, Pap T. Cell death in rheumatoid arthritis. *Apoptosis*. 2009;14:447-454.
21. Kuchen S, Seemayer CA, Rethage J, et al. The L1 retroelement-related p40 protein induces p38delta MAP kinase. *Autoimmunity*. 2004;37:57-65.
22. Lee DM, Weinblatt ME. Rheumatoid arthritis. *Lancet*. 2001;358:903-911.

23. Lin HS, Hu CY, Chan HY, et al. Anti-rheumatic activities of histone deacetylase (HDAC) inhibitors in vivo in collagen-induced arthritis in rodents. *Br J Pharmacol.* 2007;150:862-872.
24. Liu Y, Chen Y, Richardson B. Decreased DNA methyltransferase levels contribute to abnormal gene expression in "senescent" CD4(+)CD28(−) T cells. *Clin Immunol.* 2009;132:257-265.
25. Lujambio A, Ropero S, Ballestar E, et al. Genetic unmasking of an epigenetically silenced microRNA in human cancer cells. *Cancer Res.* 2007;67:1424-1429.
26. Manabe H, Nasu Y, Komiyama T, et al. Inhibition of histone deacetylase down-regulates the expression of hypoxia-induced vascular endothelial growth factor by rheumatoid synovial fibroblasts. *Inflamm Res.* 2008;57:4-10.
27. McInnes IB, Schett G. Cytokines in the pathogenesis of rheumatoid arthritis. *Nat Rev Immunol.* 2007;7:429-442.
28. Morinobu A, Wang B, Liu J, Yoshiya S, Kurosaka M, Kumagai S. Trichostatin A cooperates with Fas-mediated signal to induce apoptosis in rheumatoid arthritis synovial fibroblasts. *J Rheumatol.* 2006;33:1052-1060.
29. Muller-Ladner U, Ospelt C, Gay S, Distler O, Pap T. Cells of the synovium in rheumatoid arthritis. Synovial fibroblasts. *Arthritis Res Ther.* 2007;9:223.
30. Nagata Y, Nakasa T, Mochizuki Y, et al. Induction of apoptosis in the synovium of mice with autoantibody-mediated arthritis by the intraarticular injection of double-stranded MicroRNA-15a. *Arthritis Rheum.* 2009;60:2677-2683.
31. Nakamachi Y, Kawano S, Takenokuchi M, et al. MicroRNA-124a is a key regulator of proliferation and monocyte chemoattractant protein 1 secretion in fibroblast-like synoviocytes from patients with rheumatoid arthritis. *Arthritis Rheum.* 2009;60:1294-1304.
32. Nakamura T, Kukita T, Shobuike T, et al. Inhibition of histone deacetylase suppresses osteoclastogenesis and bone destruction by inducing IFN-beta production. *J Immunol.* 2005;175:5809-5816.
33. Nakasa T, Miyaki S, Okubo A, et al. Expression of microRNA-146 in rheumatoid arthritis synovial tissue. *Arthritis Rheum.* 2008;58:1284-1292.
34. Nasu Y, Nishida K, Miyazawa S, et al. Trichostatin A, a histone deacetylase inhibitor, suppresses synovial inflammation and subsequent cartilage destruction in a collagen antibody-induced arthritis mouse model. *Osteoarthritis Cartilage.* 2008;16:723-732.
35. Neidhart M, Rethage J, Kuchen S, et al. Retrotransposable L1 elements expressed in rheumatoid arthritis synovial tissue: association with genomic DNA hypomethylation and influence on gene expression. *Arthritis Rheum.* 2000;43:2634-2647.
36. Nile CJ, Read RC, Akil M, Duff GW, Wilson AG. Methylation status of a single CpG site in the IL6 promoter is related to IL6 messenger RNA levels and rheumatoid arthritis. *Arthritis Rheum.* 2008;58:2686-2693.
37. Nishida K, Komiyama T, Miyazawa S, et al. Histone deacetylase inhibitor suppression of autoantibody-mediated arthritis in mice via regulation of p16INK4a and p21(WAF1/Cip1) expression. *Arthritis Rheum.* 2004;50:3365-3376.
38. Pauley KM, Cha S, Chan EK. MicroRNA in autoimmunity and autoimmune diseases. *J Autoimmun.* 2009;32:189-194.
39. Pauley KM, Satoh M, Chan AL, Bubb MR, Reeves WH, Chan EK. Upregulated miR-146a expression in peripheral blood mononuclear cells from rheumatoid arthritis patients. *Arthritis Res Ther.* 2008;10:R101.
40. Pope RM. Apoptosis as a therapeutic tool in rheumatoid arthritis. *Nat Rev Immunol.* 2002;2:527-535.
41. Reedquist KA, Ludikhuize J, Tak PP. Phosphoinositide 3-kinase signalling and FoxO transcription factors in rheumatoid arthritis. *Biochem Soc Trans.* 2006;34:727-730.
42. Richardson B, Scheinbart L, Strahler J, Gross L, Hanash S, Johnson M. Evidence for impaired T cell DNA methylation in systemic lupus erythematosus and rheumatoid arthritis. *Arthritis Rheum.* 1990;33:1665-1673.
43. Saouaf SJ, Li B, Zhang G, et al. Deacetylase inhibition increases regulatory T cell function and decreases incidence and severity of collagen-induced arthritis. *Exp Mol Pathol.* 2009;87:99-104.
44. Sheedy FJ, O'Neill LA. Adding fuel to fire: microRNAs as a new class of mediators of inflammation. *Ann Rheum Dis.* 2008;67(Suppl 3):iii50-iii55.
45. Smeets TJ, Barg EC, Kraan MC, Smith MD, Breedveld FC, Tak PP. Analysis of the cell infiltrate and expression of proinflammatory cytokines and matrix metalloproteinases in arthroscopic synovial biopsies: comparison with synovial samples from patients with end stage, destructive rheumatoid arthritis. *Ann Rheum Dis.* 2003;62:635-638.
46. Smeets TJ, Kraan MC, Galjaard S, Youssef PP, Smith MD, Tak PP. Analysis of the cell infiltrate and expression of matrix metalloproteinases and granzyme B in paired synovial biopsy specimens from the cartilage-pannus junction in patients with RA. *Ann Rheum Dis.* 2001;60:561-565.
47. Sonkoly E, Stahle M, Pivarcsi A. MicroRNAs and immunity: novel players in the regulation of normal immune function and inflammation. *Semin Cancer Biol.* 2008;18:131-140.
48. Stanczyk J, Pedrioli DM, Brentano F, et al. Altered expression of MicroRNA in synovial fibroblasts and synovial tissue in rheumatoid arthritis. *Arthritis Rheum.* 2008;58:1001-1009.
49. Strietholt S, Maurer B, Peters MA, Pap T, Gay S. Epigenetic modifications in rheumatoid arthritis. *Arthritis Res Ther.* 2008;10:219.
50. Tak PP, Bresnihan B. The pathogenesis and prevention of joint damage in rheumatoid arthritis: advances from synovial biopsy and tissue analysis. *Arthritis Rheum.* 2000;43:2619-2633.
51. Tak PP, Firestein GS. NF-kappaB: a key role in inflammatory diseases. *J Clin Invest.* 2001;107:7-11.
52. Takami N, Osawa K, Miura Y, et al. Hypermethylated promoter region of DR3, the death receptor 3 gene, in rheumatoid arthritis synovial cells. *Arthritis Rheum.* 2006;54:779-787.
53. Tang Y, Luo X, Cui H, et al. MicroRNA-146A contributes to abnormal activation of the type I interferon pathway in human lupus by targeting the key signaling proteins. *Arthritis Rheum.* 2009;60:1065-1075.
54. Tao R, de Zoeten EF, Ozkaynak E, et al. Deacetylase inhibition promotes the generation and function of regulatory T cells. *Nat Med.* 2007;13:1299-1307.
55. Tas SW, Remans PH, Reedquist KA, Tak PP. Signal transduction pathways and transcription factors as therapeutic targets in inflammatory disease: towards innovative antirheumatic therapy. *Curr Pharm Des.* 2005;11:581-611.

56. Tas SW, Vervoordeldonk MJ, Tak PP. Gene therapy targeting nuclear factor-kappaB: towards clinical application in inflammatory diseases and cancer. *Curr Gene Ther.* 2009;9:160-170.
57. van der Pouw Kraan TC, Wijbrandts CA, van Baarsen LG, et al. Rheumatoid arthritis subtypes identified by genomic profiling of peripheral blood cells: assignment of a type I interferon signature in a subpopulation of patients. *Ann Rheum Dis.* 2007;66:1008-1014.
58. van Loosdregt J, Vercoulen Y, Guichelaar T, et al. Regulation of Treg functionality by acetylation-mediated Foxp3 protein stabilization. *Blood.* 2010;115:965-974.
59. Vashisht Gopal YN, Arora TS, Van Dyke MW. Tumour necrosis factor-alpha depletes histone deacetylase 1 protein through IKK2. *EMBO Rep.* 2006;7:291-296.
60. Waldburger JM, Firestein GS. Garden of therapeutic delights: new targets in rheumatic diseases. *Arthritis Res Ther.* 2009;11:206.
61. Young DA, Lakey RL, Pennington CJ, et al. Histone deacetylase inhibitors modulate metalloproteinase gene expression in chondrocytes and block cartilage resorption. *Arthritis Res Ther.* 2005;7:R503-R512.

DNA Methylation Changes in Osteoarthritis

8

Helmtrud I. Roach†

8.1 Introduction

Osteoarthritis (OA) is a common chronic disease that affects some two-thirds of the elderly population. In OA, the articular cartilage that covers the ends of diarthrodial joints, facilitating articulation and acting as a shock absorber, has become degraded.[1,29,30,57] When cartilage is degraded and ultimately lost, movement becomes difficult, painful, and everyday activities are seriously limited. Many practitioners consider the disease as an inevitable, age-related "wear and tear," often exacerbated by obesity. However, age and obesity alone do not bring about the disease.[42] Contributing factors are genetic predisposition,[44] trauma, and malalignment. Cellular events that underlie OA include premature senescence,[2,46] oxidative stress,[35,77] and epigenetic changes.[56]

In animal models of OA, the disease develops within days or weeks,[5] but in humans it often takes decades before OA becomes clinically apparent. Changes are therefore gradual, although once the disease process starts, its progression cannot be stopped. The reasons for this inevitable progression may, as will be shown, relate to the transmission of an aberrant epigenetic code to daughter cells.

This chapter reviews the current state of knowledge of the changes in the epigenetic status, specifically DNA methylation, that are associated with osteoarthritis. An overview of the cellular and molecular features that distinguish osteoarthritic cartilage from normal articular cartilage is followed by a discussion of the epigenetic changes that may underlie disease progression.

H.I. Roach†
Bone and Joint Research Group, School of Medicine, University of Southampton, Hampshire, UK
e-mail: roco@soton.ac.uk

8.2 Matrix Changes During the Development of Osteoarthritis

The shock-absorbing capacity of articular cartilage is a function of the structure of the cartilage matrix. Cartilage is a highly hydrated gel, made up of 70% water, with an organic matrix that consists of a loose mesh of type II collagen, together with types IX and XI collagens and other minor collagen types.[3] Highly negatively charged aggrecan molecules are interspersed within the collagen mesh. Aggrecan absorbs water, causing the matrix to swell, but the swelling is limited by the meshwork of collagen fibers. The collagens thus provide tensile strength and compressive stiffness, whereas the aggrecan/water structure provides elasticity and the capacity to absorb shock.

The availability of human articular cartilage as a consequence of joint replacement surgery has enabled researchers to study the stages of the disease in cartilage obtained from osteoporotic patients (who had a femoral neck fracture) or from osteoarthritic patients (for recent reviews, see refs.[1,30,32,57,75]). People who suffer from osteoarthritis do not have osteoporosis, and patients with osteoporosis do not have osteoarthritis.[14] The cartilage of the femoral heads of osteoporotic patients (#NOF; Fracture Neck of Femur) generally does not show macroscopic signs of osteoarthritis (Fig. 8.1a), whereas the cartilage of femoral heads of OA patients is eroded (Fig. 8.1b). Histological examination of cartilage from #NOF patients shows it to be normal articular cartilage (Fig. 8.2a). Superficial, intermediate, and deep zones are present and, in histological sections stained with Alcian blue/Sirius red, only the subchondral bone stains bright red (Fig. 8.2a).

Cartilage degradation in OA patients starts in the superficial zone with the loss of aggrecan in weight-bearing

H.I. Roach et al. (eds.), *Epigenetic Aspects of Chronic Diseases*,
DOI: 10.1007/978-1-84882-644-1_8, © Springer-Verlag London Limited 2011

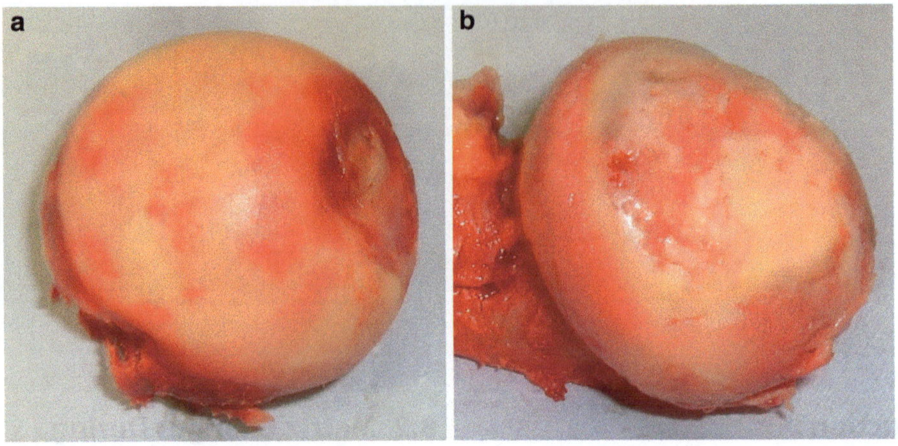

Fig. 8.1 Femoral head from a 79-year-old patient with a femoral neck fracture (**a**) and a 69-year-old osteoarthritic patient (**b**). The former is covered by smooth, near-normal aged cartilage, which represents control cartilage. In the OA patient, the cartilage is worn away, subchondral bone is visible in places, but some full thickness cartilage remains near the femoral neck (Reprinted from Roach and Tilley[57])

regions, as demonstrated by Sirius red staining in the superficial zone (Fig. 8.2b). Sirius red stains fibrous collagens (e.g., the type I collagen of bone matrix), but because the aggrecans of cartilage prevent access,[64] Sirius red does not normally stain type II collagen. Sirius red stain in cartilage therefore indicates aggrecan loss. As OA progresses, gradually all of the cartilage matrix and cartilage cells are lost, first from the weight-bearing regions, and ultimately from all other regions (Fig. 8.2c, d).

8.3 Cellular and Molecular Changes Leading to Osteoarthritis

The cartilage matrix is maintained by the articular chondrocytes, which make up about 3% in volume of the cartilage matrix. In the adult, these cells have low metabolic activity, divide rarely, and have a long life. They express the genes for cartilage matrix proteins, i.e., aggrecan, collagen types II, IX, and XI, many minor collagens, cartilage oligomeric protein, and the transcription factors for chondrocytes, i.e., Sox-5, Sox-6, and Sox-9. With the exception of a few chondrocytes in the superficial zone (Fig. 8.2e), cartilage cells do not express the matrix-degrading enzymes, such as the matrix metalloproteinases (MMPs) and aggrecanases cells in the deep zone (Fig 8.2f) are immuno-negative. The chondrocytes also do not express inflammatory cytokines, such as interleukin-1β (IL-1β) or tumor necrosis factor-α (TNF-α). However, in OA, matrix-degrading or inflammatory genes are expressed, but not by all chondrocytes. Cells that express these genes have been termed "degradative" chondrocytes.[58] In low-grade OA, degradative chondrocytes are only present in the superficial zone (shown for MMP-13 in Fig. 8.2g). As OA progresses and the superficial zone is worn away, chondrocytes in the intermediate and deep zones also become degradative (shown for ADAMTS-4 in Fig. 8.2h). The resulting proteases are secreted into the extracellular matrix (Fig. 8.2h) and chondrocyte-derived enzymes become the main source of the proteases.[28-30,55,57,60,68] In contrast to healthy chondrocytes, "degradative" cells divide to form doublets, quadruplets which, in severe OA, lead to clones (Fig. 8.2i). Clusters of the "degradative" chondrocytes clearly show that the aberrant gene expression is transmitted to the daughter cells.[49,58,68,73,76] The enzymes involved in the degradation of aggrecan are ADAMTS-4 and ADAMTS-5[45,61,66,69] and MMP-3 (stromelysin).[26,47,49] Type II collagen is degraded by MMP-13 (collagenase-3).[48,63] Degradative chondrocytes also express other genes not normally expressed by articular chondrocytes. Examples are IL-1β,[22,41,70] VEGF,[17] pleiotrophin,[52] and leptin.[37]

In a comparison of the molecular differences between control and OA chondrocytes, it is important to be aware that OA cartilage contains both normal and degradative chondrocytes. Yet this heterogeneity

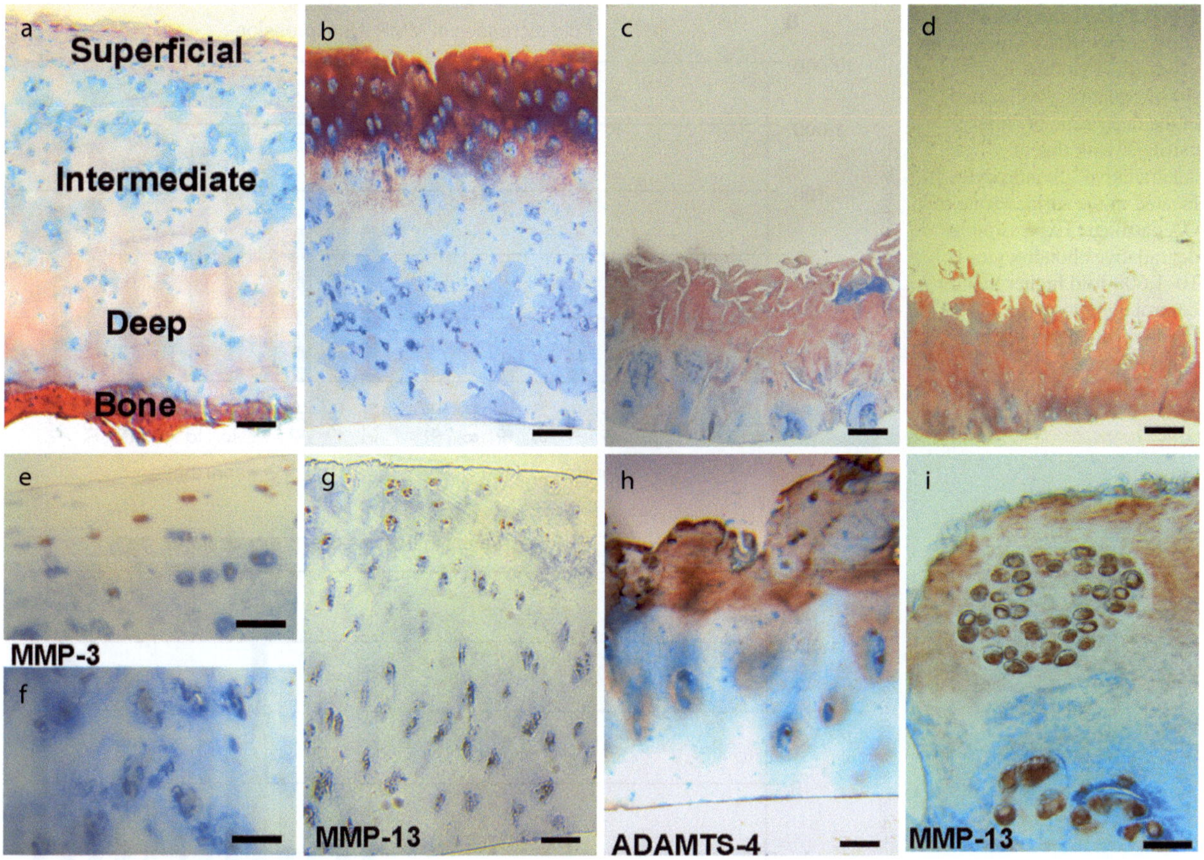

Fig. 8.2 Microscopic changes associated with increasing severity of osteoarthritis. (**a–d**) Alcian blue/Sirius red staining illustrates the changes in the cartilage matrix. All four figures are aligned so that the subchondral bone edge is at the bottom and all are of the same magnification so that the loss of cartilage thickness in severe OA (**c**, **d**) is clearly seen when compared with control or early OA cartilage (**a**, **b**). (**a**) In control cartilage, the typical zones of articular cartilage can be distinguished and only the subchondral bone stains with Sirius red. (**b**) In early OA, proteoglycans have been lost from the superficial zone, as indicated by the red stain. In severe OA (**c**) and end-stage OA (**d**), the cartilage has been eroded and deep fissures are present. (**e–i**) Immunostaining for cartilage-degrading enzymes shows that in control cartilage (**e**, **f**), only a few chondrocytes in the superficial zone produce MMP-3 (**e**), while cells in the deep zone (**f**) are immuno-negative. In OA cartilage (**g–i**), the number of immuno-positive chondrocytes increases and enzyme activity is also found in the matrix (In part reprinted from Roach and Tilley[57])

is often overlooked and differences between chondrocytes located near the surface and those in the deep zone may therefore not be recognized. It is advisable to microdissect the cartilage, i.e., to select only the surface region of the cartilage of OA patients and to contrast this with the deep zone of the cartilage of #NOF patients. This will reveal the tremendous increase in expression of, for example, MMP-13 (Fig. 8.3a) or IL-1β (Fig.8.3b) in degradative chondrocytes, compared with their expression in control chondrocytes.

8.4 Contribution of Genetics

OA is polygenetic, and susceptibility is the likely result of the interaction of many genes. Linkage studies have implicated IL-1B, matrilin-3, IL-4 receptor alpha, secreted frizzled-related protein 3 (SFRP-3), ADAM12, and asporin (ASPN).[44] Of these, IL-1, IL-4Rα, FRZB, and ASPN function in chondrocyte differentiation. Genome-wide association studies have also implicated the vitamin D receptor, the estrogen receptor alpha (ERα), the transforming growth factor-beta (TGF-β),

Fig. 8.3 Real-time RT-PCR comparing mRNA expression of MMP-13 (a) and IL-1β (b) of chondrocytes located in the deep zone of control cartilage with that of "degradative" chondrocytes located in the surface zone of OA cartilage. Expression in degradative chondrocytes is 10–1,000-fold higher than in control chondrocytes (b) (Reprinted from Hashimoto et al.[34])

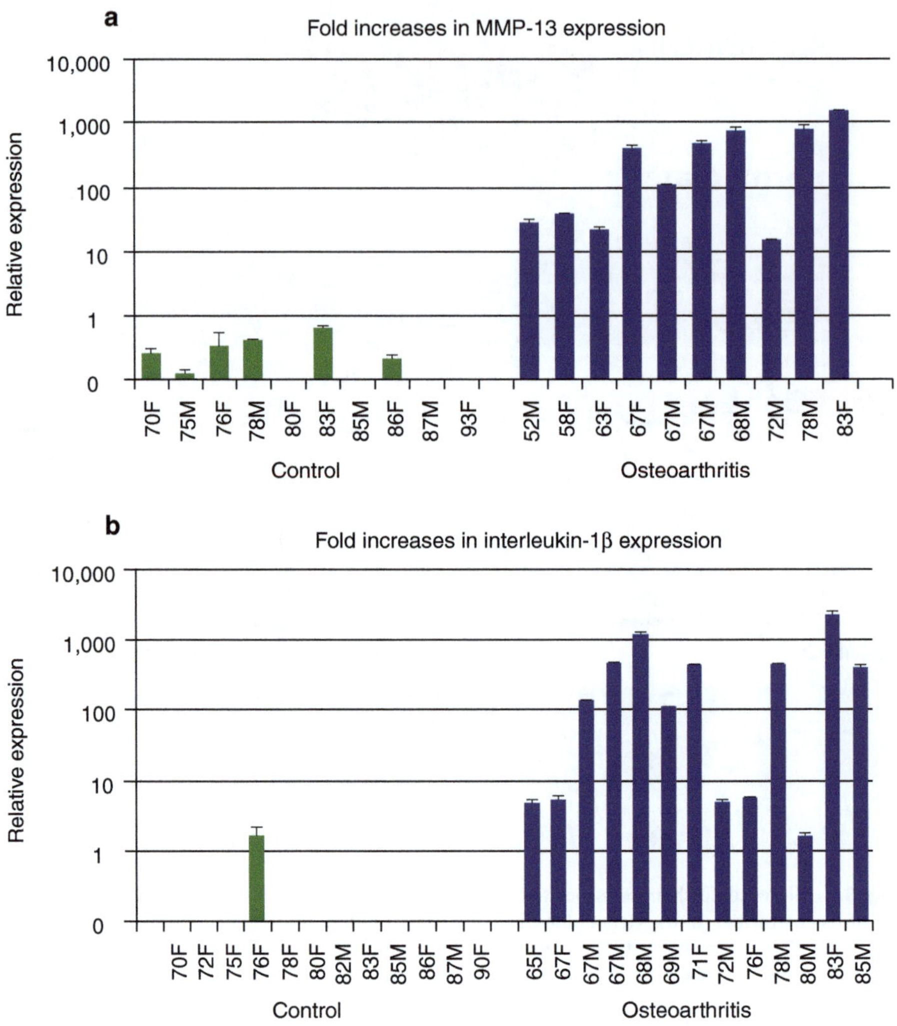

IGF-1, cartilage matrix protein, COL9A1, COL11A1, and ANK.[65] The importance of a single gene in the disease process is relatively modest, but the risk of OA is markedly increased when polymorphisms occur in several genes in combination.[71] Identical twin studies have shown that in women genetic factors appear to contribute 39–65% to hand and knee OA, 60% to hip OA, and up to 70% to spine OA.[65]

8.4.1 Why Suspect Epigenetic Regulation in OA?

The incomplete concordance in monozygotic twins for OA suggests a role for epigenetics. Epigenetic changes induced by environmental factors may in fact determine whether an individual with a given genome develops the disease. Monozygotic twin studies have shown that the differences in epigenetic marks increase with age.[25,74] Another reason for inferring epigenetic regulation in OA is the change in the repertoire of gene expression, with many non-chondrocytic genes being activated aberrantly, whereas typical chondrocytic genes are silenced. Moreover, once an articular chondrocyte has acquired the aberrant pattern of gene expression, the cell continues to express aberrant genes and aberrant expression is stably transmitted to daughter cells.

Thus, one would expect that non-chondrocytic genes are silenced by epigenetic mechanisms in control chondrocytes. If this is so, DNA methylation, combined with histone de-acetylation should be high in the promoter regions. There should be methylation at histone 3 lysine 9 (H3K9) and H3K27, interaction with heterochromatin proteins, and the chromatin structure should be closed.[24,27] However, genes that are aberrantly induced during the disease process must have undergone "un-silencing," i.e., loss of DNA methylation in the promoter regions, acetylated histones, and methylated lysine 4 on H3 (H3K4), among others. As the DNA methylation status in the promoter region is an indication of whether a gene is active or silenced, it constitutes a good index of whether epigenetic changes have occurred.

8.5 Aberrant Gene Expression and DNA Hypo-methylation

Early studies merely determined the degree of DNA methylation in arthritic diseases. For example, hypermethylation of T cells and peripheral blood mononuclear cells is characteristic of inflammatory arthritis.[39,53] Yet in OA the overall level of DNA methylation was no different from that of controls.[62] To demonstrate whether epigenetic changes play a role in a given disease, it is necessary to determine the DNA methylation status of specific CpG sites in the promoters of differentially expressed genes.

When investigating whether CpG methylation correlates with differential gene expression, many investigators deliberately select genes with CpG island promoters on the assumption that such promoters play a greater role in epigenetic regulation than sparse CpG promoters. An example is the study by Ezura et al.[21] which determined the DNA methylation status of 11 CpG islands in chondrocytic genes before and after chondrogenesis. Of the 11 CpG islands, 10 were found to be hypo-methylated, regardless of the expression levels of the genes. Similar results were obtained in the Human Epigenome project (www.epigenome.org), in which the methylation status was determined for 1.9 million CpG sites in 873 genes located on chromosomes 6, 20, and 22. This was carried out in 12 different tissues, and methylation status was correlated to expression. Eighty-eight percent of CpG island promoters were found to be non-methylated irrespective of expression.[9,16] It is, therefore, the author's view that one is more likely to find differential DNA methylation in genes with sparse CpG promoters.

8.5.1 De-methylation at Specific CpG Sites in the Promoters of Proteases

Many cartilage-degrading enzymes are aberrantly activated in OA chondrocytes and are, therefore, candidate genes for investigating the DNA methylation status of the CpG sites in the relevant promoters. MMP-3, MMP-9, MMP-13, and ADAMTS-4 are proteases that are typically expressed de novo in OA chondrocytes (see Fig. 8.4 for structure of the *MMP-13* and *IL1B* promoters). These proteases have promoters with relatively few CpG sites. Ten CpG sites are located in the 600-bp promoter sequence of MMP-13 (Fig. 8.4a); ten CpG sites are in the 870 promoter region of MMP-9 (Fig. 8.4b); seven CpG sites are in the 2,000-bp 5'-flanking region of MMP-3 and MMP-13 and in the 900-bp region for ADAMTS-4. The methylation status of individual CpG sites varied considerably.[11,58] Some sites were methylated, whereas others were largely

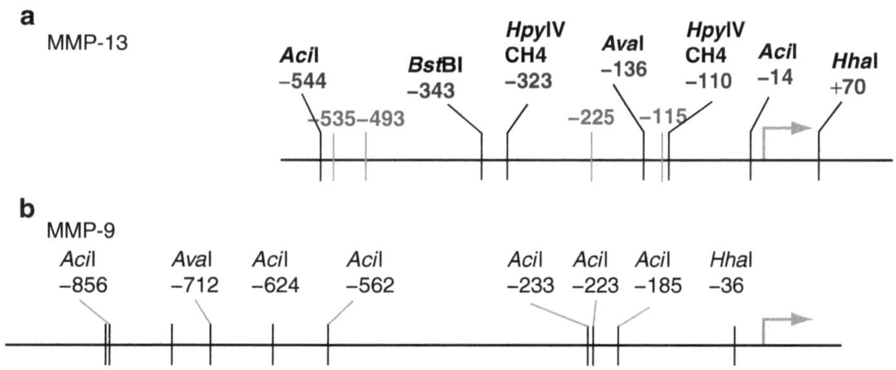

Fig. 8.4 Diagram of the MMP-13 (**a**) and MMP-9 (**b**) promoter illustrating the location of CpG sites and associated methylation-sensitive restriction enzymes (Adapted from Fig. 2 in Roach et al.[58])

un-methylated in both control and OA samples. For each protease, there was at least one CpG site where loss of DNA methylation had occurred in an OA sample. Moreover, the overall percentage of methylated CpG sites was only 52% in OA sample, reduced from 80% in controls.[58] This suggests that aberrant expression loss is associated with loss of DNA methylation at specific CpG sites in the sparse CpG promoters of cartilage-degrading proteases.

8.6 Non-epigenetic and Epigenetic Regulation of Interleukin-1β Expression In Vitro

An experimental model that is extensively used to mimic the changes in gene expression that occur in OA involves monolayer cultures of healthy chondrocytes that have been treated with either IL-1β[59] or TNFα in combination with oncostatin M (OSM).[6] The cytokines induce in vitro expression of the proteolytic enzymes and of pro-inflammatory factors, such as IL-1β. The question then arises whether this aberrant induction involves epigenetic changes or non-epigenetic regulation by transcription factors. When mRNA expression of *IL1B* was determined after short-term treatment with exogenous IL-1β/OSM, expression was induced within 24 h, but was readily reversible after cytokine withdrawal (Fig. 8.5a). When however cells were treated twice weekly for 3 weeks, expression persisted after cytokine withdrawal (Fig. 8.5b).[34] Short-term induction therefore is readily reversible and, therefore, probably does not involve CpG methylation changes. In contrast, long-term induction, leads to persistent expression that is maintained even after passaging, and therefore is consistent with stable heritable changes. With the aid of bisulfite sequencing it was shown that of the 21 CpG sites in the 1,300-bp sequence of the IL1B promoter upstream of exon 1, 16 distal CpG sites are methylated in all cultures.[40]

Therefore, epigenetic regulation would not seem to involve the region between −300 to −590 bp. In addition, the *Ava*I site at −511 bp and the two CpG sites (−20 and +13) that encompass the transcription start site were non-methylated in all groups. By contrast, the two CpG sites at −299 and −256 bp that were methylated in control samples had become de-methylated in cytokine-treated chondrocytes. This identifies the two CpG sites at −299 and −256 bp as critical for epigenetic regulation of *IL1B*.

8.6.1 Experimental De-methylation Increases Expression of IL1B

If DNA de-methylation underpins aberrant *IL1B* expression in chondrocytes, then experimentally induced de-methylation should also lead to increased

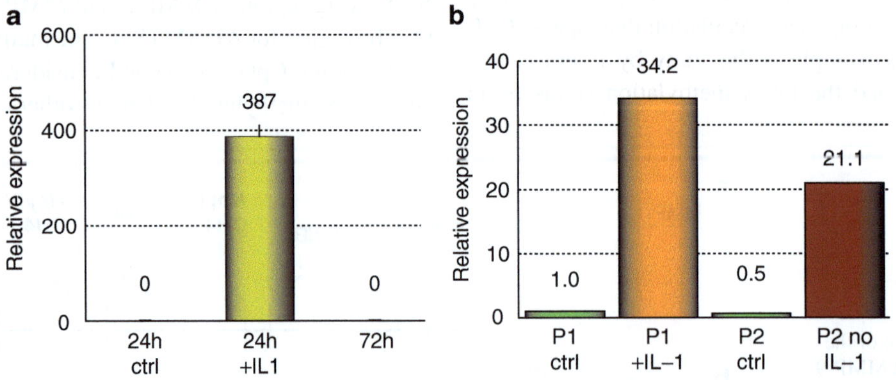

Fig. 8.5 Effects of cytokine withdrawal following short-term (**a**) or long-term (**b**) treatment of control chondrocytes with IL-1β/OSM on the mRNA expression of *IL1B*. (**a**) Treatment for 24 h results in a considerable increase in expression, but this is lost again upon cytokine withdrawal. (**b**) Long-term treatment for 3 weeks also increases expression. After passage and culture for a further 3 weeks without cytokines, expression is maintained even after cytokine withdrawal (Reprinted from Hashimoto et al.[34])

expression. Hashimoto et al[34] cultured normal chondrocytes with 5-aza-dC and showed that the expression of *IL1B* increased ~5 fold compared with control cultures (Fig. 8.6a). As a check that 5-aza-dC treatment actually led to loss of DNA methylation in the *IL1B* promoter, the percentage of DNA methylation was also quantified: The CpG site at −299 bp was methylated in 60% uncultured chondrocytes. Culture alone reduced the percentage methylation to ~45%, but addition of 5-aza-dC reduced methylation further, i.e., to ~33% (Fig. 8.6b). These results support the notion of a cause–effect relationship between DNA de-methylation and aberrant transcription of *IL1B*.

8.7 DNA De-methylation in a Young OA Patient

Although OA occurs predominantly in the elderly, degenerative cartilage changes can also occur in the young, as in developmental dysplasia of the hip (DDH), a condition where the femoral head does not develop correctly within the acetabulum. If this defect is not corrected in infancy, or if treatment fails to correct the dysplasia, secondary osteoarthritis results. Da Silva et al.[12] showed that the cartilage from a 23-year-old female with DDH had histological features similar to those in the cartilage of aged individuals with OA, including loss of aggrecan from the surface zone and extensive fissures (Fig. 8.7a). Loss of DNA methylation also had occurred at the same CpG sites where de-methylation had been observed in OA (shown for MMP-13 and MMP-9 in Fig. 8.7b; for the results for MMP-3 and ADAMTS-4, see da Silva et al.[12]). The DNA methylation pattern in the DDH patient was similar, but not identical to that in older people with OA. Thus, loss of CpG methylation in the promoters of proteases is a consistent feature of OA, irrespective of age.

8.8 Activation of Leptin

Leptin, a hormone whose major role relates to food intake and energy consumption, is aberrantly expressed in OA,[37] inducing nitric oxide synthesis, some pro-inflammatory cytokines, and MMP-13.[15,37] Iliopoulos et al.[37] studied DNA methylation in 32 CpG sites in the leptin promoter. In normal cartilage, 22/32 CpG sites were methylated and leptin expression was negligible. In minimally damaged cartilage, five CpG sites had lost DNA methylation and leptin expression had increased modestly. In cartilage from patients with extensive OA, only 2/32 CpG sites had remained methylated and leptin expression had increased some 60-fold. When normal chondrocytes were treated with 5-aza-deoxycytidine, the DNA methylation inhibitor in leptin expression had increased and DNA methylation was lost. This confirms that DNA methylation status is inversely related to leptin expression.

The preceding studies demonstrate that not all CpG sites have to be methylated to silence a gene, even though the overall degree of CpG methylation in non-expressing cells is generally higher than in expressing cells. The converse is also true: Not all CpG sites have to be de-methylated for expression to occur. It is loss of CpG methylation at specific sites that is important, with a decrease in the number of methylated CpG sites probably also contributing.

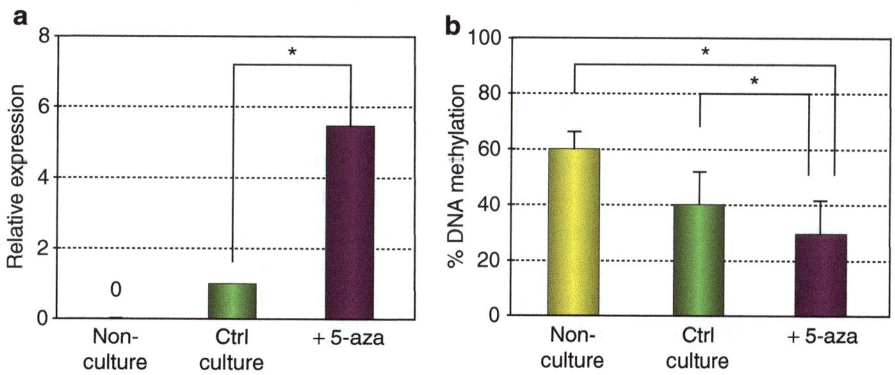

Fig. 8.6 Effects of 5-aza-deoxycytidine on (**a**) the expression of *IL1B* and (**b**) the % DNA methylation at −299 bp in the IL1B promoter (Reprinted from Hashimoto et al.[34])

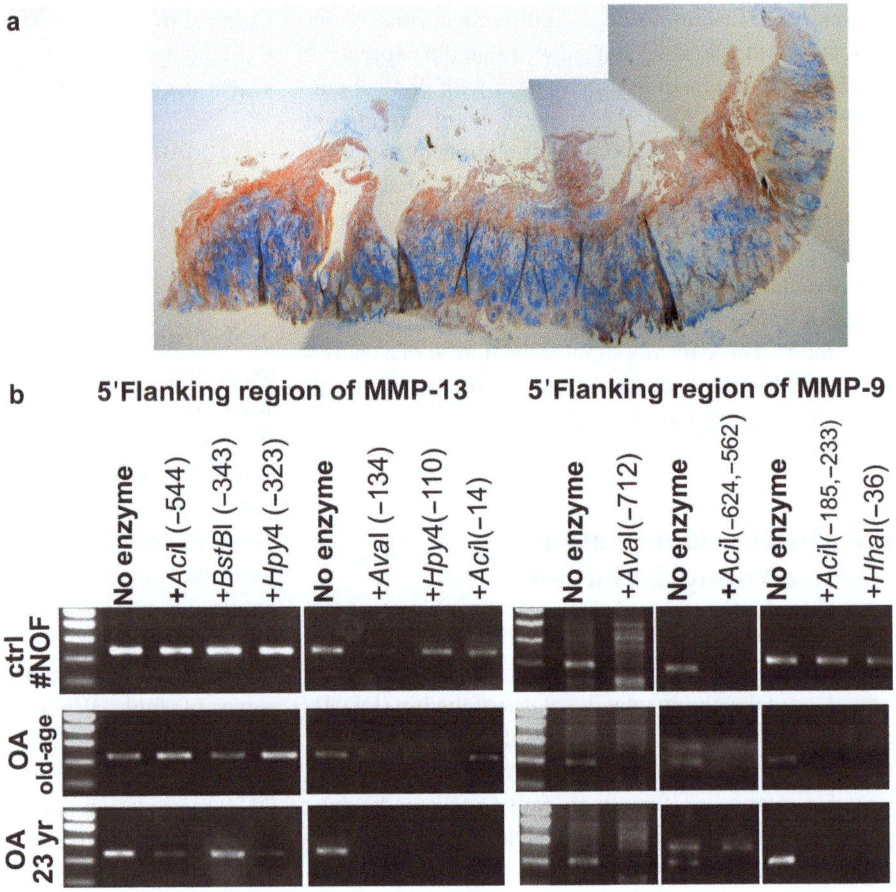

Fig. 8.7 Osteoarthritis in a young patient is also associated with loss of DNA methylation. (**a**) A section through the articular cartilage of a 23-year-old patient whose developmental dysplasia of the hip had not been successfully corrected during infancy. (**b**) DNA methylation status of CpG sites in the promoters of MMP-13 and MMP-9, assessed by the methylation-sensitive restriction enzyme method. Presence of a band indicates presence of CpG methylation. In the 23-year old, even more CpG sites are de-methylated compared with old-age OA (Reprinted from da Silva et al.[12])

8.9 Which Factors Cause DNA De-methylation?

Inflammatory cytokines, especially IL-1β, are known to play a role in OA pathology. Hashimoto et al.[34] investigated whether inflammatory cytokines cause loss of DNA methylation, thus leading to the activation of aberrant genes. When healthy chondrocytes were cultured for 4–5 weeks with twice-weekly additions of exogenous IL-1β or TNF-α/OSM (Fig. 8.8), IL-1β expression increased 15- to 197-fold compared with controls, whereas the combination of TNF-α/OSM increased expression of *IL1B* by 300- to 1,750-fold. Addition of the cytokines, moreover, caused considerable loss of DNA methylation, with IL-1β addition reducing methylation to less than 25% and TNFα/OSM treatment leaving hardly any cells methylated at the CpG site at −299 bp. Interestingly, the two crucial CpG sites (−256 and −299 bp) are located around an NF-κB binding site, and NF-κB may be involved in de-methylation.[40]

If the above findings also apply to other genes and occur in vivo, then OA progression may involve these events: An inflammatory episode in the synovium, perhaps following mechanical stress, activates synovial macrophages to produce IL-1β and TNFα which diffuse into the articular cartilage, where they induce a phenotypic change to "degradative" chondrocytes. This change includes loss of DNA methylation, aberrant expression of proteases and *IL1B* in the chondrocytes of the superficial layer.[31] *IL1B* will now be included in the expression repertoire of the "degradative" chondrocytes and will be expressed (and translated to

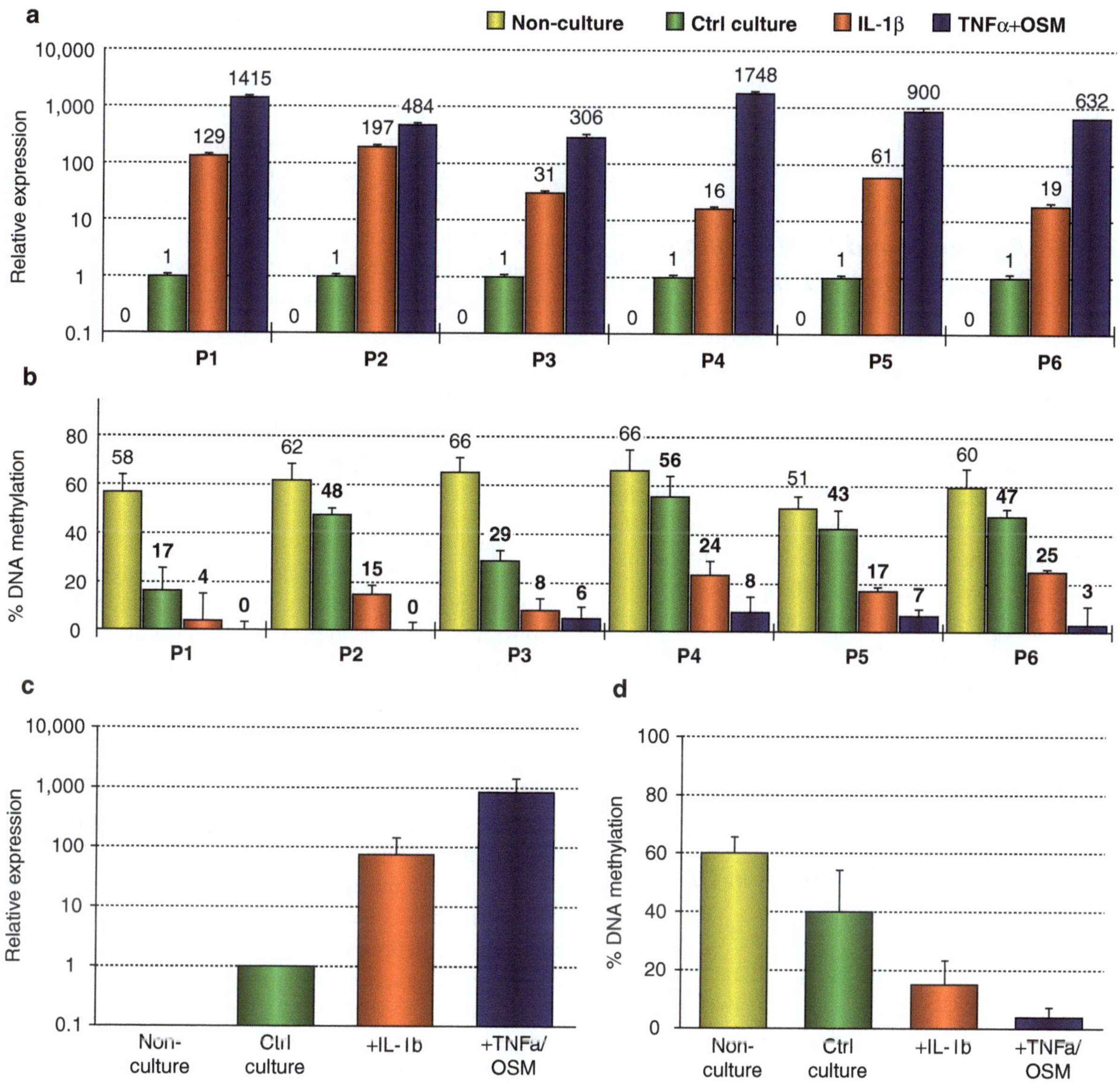

Fig. 8.8 (**a**) Relative mRNA expression and (**b**) % DNA methylation at −299 bp in six patients (P1–P6) after long-term cultures treated as indicated in the legend. A clear inverse relationship between expression (**c**) and % DNA methylation (**d**) can be seen (Reprinted from Hashimoto et al.[34])

IL-1β) even after synovial inflammation has abated. The chondrocyte-derived cytokine then diffuses to adjacent cells where it induces the altered "degradative" phenotype. This scenario may explain why protease inhibitors, such as TIMPs, have little effect when injected into the joint, and why, once degradative processes have been operative, OA progression cannot be halted.

8.10 Are Chondrocytic Genes Silenced in OA by DNA Hyper-methylation?

In addition to aberrantly expressed genes, many chondrocytic genes are down-regulated or silenced in "degradative" OA chondrocytes in vivo or by inflammatory cytokines in vitro. In part, this is because aged chondrocytes generally have a low level of metabolic activity. On the other hand, epigenetic silencing may also have taken place. To date, very few studies have examined the latter possibility.

8.10.1 Aggrecan and Type II Collagen: No Silencing by DNA Methylation

The major components of the cartilage matrix are aggrecan and type II collagen. These matrix components turn over slowly, but continuously in healthy chondrocytes, but not in OA. It was of interest therefore to determine whether DNA hyper-methylation is the process that silences the *aggrecan* or *COL2A1* genes in OA chondrocytes. Pöschl et al.[51] found that all of 33 CpG sites located in a 340-bp CpG island were un-methylated, irrespective of whether the cartilage was from controls or OA patients. Therefore, silencing by increased DNA methylation is not the mechanism that down-regulates aggrecan in OA. Similarly, there is no differential methylation in the *COL2A1* promoter in expressing versus non-expressing cells.[78]

8.10.2 p21WAF1/CIP1 Gene in OA

p21WAF1/CIP1, an inhibitor of proliferation, is another example of the lack of association between decreased expression and DNA methylation. This gene is expressed in normal slowly proliferating chondrocytes. In OA, *p21WAF1/CIP1* is down-regulated; this is associated with the increase in cell division in OA, yet the extent of DNA methylation of the *p21WAF1/CIP1* promoter is the same in normal and OA chondrocytes.[62] The promoter contains a CpG island, lending further support to the notion that silencing of genes with CpG island promoters in non-cancer cells cannot be explained by DNA methylation.

8.10.3 Osteogenic Protein-1 in Aged Chondrocytes

OP-1, also known as bone morphogenetic protein-7, is an anabolic factor that stimulates cartilage matrix synthesis. Hence, loss of expression would be detrimental to the maintenance of articular cartilage and may contribute to OA. Loeser et al.[43] showed that some CpG sites were un-methylated in young people, but were methylated in an age-dependent manner in older individuals. Treatment with 5-azacytidine increased expression of chondrocytic genes approximately twofold; this supports the notion that DNA methylation leads to partial silencing of this anabolic gene with age. It is not known whether this is also observed in OA chondrocytes.

8.10.4 Type IX Collagen: Epigenetic Silencing in a Sparse CpG Promoter

Collagen type IX is present on the surface of type II collagen and binds covalently to two collagen II molecules and to COMP (Fig. 8.9a[19]). Collagen type IX constitutes only 1% to the total collagen of mature cartilage, but is important for the integrity and stability of articular cartilage.[20] Mice that lack *COL9A1* develop normally, but display osteoarthritis-like cartilage degradation when older.[4,8,23,36] Gradual reduction in *COL9A1* expression in humans over decades would reduce collagen IX in the matrix; this in turn would affect matrix integrity and render the cartilage more susceptible to mechanical damage. Reduced collagen IX expression could thus contribute to the pathoetiology of osteoarthritis.

Imagawa and Roach[38] compared the expression of *COL9A1* in OA chondrocytes with that in control chondrocytes. Considerable variation between patients notwithstanding, *COL9A1* was highly expressed in most control samples (Fig. 8.9b). However, expression was 500-fold lower in samples from OA patients. In addition, the percentage of DNA methylation at all CpG sites (Fig. 8.9c) was higher in OA than control samples, with the greatest differences found at CpG sites −632, −614, and −599 bp (Fig. 8.9d). This is the first demonstration that down-regulated expression of a chondrocytic gene in osteoarthritis is associated with silencing by hyper-methylation.

8.11 Summary and Conclusions

Methylation of genomic DNA represents a significant mechanism for regulating tissue-specific gene

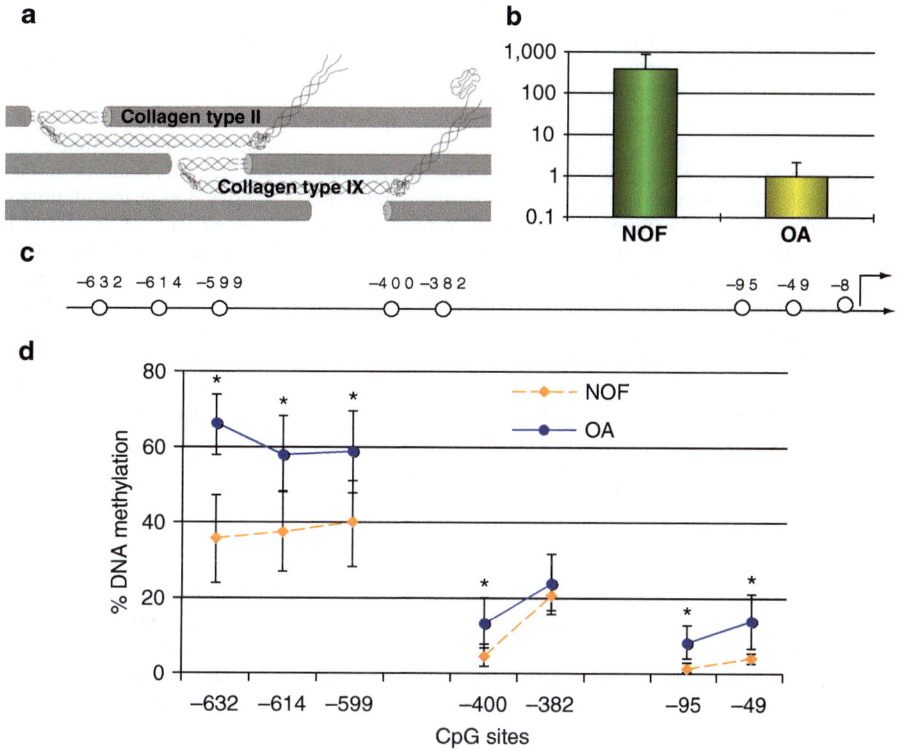

Fig. 8.9 Hyper-methylation underlies the silencing of type IX collagen in osteoarthritis. (**a**) Type IX collagen fibrils cross-link with type II collagen microfibrils which is important for integrity of the collagen network. (**b**) Fold decreases in COL9A1 expression in OA cartilage. (**c**) Structure of the *COL9A1* promoter, showing CpG sites as circles. (**d**) The % DNA methylation in the *COL9A1* promoter was quantified after bisulfite modification with a pyrosequencer (Biotage). CpG methylation has increased in OA patients compared with control (NOF patients), especially at the CpG sites at –632, –614, and –599 bp. $N = 10$ for NOF and 12 for OA. *$P < 0.01$ (**a**: Adapted from Eyre[19]; **b–d**: Reproduced from Imagawa and Roach[38])

expression. Aberrant methylation patterns play a role in cancer[7,13,18,33,67] and other non-Mendelian diseases.[50] Changes in DNA methylation as discussed in this Chapter can explain the changes in gene expression observed in OA. Loss of DNA methylation results in aberrant expression of IL-1β, MMP-3, MMP-9, MMP-13, ADAMTS-4, and leptin in OA, whereas increased methylation is associated with decreased expression of *COL9A1*. Changes in DNA methylation probably also underlie the aberrant expression or silencing of other genes. However, changes in DNA methylation will not always explain permanently altered gene expression, as shown by the lack of differential DNA methylation in the promoters of aggrecan, collagen type II, and *p21WAF1/CIP1*.

The molecular steps by which CpG methylation status is permanently changed are not known. Stochastic changes as a result of accumulated errors in replicating the DNA methylation pattern during cell division may contribute, but are unlikely to be the only explanation. Inflammatory cytokines play a role in de-methylation of specific CpG sites in the *IL1B* promoter, but whether this is generally true is not known. Oxidative stress modulates DNA methylation in cancer cells,[10] perhaps by preventing methyl-CpG from binding proteins.[72]

The pathological consequences of changes in epigenetic status cannot readily be reversed. Therefore, the consequences of epigenetic changes are more disastrous than the transient changes in expression that occur in response to transcription factors or cytokines. If epigenetic changes could be prevented or reversed, OA would be slowed, even halted. This would constitute a major therapeutic benefit to patients with arthritis[54] whose numbers continue to increase given the lengthened lifespan throughout the world.

References

1. Aigner T, Sachse A, Gebhard PM, Roach HI. Osteoarthritis: pathobiology-targets and ways for therapeutic intervention. *Adv Drug Deliv Rev*. 2006;58:128-149.
2. Aigner T, Soder S, Gebhard PM, McAlinden A, Haag J. Mechanisms of disease: role of chondrocytes in the pathogenesis of osteoarthritis–structure, chaos and senescence. *Nat Clin Pract Rheumatol*. 2007;3:391-399.
3. Aigner T, Stove J. Collagens–major component of the physiological cartilage matrix, major target of cartilage degeneration, major tool in cartilage repair. *Adv Drug Deliv Rev*. 2003;55: 1569-1593.

4. Allen KD, Griffin TM, Rodriguiz RM, et al. Decreased physical function and increased pain sensitivity in mice deficient for type IX collagen. *Arthritis Rheum.* 2009;60:2684-2693.
5. Ameye LG, Young MF. Animal models of osteoarthritis: lessons learned while seeking the "Holy Grail". *Curr Opin Rheumatol.* 2006;18:537-547.
6. Barksby HE, Hui W, Wappler I, et al. Interleukin-1 in combination with oncostatin M up-regulates multiple genes in chondrocytes: implications for cartilage destruction and repair. *Arthritis Rheum.* 2006;54:540-550.
7. Baylin SB, Esteller M, Rountree MR, Bachman KE, Schuebel K, Herman JG. Aberrant patterns of DNA methylation, chromatin formation and gene expression in cancer. *Hum Mol Genet.* 2001;10:687-692.
8. Blumbach K, Niehoff A, Paulsson M, Zaucke F. Ablation of collagen IX and COMP disrupts epiphyseal cartilage architecture. *Matrix Biol.* 2008;27:306-318.
9. Brena RM, Huang TH, Plass C. Toward a human epigenome. *Nat Genet.* 2006;38:1359-1360.
10. Campos AC, Molognoni F, Melo FH, et al. Oxidative stress modulates DNA methylation during melanocyte anchorage blockade associated with malignant transformation. *Neoplasia.* 2007;9:1111-1121.
11. Cheung KS, Hashimoto K, Yamada N, Roach HI. Expression of ADAMTS-4 by chondrocytes in the surface zone of human osteoarthritic cartilage is regulated by epigenetic DNA de-methylation. *Rheumatol Int.* 2009;29:525-534.
12. da Silva MA, Yamada N, Clarke NM, Roach HI. Cellular and epigenetic features of a young healthy and a young osteoarthritic cartilage compared with aged control and OA cartilage. *J Orthop Res.* 2009;27:593-601.
13. Davis CD, Uthus EO. DNA methylation, cancer susceptibility, and nutrient interactions. *Exp Biol Med (Maywood).* 2004;229:988-995.
14. Dequeker J, Aerssens J, Luyten FP. Osteoarthritis and osteoporosis: clinical and research evidence of inverse relationship. *Aging Clin Exp Res.* 2003;15:426-439.
15. Dumond H, Presle N, Terlain B, et al. Evidence for a key role of leptin in osteoarthritis. *Arthritis Rheum.* 2003;48:3118-3129.
16. Eckhardt F, Lewin J, Cortese R, et al. DNA methylation profiling of human chromosomes 6, 20 and 22. *Nat Genet.* 2006;38:1378-1385.
17. Enomoto H, Inoki I, Komiya K, et al. Vascular endothelial growth factor isoforms and their receptors are expressed in human osteoarthritic cartilage. *Am J Pathol.* 2003;162:171-181.
18. Esteller M. Epigenetic gene silencing in cancer: the DNA hypermethylome. *Hum Mol Genet.* 2007;16 Spec No 1:R50-R59.
19. Eyre D. Collagen of articular cartilage. *Arthritis Res.* 2002;4:30-35.
20. Eyre DR. Collagens and cartilage matrix homeostasis. *Clin Orthop Relat Res.* 2004;472S:S118-S122.
21. Ezura Y, Sekiya I, Koga H, Muneta T, Noda M. Methylation status of CpG islands in the promoter regions of signature genes during chondrogenesis of human synovium-derived mesenchymal stem cells. *Arthritis Rheum.* 2009;60:1416-1426.
22. Fan Z, Soder S, Oehler S, Fundel K, Aigner T. Activation of interleukin-1 signaling cascades in normal and osteoarthritic articular cartilage. *Am J Pathol.* 2007;171:938-946.
23. Fassler R, Schnegelsberg PN, Dausman J, et al. Mice lacking alpha 1 (IX) collagen develop noninflammatory degenerative joint disease. *Proc Natl Acad Sci USA.* 1994;91:5070-5074.
24. Feinberg AP. Phenotypic plasticity and the epigenetics of human disease. *Nature.* 2007;447:433-440.
25. Fraga MF, Ballestar E, Paz MF, et al. Epigenetic differences arise during the lifetime of monozygotic twins. *Proc Natl Acad Sci USA.* 2005;102:10604-10609.
26. Freemont AJ, Hampson V, Tilman R, Goupille P, Taiwo Y, Hoyland JA. Gene expression of matrix metalloproteinases 1, 3, and 9 by chondrocytes in osteoarthritic human knee articular cartilage is zone and grade specific. *Ann Rheum Dis.* 1997;56:542-549.
27. Gan Q, Yoshida T, McDonald OG, Owens GK. Epigenetic mechanisms contribute to pluripotency and cell lineage determination of embryonic stem cells. *Stem Cells.* 2007;25:2-9.
28. Goldring MB. The role of the chondrocyte in osteoarthritis. *Arthritis Rheum.* 2000;43:1916-1926.
29. Goldring MB. Update on the biology of the chondrocyte and new approaches to treating cartilage diseases. *Best Pract Res Clin Rheumatol.* 2006;20:1003-1025.
30. Goldring MB, Goldring SR. Osteoarthritis. *J Cell Physiol.* 2007;213:626-634.
31. Goldring MB, Otero M, Tsuchimochi K, Ijiri K, Li Y. Defining the roles of inflammatory and anabolic cytokines in cartilage metabolism. *Ann Rheum Dis.* 2008;67(Suppl 3):iii75-iii82.
32. Goldring SR, Goldring MB. Clinical aspects, pathology and pathophysiology of osteoarthritis. *J Musculoskelet Neuronal Interact.* 2006;6:376-378.
33. Gronbaek K, Hother C, Jones PA. Epigenetic changes in cancer. *APMIS.* 2007;115:1039-1059.
34. Hashimoto K, Oreffo RO, Gibson MB, Goldring MB, Roach HI. DNA demethylation at specific CpG sites in the IL1B promoter in response to inflammatory cytokines in human articular chondrocytes. *Arthritis Rheum.* 2009;60:3303-3313.
35. Henrotin Y, Kurz B, Aigner T. Oxygen and reactive oxygen species in cartilage degradation: friends or foes? *Osteoarthritis Cartilage.* 2005;13:643-654.
36. Hu K, Xu L, Cao L, et al. Pathogenesis of osteoarthritis-like changes in the joints of mice deficient in type IX collagen. *Arthritis Rheum.* 2006;54:2891-2900.
37. Iliopoulos D, Malizos KN, Tsezou A. Epigenetic regulation of leptin affects MMP-13 expression in osteoarthritic chondrocytes: possible molecular target for osteoarthritis therapeutic intervention. *Ann Rheum Dis.* 2007;66:1616-1621.
38. Imagawa K, Roach HI (2009) Reduced expression of collagen type IX in osteoarthritic chondrocytes is associated with epigenetic silencing by DNA hypermethylation. In: Transactions of the 55th Annual Meeting of the Orthopaedic Research Society; March 2009; New Orleans.
39. Kim YI, Logan JW, Mason JB, Roubenoff R. DNA hypomethylation in inflammatory arthritis: reversal with methotrexate. *J Lab Clin Med.* 1996;128:165-172.
40. Kirillov A, Kistler B, Mostoslavsky R, Cedar H, Wirth T, Bergman Y. A role for nuclear NF-kappaB in B-cell-specific demethylation of the Igkappa locus. *Nat Genet.* 1996;13:435-441.

41. Kobayashi M, Squires GR, Mousa A, et al. Role of interleukin-1 and tumor necrosis factor alpha in matrix degradation of human osteoarthritic cartilage. *Arthritis Rheum.* 2005;52: 128-135.
42. Loeser RF. Aging and osteoarthritis: the role of chondrocyte senescence and aging changes in the cartilage matrix. *Osteoarthritis Cartilage.* 2009;17:971-979.
43. Loeser RF, Im HJ, Richardson B, Lu Q, Chubinskaya S. Methylation of the OP-1 promoter: potential role in the age-related decline in OP-1 expression in cartilage. *Osteoarthritis Cartilage.* 2009;17:513-517.
44. Loughlin J. The genetic epidemiology of human primary osteoarthritis: current status. *Expert Rev Mol Med.* 2005;7: 1-12.
45. Malfait AM, Liu RQ, Ijiri K, Komiya S, Tortorella MD. Inhibition of ADAM-TS4 and ADAM-TS5 prevents aggrecan degradation in osteoarthritic cartilage. *J Biol Chem.* 2002;277:22201-22208.
46. Martin JA, Brown T, Heiner A, Buckwalter JA. Post-traumatic osteoarthritis: the role of accelerated chondrocyte senescence. *Biorheology.* 2004;41:479-491.
47. Mehraban F, Lark MW, Ahmed FN, Xu F, Moskowitz RW. Increased secretion and activity of matrix metalloproteinase-3 in synovial tissues and chondrocytes from experimental osteoarthritis. *Osteoarthritis Cartilage.* 1998;6: 286-294.
48. Neuhold LA, Killar L, Zhao W, et al. Postnatal expression in hyaline cartilage of constitutively active human collagenase-3 (MMP-13) induces osteoarthritis in mice. *J Clin Invest.* 2001;107:35-44.
49. Okada Y, Shinmei M, Tanaka O, et al. Localization of matrix metalloproteinase 3 (stromelysin) in osteoarthritic cartilage and synovium. *Lab Invest.* 1992;66:680-690.
50. Petronis A. Human morbid genetics revisited: relevance of epigenetics. *Trends Genet.* 2001;17:142-146.
51. Poschl E, Fidler A, Schmidt B, Kallipolitou A, Schmid E, Aigner T. DNA methylation is not likely to be responsible for aggrecan down regulation in aged or osteoarthritic cartilage. *Ann Rheum Dis.* 2005;64:477-480.
52. Pufe T, Bartscher M, Petersen W, Tillmann B, Mentlein R. Pleiotrophin, an embryonic differentiation and growth factor, is expressed in osteoarthritis. *Osteoarthritis Cartilage.* 2003;11:260-264.
53. Richardson B, Scheinbart L, Strahler J, Gross L, Hanash S, Johnson M. Evidence for impaired T cell DNA methylation in systemic lupus erythematosus and rheumatoid arthritis. *Arthritis Rheum.* 1990;33:1665-1673.
54. Roach HI. Potential directions for drug development for osteoarthritis. *Expert Opin Drug Discovery.* 2008;3: 475-487.
55. Roach HI. The complex pathology of osteoarthritis: even mitochondria are involved. *Arthritis Rheum.* 2008;58: 2217-2218.
56. Roach HI, Aigner T. DNA methylation in osteoarthritic chondrocytes: a new molecular target. *Osteoarthritis Cartilage.* 2007;15:128-137.
57. Roach HI, Tilley S. The pathogenesis of osteoarthritis. In: Bronner F, Farach-Carson MC, eds. *Bone and Osteoarthritis*, Topics in Bone Biology, vol. 4. London: Springer; 2007: 1-18.
58. Roach HI, Yamada N, Cheung KS, et al. Association between the abnormal expression of matrix-degrading enzymes by human osteoarthritic chondrocytes and demethylation of specific CpG sites in the promoter regions. *Arthritis Rheum.* 2005;52:3110-3124.
59. Saas J, Haag J, Rueger D, et al. IL-1beta, but not BMP-7 leads to a dramatic change in the gene expression pattern of human adult articular chondrocytes–portraying the gene expression pattern in two donors. *Cytokine.* 2006;36: 90-99.
60. Sandell LJ, Aigner T. Articular cartilage and changes in arthritis. An introduction: cell biology of osteoarthritis. *Arthritis Res.* 2001;3:107-113.
61. Sandy JD, Verscharen C. Analysis of aggrecan in human knee cartilage and synovial fluid indicates that aggrecanase (ADAMTS) activity is responsible for the catabolic turnover and loss of whole aggrecan whereas other protease activity is required for C-terminal processing in vivo. *Biochem J.* 2001;358:615-626.
62. Sesselmann S, Soder S, Voigt R, Haag J, Grogan SP, Aigner T. DNA methylation is not responsible for p21WAF1/CIP1 down-regulation in osteoarthritic chondrocytes. *Osteoarthritis Cartilage.* 2009;17:507-512.
63. Shlopov BV, Gumanovskaya ML, Hasty KA. Autocrine regulation of collagenase 3 (matrix metalloproteinase 13) during osteoarthritis. *Arthritis Rheum.* 2000;43:195-205.
64. Smith JO, Oreffo RO, Clarke NM, Roach HI. Changes in the antiangiogenic properties of articular cartilage in osteoarthritis. *J Orthop Sci.* 2003;8:849-857.
65. Spector TD, Macgregor AJ. Risk factors for osteoarthritis: genetics. *Osteoarthritis Cartilage.* 2004;12(Suppl A):S39-S44.
66. Stanton H, Rogerson FM, East CJ, et al. ADAMTS5 is the major aggrecanase in mouse cartilage in vivo and in vitro. *Nature.* 2005;434:648-652.
67. Szyf M. Targeting DNA methylation in cancer. *Ageing Res Rev.* 2003;2:299-328.
68. Tetlow LC, Adlam DJ, Woolley DE. Matrix metalloproteinase and proinflammatory cytokine production by chondrocytes of human osteoarthritic cartilage: associations with degenerative changes. *Arthritis Rheum.* 2001;44:585-594.
69. Tortorella MD, Malfait AM, Deccico C, Arner E. The role of ADAM-TS4 (aggrecanase-1) and ADAM-TS5 (aggrecanase-2) in a model of cartilage degradation. *Osteoarthritis Cartilage.* 2001;9:539-552.
70. Towle CA, Hung HH, Bonassar LJ, Treadwell BV, Mangham DC. Detection of interleukin-1 in the cartilage of patients with osteoarthritis: a possible autocrine/paracrine role in pathogenesis. *Osteoarthritis Cartilage.* 1997;5:293-300.
71. Valdes AM, Doherty M, Spector TD. The additive effect of individual genes in predicting risk of knee osteoarthritis. *Ann Rheum Dis.* 2008;67:124-127.
72. Valinluck V, Tsai HH, Rogstad DK, Burdzy A, Bird A, Sowers LC. Oxidative damage to methyl-CpG sequences inhibits the binding of the methyl-CpG binding domain (MBD) of methyl-CpG binding protein 2 (MeCP2). *Nucleic Acids Res.* 2004;32:4100-4108.
73. Walter H, Kawashima A, Nebelung W, Neumann W, Roessner A. Immunohistochemical analysis of several proteolytic enzymes as parameters of cartilage degradation. *Pathol Res Pract.* 1998;194:73-81.

74. Wong AH, Gottesman II, Petronis A. Phenotypic differences in genetically identical organisms: the epigenetic perspective. *Hum Mol Genet.* 2005;14 Spec No 1:R11-R18.
75. Wu CW, Kalunian KC. New developments in osteoarthritis. *Clin Geriatr Med.* 2005;21:589-601.
76. Wu W, Billinghurst RC, Pidoux I, et al. Sites of collagenase cleavage and denaturation of type II collagen in aging and osteoarthritic articular cartilage and their relationship to the distribution of matrix metalloproteinase 1 and matrix metalloproteinase 13. *Arthritis Rheum.* 2002;46: 2087-2094.
77. Yudoh K, Nguyen T, Nakamura H, Hongo-Masuko K, Kato T, Nishioka K. Potential involvement of oxidative stress in cartilage senescence and development of osteoarthritis: oxidative stress induces chondrocyte telomere instability and downregulation of chondrocyte function. *Arthritis Res Ther.* 2005;7:R380-R391.
78. Zimmermann P, Boeuf S, Dickhut A, Boehmer S, Olek S, Richter W. Correlation of COL10A1 induction during chondrogenesis of mesenchymal stem cells with demethylation of two CpG sites in the COL10A1 promoter. *Arthritis Rheum.* 2008;58:2743-2753.

9 Epigenetics and Type 2 Diabetes

Charlotte Ling, Tina Rönn, and Marloes Dekker Nitert

9.1 Introduction

Type 2 diabetes mellitus (T2D) is a chronic multifactorial disease characterized by hyperglycemia, a result of impaired pancreatic beta cell function and insulin resistance by liver and peripheral target tissues, e.g., skeletal muscle and adipose tissue. In insulin resistance, cells in the body show reduced response to insulin. This in turn decreases glucose clearance from the blood and leads to increased glucose production by the liver. The pancreatic beta cells compensate by producing more insulin, but loss of this balance leads to hyperglycemia (Fig. 9.1). High levels of glucose can damage organs and lead to complications of the cardiovascular system, the eyes, neurons, and kidneys.

T2D incidence varies widely, but is rapidly increasing worldwide, mainly due to the increasing age of many populations, urbanization, and the increasing prevalence of obesity and physical inactivity. Between 2000 to 2030, the number of diabetics throughout the world is expected to double, to a total of some 366 million individuals.[91]

T2D has long been viewed as the outcome of a complex interaction between the genome and environmental factors. The importance of genetics is brought out by the fact that monozygotic twins have a T2D concordance rate of approximately 70%, compared with only 20–30% in dizygotic twins.[56] Offspring of one T2D parent have a lifetime risk of developing the disease of about 40%, whereas if both parents have T2D, the risk becomes 70%.[21] Many genetic variants have been associated with the disease, but causality has proven elusive.[42] Although, recent genome-wide studies have discovered approximately 40 previously unknown common genetic variants associated with T2D, there has been less success in identifying epigenetic factors of importance for the disease. While aging, obesity, and physical inactivity increase susceptibility to T2D, these factors may also change the epigenetic pattern in T2D target tissues and subsequently affect

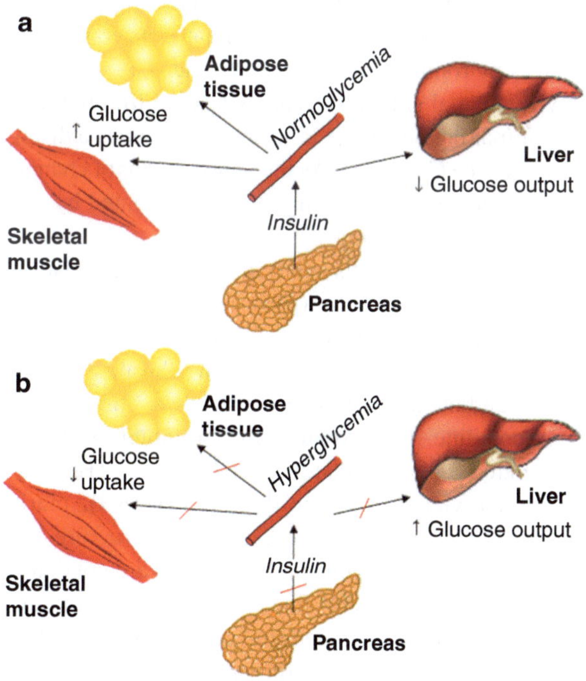

Fig. 9.1 Pathophysiology of hyperglycemia in T2D. (**a**) In non-diabetic subjects, insulin suppresses glucose production from the liver and stimulates glucose uptake into skeletal muscle and adipose tissue. (**b**) T2D patients display defects in insulin secretion in pancreas and insulin action in target tissues. As a result, glucose uptake decreases and hepatic glucose production increases, resulting in hyperglycemia.

C. Ling (✉)
Department of Clinical Sciences, Lund University, Malmö, Sweden
e-mail: charlotte.ling@med.lu.se

H.I. Roach et al. (eds.), *Epigenetic Aspects of Chronic Diseases*,
DOI: 10.1007/978-1-84882-644-1_9, © Springer-Verlag London Limited 2011

gene expression and metabolism. Epigenetic mechanisms may therefore play an important role in the pathogenesis of T2D and its complications. Further understanding of the epigenetic changes induced by these risk factors may help develop tools that predict, diagnose, and treat the disease. With the help from more genome-wide epigenetic studies, it may become possible to further dissect the role of DNA methylation and histone modifications in the development of T2D.

9.2 The Role of Epigenetics

In 1992, Hales and Barker proposed that environmental factors experienced in early life may enhance the risk of T2D in later life.[23] In particular, under-nutrition and low birth weight have shown a relation to adult T2D, insulin resistance and impaired insulin secretion.[36] Inadequate nutrition, by inducing chronic alterations in metabolism, hormone levels, and cell numbers, contributes to the risk of T2D.[4] Developmental plasticity makes it possible for the early human embryo to adapt to its environment at any given time, but when the environmental situation changes later in life, the benefit of making better use of nutrients becomes a disadvantage (see Chap. 13). As the genome cannot change, environmental programming may be mediated by epigenetic reprogramming.

Pancreatic beta cells synthesize and secrete insulin. The regulation of insulin (*INS*) gene expression is not fully understood, but there is evidence for epigenetic involvement both from studies on the chromatin structure[9,53] and the level of DNA methylation.[30] In a mouse beta cell line, the proximal *Ins* promoter is hyperacetylated at lysine residues of histone 3 (H3) and hypermethylated at lysine 4 of H3 (H3K4), marks associated with an open euchromatin structure and actively transcribed genes. These marks are not detected in the non-beta cell lines. Embryonic stem cells have an intermediate pattern, consistent with their potential to differentiate into an insulin-expressing cell.[9] Furthermore, in human pancreatic islets, the *INS* gene displays a chromatin pattern typical of active genes, including hyperacetylation of histone 4 (H4) and dimethylation of H3K4 (H3K4me2). These patterns of histone modifications are not present in other cell types, which instead display elevated levels of inactive marks.[53] CpG sites in both the mouse *Ins2* and human *INS* promoter are demethylated in insulin-producing beta cells and methylation of these sites suppresses insulin gene expression.[30]

Important evidence for a role of epigenetic factors in the pathogenesis of T2D comes from a data-mining analysis of more than 12 million Medline records.[93] The study found that methylation and chromatin are top hits, implicitly related to T2D. Common phenotypes involved in the onset and pathology of T2D, which are shared by diseases associated with changes in DNA methylation, were also identified. Examples are aberrant expression of X-linked genes, oncogenesis, onset of Huntington's disease, in all of which the probability of disease increases with age. Similarly, the onset of T2D tends to occur later in life, with the severity increasing over time.

Although there is support for the role for epigenetics in the pathogenesis of T2D, conclusive studies from human T2D tissues are limited. In one study, *S*-adenosylmethionine, the main physiological donor of methyl groups, was decreased in the erythrocytes of patients with T2D. Moreover, a decrease of the methyl donor was associated with the progression of the disease.[64] Indeed, treatment with *S*-adenosylmethionine improves insulin sensitivity in a rat model of insulin resistance and T2D, possibly because of an increase in skeletal muscle mitochondrial DNA density.[27] Another functional study evaluating epigenetics in human T2D tissue concerns Peroxisome proliferator-activated receptor gamma coactivator 1 alpha (also known as PGC-1α, and encoded by *PPARGC1A*), a transcriptional coactivator of mitochondrial genes involved in normal ATP-production and insulin secretion from the pancreatic beta cells. The study showed that the level of DNA methylation is increased in a promoter region of *PPARGC1A* in pancreatic islets from patients with T2D, as compared with islets from healthy human donors.[34] This increase in DNA methylation correlates with a decrease in *PPARGC1A* mRNA expression; moreover, the *PPARGC1A* expression is positively correlated with glucose-stimulated insulin secretion.[38] Furthermore, in skeletal muscle from patients with T2D, an increase in DNA methylation parallels a decrease in *PPARGC1A* mRNA expression and mitochondrial content, with a high proportion of non-CpG methylation in the region of the promoter of *PPARGCIA*.[5]

9.3 Aging, Type 2 Diabetes, and Epigenetic Changes

Oxidative phosphorylation (OXPHOS) in the mitochondria results in the production of ATP, which is the main cellular source of energy. Oxidative capacity and mitochondrial function decline with age and are of great importance in understanding the pathogenesis of T2D.[28,62] A number of OXPHOS genes are downregulated in skeletal muscle from patients with T2D.[52,61] One of these, *COX7A1*, has more promoter DNA methylation in skeletal muscle of elderly compared with that of young individuals. In muscle cells of the elderly individuals, *COX7A1* mRNA expression is decreased and the transcript level is correlated positively with in vivo glucose uptake (Fig. 9.2).[70] The expression of another OXPHOS gene, *NDUFB6*, is also affected by DNA methylation in human skeletal muscle (Fig. 9.2). The DNA methylation site associated with *NDUFB6* mRNA expression is introduced by a single nucleotide polymorphism (SNP), rs629566, which introduces both a CG-dinucleotide and a putative transcription factor-binding site for neural retina leucine zipper (NRL) in the promoter of *NDUFB6*. Young carriers of this SNP display no DNA methylation in the CpG-site introduced by the SNP nor in three additional CpG sites located at positions –634, –663, and –676, and they have higher mRNA expression than carriers of the common allele which does not introduce a methylation site. In contrast, elderly individuals who carry the SNP that introduces the methylation site show an increase in promoter DNA methylation (58% ± 16%), a decrease in *NDUFB6* mRNA expression, and a reduced in vivo metabolism.[37] These findings suggest that age increases the dysregulation of DNA methylation in human skeletal muscle. This change affects gene expression and T2D phenotypes, e.g., insulin sensitivity.

A change in DNA methylation with increasing age is also found in animal models. A decrease in the activity of hepatic glucokinase (GCK), a key enzyme in glucose utilization, is associated with insulin resistance and T2D. In rat hepatocytes, *Gck* expression and activity decline with age, with DNA methylation increasing concurrently. Moreover, culturing rat hepatocytes from elderly rats with 5-aza-2′-deoxycytidine restored *Gck* expression and decreased DNA methylation.[26]

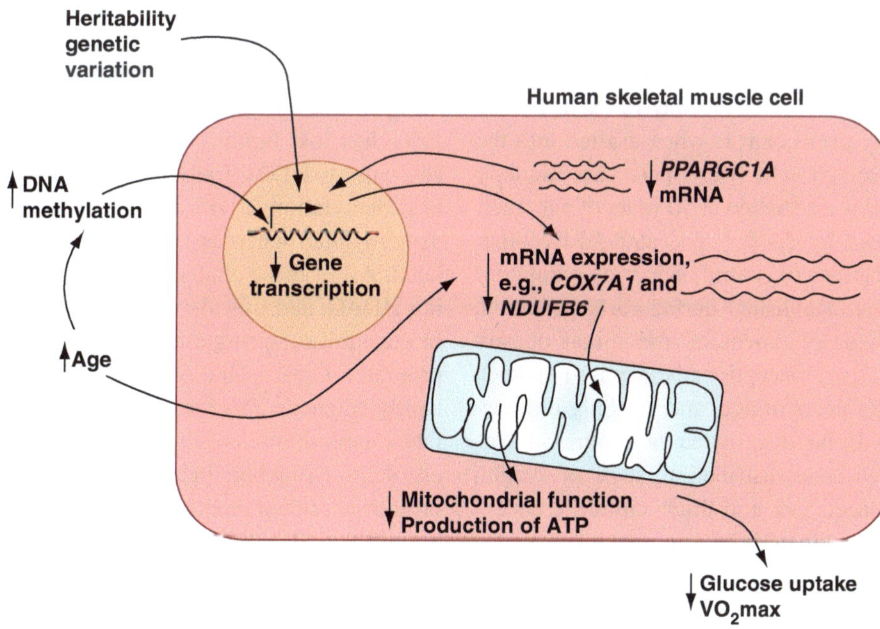

Fig. 9.2 Schematic drawing of how epigenetic mechanisms might be involved in the regulation of OXPHOS genes in human skeletal muscle and its effect on in vivo metabolism during aging. Age increases DNA methylation and decreases mRNA expression of *COX7A1* and *NDUFB6*, respectively. Furthermore, genetic variation (SNPs) is associated with *COX7A1* and *NDUFB6* mRNA levels. The expression of both genes correlates with *PPARGC1A* mRNA expression. This could affect in vivo metabolism, as the mRNA expression of both *COX7A1* and *NDUFB6* is positively associated with insulin-stimulated glucose uptake and VO_{2max} (Modified from Rönn et al.[70] With kind permission from Springer Science+Business Media)

Pancreatic beta cell proliferation declines with age in both rodents and humans.[10] A link between the *CDKN2A* locus and T2D susceptibility has been ascertained.[73,74,97] The murine *Cdkn2a* locus encodes Ink4a and Arf, which are negative cell cycle regulators that limit proliferation of beta cells. Ezh2 is a histone methyltransferase that represses *Ink4a* and *Arf* in islet beta cells and is positively correlated with beta cell proliferation. As Ezh2 levels decline in aging beta cells, H3 trimethylation decreases at the *Ink4a/Arf* locus, concurrent with an increase in *Ink4a* and *Arf* expression.[10] Whether this histone methyltransferase also has a regulatory role in human pancreatic islets and in the pathogenesis of T2D must still be determined.

9.4 The Role of Nutrition and Obesity in Epigenetics of Type 2 Diabetes

Obesity and diet are important factors in the susceptibility to T2D. Yet evidence on the effects of nutrition on epigenetic changes related to T2D has, until recently, been mostly circumstantial and not experimentally confirmed. Detailed epidemiological studies have been carried out on well-defined groups of individuals exposed to famine at different time-points. Ravelli and colleagues[67] showed that boys whose mothers were exposed to famine in early and mid-gestation during the Dutch Hunger Winter (Nov 1944–May 1945) had twice the rate of obesity over controls when drafted into the army.[67] When women who were periconceptionally exposed to famine were studied at 50 years of age, their BMI had increased by 7.4%.[66] The authors hypothesized that perturbations of central endocrine regulatory systems, that were established during early gestation, contribute to the development of abdominal obesity later in life. The periconceptionally exposed women were studied at 55 years of age, and not only did they exhibit higher BMI, but also increases in other indices of body mass and mass distribution, such as weight, waist circumference, and mid-thigh circumference.[79] Exposure to famine in mid- to late gestation is also associated with a 0.4–0.5 mM increase in the 120 min blood glucose concentration in the oral glucose tolerance test (OGTT). Furthermore, prenatal famine exposure is related to increases in fasting pro-insulin and insulin concentrations at 120 min in the OGTT, suggesting an association with insulin resistance.[65] In fact, low birth weight babies whose mothers had low body weight have the highest blood glucose concentrations at 120 min. This association is particularly true in those that develop into obese adults. Offspring of mothers with diet-treated gestational diabetes or type 1 diabetes (T1D) have an increased risk of the metabolic syndrome, i.e., high plasma triglycerides, central obesity, and high blood pressure, symptoms associated with increased risk of cardiovascular disease and T2D.[11] The risk increased with a rise in the mother's fasting blood glucose and the 120 min blood glucose levels after a glucose challenge in the OGTT setting.[11]

These epidemiological studies have now been expanded to assess changes in epigenetic marks. In the individuals exposed to the Dutch Hunger Winter, changes in DNA methylation were present 6 decades later. Periconceptional exposure to famine was found to be associated with decreases in DNA methylation in the promoter of the *IGF2* and *INSIGF* genes, whereas DNA methylation was increased in the promoters of *IL10*, *LEP*, *ABCA1*, *GNASAS*, and *MEG3*. Exposure during late pregnancy was only associated with increased methylation of the *GNASAS* promoter region.[24,87]

Offspring in a rat model of intrauterine growth retardation have a reduced beta cell mass and a lower level of *Pdx1* expression.[59] *Pdx1* is the pancreatic and duodenal homeobox 1 transcription factor that is essential for the development and function of beta cells. In the growth-retarded fetus, the proximal promoter of *Pdx1* has lost binding of transcription factor USF1, and recruits HDAC1 and corepressor Sin3A, resulting in deacetylation of H3 and H4. After birth, H3K4 is demethylated whereas H4 lysine 9 (H4K9) is methylated. All of these changes can be reversed by inhibiting HDAC, and this also normalizes *Pdx1* expression. In the adult offspring that develops diabetes, the *Pdx1* promoter CpG island is methylated, which permanently silences *Pdx1* expression. These changes in DNA methylation are linked to higher expression levels of Dnmt1 and 3a. In another model,[8] rats were fed a protein-restricted diet during pregnancy. The offspring in both the F1 and F2 generations had lower levels of methylation of the hepatic *Pparα* and glucocorticoid receptor ($GR1_{10}$) promoters. In both generations, these lower levels were accompanied by tendencies for increased expression of *Pparα*, $GR1_{10}$, acyl-CoA oxidase and *Pepck*. This study provides evidence for nutritionally induced epigenetic changes being transmitted transgenerationally. In a more detailed study by the same group, protein restriction during pregnancy led to a decrease in DNA methylation and an increase in expression of the hepatic $GR1_{10}$. *Dnmt1* expression was decreased in the offspring of the rats fed with a protein-restricted diet.[32] Additionally, levels of activating histone marks were raised, whereas levels of inactivating marks were lowered in the $GR1_{10}$ promoter. Folic acid supplementation of the protein-restricted diet during pregnancy prevented the decrease in DNA methylation of the hepatic *Pparα* promoter.[33] When folic acid instead was supplemented during juvenile development, it induced *Pparα* and $GR1_{10}$ promoter methylation in the liver and decreased insulin receptor promoter methylation in the liver and adipose tissue, irrespective of the maternal diet during pregnancy. These changes were reflected in the mRNA expression levels.[8]

In sheep, dietary restriction of methionine, folate, and vitamin B12 periconceptionally led to offspring that were heavier, fatter, insulin-resistant, and had an altered immune response to allergenic challenges. Moreover, male sheep were hypertensive. This may have been due to a change in methylation status in 4% of 1,400 CpG islands in fetal liver, over half of which were specific for males.[76]

In fetuses of Japanese macaques fed with a high-fat diet for a period of 1–4 years, the levels of hepatic acetylated H3 lysine 14 (H3K14) were increased, but acetylation and methylation levels of other histone marks were not changed. HDAC1 mRNA and protein expressions were decreased whereas expressions of genes involved in metabolism, oxidative stress, and circadian rhythm were increased. Liver triglycerides and histological correlates of non-alcoholic fatty liver disease were increased in the fetuses.[2]

In a mouse model of non-alcoholic fatty liver disease, in which the mice were fed a methyl-deficient diet, overall liver DNA methylation was decreased, especially at repetitive sequences at major and minor satellites. This was associated with significantly decreased protein expression of *Dnmt1*, an increase in the level of H3 trimethylated lysine 9 (H3K9me3), and a decrease in H4 trimethylated lysine 20 (H4K20me3).[63] In livers from mice that had been fasted, the promoter of the fatty acid synthase (*Fasn*) gene was deacetylated by HDAC9 at the binding site for the transcription factor USF1. In the fed state or upon refeeding, USF1 was bound to the promoter driving expression of *Fasn*. *Fasn* mRNA expression levels were accordingly high in the fed state and low in the fasted state.[92] Short-duration fasting leads to gluconeogenesis in the liver in response to increases in glucagon levels resulting from increased proteolysis of skeletal muscle. However, long-duration fasting leads to protein-sparing action by hepatic ketone production, a feature characteristic for T1D. Glucagon triggers dephosphorylation and nuclear translocation of the CREB-regulated transcription coactivator 2 (CRTC2 or TORC2). At the same time, a decrease in insulin signaling augments gluconeogenic gene expression through dephosphorylation and nuclear shuttling of forkhead box O1 (FOXO1). After glucagon induction, CRTC2 associates with the histone acetyltransferase p300, which in turn stimulates gluconeogenic gene expression. During late fasting, the promoter of CRTC2 is deacetylated by means of the nutrient-sensing deacetylase sirtuin 1 (SIRT1), whereas FOXO1 supports gluconeogenic gene expression.[39] It is not known whether these processes are altered in T2D.

Leptin, encoded by the *LEP* gene, is a hormone that regulates energy uptake and expenditure and is primarily expressed in differentiated adipocytes of white adipose tissue. DNA methylation in the *Lep* promoter is modulated by high-fat diet–induced obesity in rats.[50] In adipocyte differentiation, DNA methylation of the *LEP* promoter decreases from preadipocytes to adipocytes. This is associated with *LEP* expression in adipocytes.[47,57,80,96] Furthermore, the density of methylation of *LEP* promoters in fat cells is lower than in peripheral blood leukocytes, although it also varies in leukocytes. This suggests that the epigenetic state of the *LEP* promoter is not stable and subject to fluctuations.[80] In human and mouse sperm cells, the *LEP* promoter is unmethylated; this suggests that epigenetic reprogramming occurs during spermatogenesis.[80] The *Pparg* gene promoter, a key regulator of adipogenesis, is methylated in mouse preadipocytes and, as the preadipocytes differentiate into adipocytes, methylation decreases and gene expression increases. In the db/db mouse, an obese diabetic mouse model with a mutation in the leptin receptor, the promoter of *Pparg* is more highly methylated and its expression is decreased compared to controls. When cultured cells of the db/db mouse were treated with 5-aza-deoxycytidine, the DNA methylase inhibitor, promoter methylation decreased and mRNA expression of *Pparg* increased.[15] Similarly, DNA methylation of specific

CpG sites in the mouse *Glut4* promoter decreased as preadipocytes differentiated to adipocytes.[95]

Genetic disruption of the interaction between nuclear receptor corepressor 1 (Ncor1) and HDAC3 resulted in mice whose circadian rhythm was abnormal and whose lipid consumption was increased. The resulting phenotype was lean, insulin-sensitive, and obesity-resistant.[3] Genetic studies have shown that a polymorphism in the melatonin receptor 1b increases the risk for T2D, but it is not yet known whether genes that regulate circadian rhythm and metabolism are epigenetically changed in humans with T2D.[40,71]

Class III HDACs respond to metabolic factors and the intracellular $NAD^+/NADH$ ratio because they require NAD^+ hydrolysis for their action.[12,35,94] In mammalian cells, mitochondrial-derived cytoplasmic citrate (derived from glucose in the tricarboxylic acid cycle) serves as substrate for ATP-citrate lyase (ACL) to generate acetyl-CoA for lipogenesis and histone acetylation by HATs. Inasmuch as both citrate and acetate freely diffuse through the pore complex of the nucleus where they can interact with ACL, acetyl-CoA is produced in the nucleus, thereby supplying substrate for HAT-mediated acetylation. In the differentiation of preadipocytes to adipocytes, silencing of ACL resulted in adipocytes with smaller lipid droplets, the size of which could be partially restored by incubating cells in acetate. This suggests that ACL-dependent changes in histone acetylation are needed for glucose uptake and metabolism preceding fat storage in adipocytes. Indeed, in ACL-silenced cells, expressions of the glucose transporter GLUT4, of the glycolytic regulators hexokinase 2, phosphofructokinase-1, and lactate dehydrogenase A, and of genes involved in conversion of glucose to fatty acids and nonessential amino acids were diminished. The reduction in GLUT4 expression was coupled to specific reduction of H3 and H4 acetylation at its promoter.[90] This finding again indicates the importance of epigenetic modifications in the regulation of metabolism.

In mice, a high-fat diet decreases the mRNA expression of *Ppargc1a* in skeletal muscle.[78] Incubation of myotubes with long-chain saturated fatty acids decreases *Ppargc1a* mRNA expression and severely decreases its promoter activity. These decreases are reversed when HDAC activity is inhibited.[13]

In diabetic rats with diabetic nephropathy, intermittent fasting prevents both phosphorylation of H3 and a diminished expression of Sir2, a NAD^+-dependent histone deacetylase.[86] This in turn improves blood urea nitrogen, creatinine, albumin, and HDL cholesterol.

A role for the H3K9-specific jumonji histone demethylase 2a (Jhdm2a) in the regulation of metabolic gene expression and weight control has been shown in mice.[85] Disruption of this enzymatic activity caused obesity and hyperlipidemia through interference with β-adrenergic-stimulated glycerol release. It also affected oxygen consumption in brown fat and reduced fat oxidation and glycerol release in skeletal muscle. In skeletal muscle, *Jhdm2a* knockout downregulated genes involved in metabolic processes. One of the targets for Jhdm2a in skeletal muscle is PPARα, an important regulator of fatty acid metabolism. In brown fat, β-adrenergic stimulation induces binding of Jhdm2a to a PPAR-responsive element in the uncoupling protein gene *Ucp1*. This reduces H3K9me2 at this responsive element, thereby facilitating the recruitment of Pparγ and RXRα, of the transcriptional coactivator Pgc-1α and of CBP/p300 and Src1 to the Ppar-responsive element of *Ucp1*. This results in an increased glycerol release and oxygen consumption in the brown fat.

9.5 Exercise and Epigenetics

Physical inactivity is a risk factor for T2D.[18] However, only a limited number of studies have investigated the role of epigenetics in exercise. Skeletal muscle cells take up glucose through an insulin-dependent translocation of the glucose transporter *GLUT4*. In acute exercise, transcription of *GLUT4* increases, as does *GLUT4* protein expression.[20,29,55] The promoter of *GLUT4* contains a transcription factor–binding site for the myocyte enhancer factor 2 (MEF2), which is critical for regulation of *GLUT4* expression.[54] In the resting state, MEF2 interacts with HDAC5, which represses MEF2 action and *GLUT4* expression via deacetylation of histones at the *GLUT4* promoter. The interaction between MEF2 and HDAC5 is disrupted by exercise, possibly by adenosine 5′-monophosphate (AMP)–activated protein kinase (AMPK)[43] or CaMKII.[77] This induces a signal that causes HDAC5 to translocate to the cytosol[46] (Fig. 9.3). After acute exercise, the association between MEF2 and HDAC5 decreases by 24%,

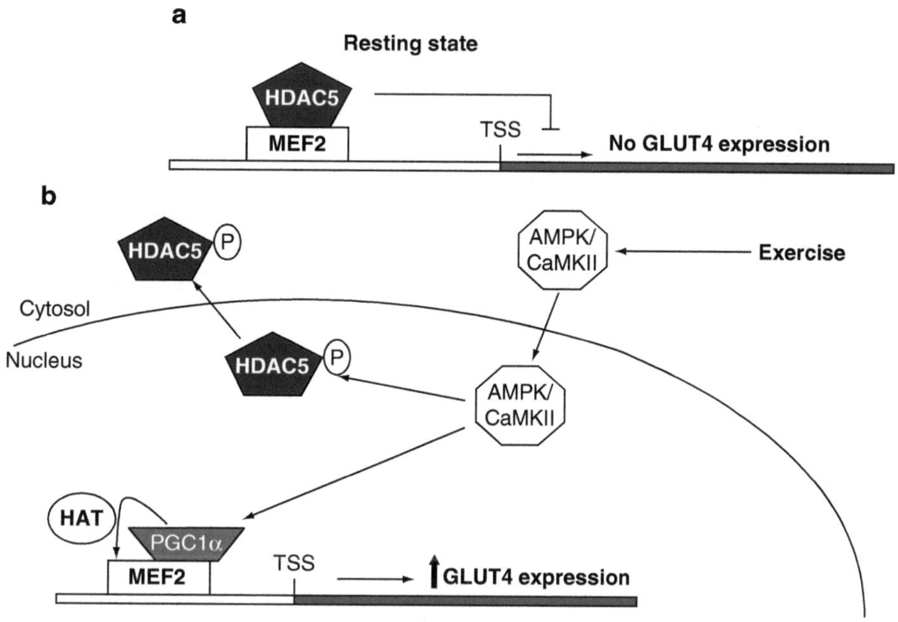

Fig. 9.3 The effect of exercise on mRNA expression of *GLUT4*. (**a**) In the resting state, HDAC5 is associated with MEF2, which inhibits *GLUT4* mRNA expression. (**b**) With exercise, AMPK is activated and relocates into the nucleus. This leads to phosphorylation and removal of HDAC5 from the nucleus and enables PGC-1α to bind to MEF2 and attract HATs to MEF2. This in turn stimulates MEF2 activity and results in increased *GLUT4* mRNA expression (Modified from McGee and Hargreaves [45])

and the amount of HDAC5 in the nucleus is down by 54%.[44] *GLUT4* is expressed in response to exercise when MEF2 associates with HATs. This involves the transcriptional coactivator PGC-1α for subsequent recruitment of HATs to the *GLUT4* promoter.[44] AMPK is activated by exercise.[98] Stimulation of skeletal muscle with AMPK activators or the diabetes drug metformin, which stimulates AMPK, increases PGC-1α protein expression in skeletal muscle.[81,82]

In rat liver, the HDAC Sirt1 deacetylates and functionally activates the PGC-1α protein.[69] In rat skeletal muscle, both PGC-1α and Sirt1 expressions are higher in the slower oxidative red muscle fiber than in the fast glycolytic white muscle fiber. Both acute exercise and endurance exercise increase PGC-1α and Sirt1 protein expressions in the red oxidative fibers. Sirt1 does also affect other metabolic genes in the skeletal muscle cell, possibly contributing to the adaptation of the muscle cell to endurance exercise.[83] Given exercise affects PGC-1α expression in muscle via AMPK and Sirt1 and also because metformin stimulates PGC-1α expression in muscle via AMPK, stimulation of PGC-1α expression may be the mechanism that prevents insulin resistance, partly because PGC-1α modulates mitochondrial biogenesis.

In inactive rats, slow oxidative muscle fibers shift to fast glycolytic fibers. This coincides with changes in H3 acetylation and H3K4me3 of myosin heavy chain genes specific for most fiber types.[58] If exercise changes the redox state of skeletal muscle cells, this would affect the deacetylase activity of Sirt1 and alter gene expression.[16]

9.6 Diabetic Complications and Epigenetic Changes

T2D can lead to secondary complications, i.e., macroangiopathy in large blood vessels of the heart, brain, and legs and microangiopathy in small blood vessels of the kidneys, eyes, and nerves. Cardiovascular complications resulting in heart attack or stroke are the main cause of death in diabetic patients.[51] Duration of the disease and degree of glycemic control are major risk factors for diabetic complications. Yet intensive glucose control for 3–5 years in patients with diabetes did not reduce the risk of macrovascular complications.[17,60,88] This may be due to the long-term, epigenetically induced changes that result from hyperglycemia and persist for more than 5 years.

Vascular inflammation and increased expression of inflammatory genes are major events in the progression of diabetic complications. The transcription factor nuclear factor κ-B (NF-κB) regulates expression of genes involved in inflammatory diseases, including diabetic complications and atherosclerosis.[48] Hyperglycemia induces NF-κB activity and expression of proinflammatory cytokines in monocytes.[25,75] This involves interaction between NF-κB and histone acetyltransferases (e.g., CBP/p300), resulting in hyperacetylation of histones at promoters of target genes, including the tumor necrosis factor-α (*TNF-α*) and *COX-2*.[48] Moreover, the H3K4 methyltransferase SET7/9, stimulates H3K4 methylation and the recruitment of NF-κB p65 to promoters of proinflammatory genes.[31] NF-κB may therefore require histone modifications to induce inflammatory gene expression during hyperglycemia.

The time-averaged mean level of glycemia, measured as HbA1c, explains only in part the variability of risk in developing diabetic complications and the persistence of vascular complications, notwithstanding improved glucose control. Transient hyperglycemia may therefore induce sustained epigenetic changes that act on NF-κB-regulated gene expression and increase the risk for long-term vascular complications.[1,14] These epigenetic changes involve recruitment of SET7 and H3K4 monomethylation in the promoter of the *NF-κB* subunit p65 and subsequent p65 expression and NF-κB activity in aortic endothelial cells. These changes persist for 6 days during subsequent culture of aortic endothelial cells at normal glucose levels; this supports the inference that transient hyperglycemia generates persistent epigenetic marks. Overexpression of genes that reduce mitochondrial superoxide production, e.g., *UCP1*, *MnSOD* or *GLO1*, prevents the epigenetic changes induced by transient hyperglycemia.[14] Moreover, transient hyperglycemia causes LSD1, a histone demethylase, to be recruited to the *NF-κB* p65 promoter, thereby lowering its H3K9 methylation.[7] Indeed, epigenetic modifications induced by transient hyperglycemia may be the explanation for "a hyperglycemic memory."[41,72]

When the chromatin mark H3K9me2, generally associated with reduced gene expression, was mapped in blood lymphocytes and monocytes from patients with T1D and healthy controls, genes involved in inflammation-related pathways, including *NF-κB* and *IL-6,* showed altered H3K9me2 in the cells from the diabetic patients.[49]

Vascular smooth muscle cells from diabetic *db/db* mice have decreased levels of H3K9me3 (an inactive mark) and elevated levels of H3K4me2 (an active mark) at the promoters of inflammatory genes, e.g., *IL-6* and *Mcp-1*, in parallel with decreased levels of the H3K9me3 methyltransferase Suv39h1 and a histone demethylase, the lysine-specific demethylase 1 (Lsd1).[68,89] Interestingly, overexpression of Suv39h1 in vascular smooth muscle cells from diabetic *db/db* mice reversed the diabetic phenotype moreover, gene silencing of *SUV39H1* in normal human vascular smooth muscle cells increased the expression of inflammatory genes.[89]

This opens the possibility that "epigenetic drugs," e.g., HDAC inhibitors, can be used to treat diabetic complications.[6,22,84] Interestingly, myocardial infarction and ischemia induce HDAC activity, in parallel leading to a decrease in histone acetylation of H3 and H4 in the heart.[19] When myocardial infarction was treated with chemical inhibitors of HDAC, the infarct area and cell death were both reduced.[19]

9.7 Outlook

Based on current knowledge, it is evident that epigenetic mechanisms play an important role in the pathogenesis of T2D and its complications. However, we are still only beginning to comprehend which and how epigenetic factors affect T2D. The use of genome-wide technologies to study epigenetic changes in T2D target tissues, both in risk groups before the disease has developed and in patients with T2D, will help identify more candidate genes that are regulated by epigenetic factors. It will also help understand whether epigenetic regulation is a cause or a consequence of the disease.

References

1. The relationship of glycemic exposure (HbA1c) to the risk of development and progression of retinopathy in the diabetes control and complications trial. *Diabetes*. 1995;44(8): 968-983.
2. Aagaard-Tillery KM, Grove K, Bishop J, et al. Developmental origins of disease and determinants of chromatin structure: maternal diet modifies the primate fetal epigenome. *J Mol Endocrinol*. 2008;41(2):91-102.
3. Alenghat T, Meyers K, Mullican SE, et al. Nuclear receptor corepressor and histone deacetylase 3 govern circadian metabolic physiology. *Nature*. 2008;456(7224):997-1000.
4. Barker DJ. The developmental origins of insulin resistance. *Horm Res*. 2005;64(Suppl 3):2-7.
5. Barres R, Osler ME, Yan J, et al. Non-CpG methylation of the PGC-1alpha promoter through DNMT3B controls mitochondrial density. *Cell Metab*. 2009;10(3):189-198.

6. Bieliauskas AV, Pflum MK. Isoform-selective histone deacetylase inhibitors. *Chem Soc Rev*. 2008;37(7):1402-1413.
7. Brasacchio D, Okabe J, Tikellis C, et al. Hyperglycemia induces a dynamic cooperativity of histone methylase and demethylase enzymes associated with gene-activating epigenetic marks that coexist on the lysine tail. *Diabetes*. 2009;58(5):1229-1236.
8. Burdge GC, Lillycrop KA, Phillips ES, Slater-Jefferies JL, Jackson AA, Hanson MA. Folic acid supplementation during the juvenile-pubertal period in rats modifies the phenotype and epigenotype induced by prenatal nutrition. *J Nutr*. 2009;139(6):1054-1060.
9. Chakrabarti SK, Francis J, Ziesmann SM, Garmey JC, Mirmira RG. Covalent histone modifications underlie the developmental regulation of insulin gene transcription in pancreatic beta cells. *J Biol Chem*. 2003;278(26): 23617-23623.
10. Chen H, Gu X, Su IH, et al. Polycomb protein Ezh2 regulates pancreatic beta-cell Ink4a/Arf expression and regeneration in diabetes mellitus. *Genes Dev*. 2009;23(8): 975-985.
11. Clausen TD, Mathiesen ER, Hansen T, et al. Overweight and the metabolic syndrome in adult offspring of women with diet-treated gestational diabetes mellitus or type 1 diabetes. *J Clin Endocrinol Metab*. 2009;94(7):2464-2470.
12. Cohen HY, Miller C, Bitterman KJ, et al. Calorie restriction promotes mammalian cell survival by inducing the SIRT1 deacetylase. *Science*. 2004;305(5682):390-392.
13. Crunkhorn S, Dearie F, Mantzoros C, et al. Peroxisome proliferator activator receptor gamma coactivator-1 expression is reduced in obesity: potential pathogenic role of saturated fatty acids and p38 mitogen-activated protein kinase activation. *J Biol Chem*. 2007;282(21):15439-15450.
14. El-Osta A, Brasacchio D, Yao D, et al. Transient high glucose causes persistent epigenetic changes and altered gene expression during subsequent normoglycemia. *J Exp Med*. 2008;205(10):2409-2417.
15. Fujiki K, Kano F, Shiota K, Murata M. Expression of the peroxisome proliferator activated receptor gamma gene is repressed by DNA methylation in visceral adipose tissue of mouse models of diabetes. *BMC Biol*. 2009;7:38.
16. Fulco M, Schiltz RL, Iezzi S, et al. Sir2 regulates skeletal muscle differentiation as a potential sensor of the redox state. *Mol Cell*. 2003;12(1):51-62.
17. Gerstein HC, Miller ME, Byington RP, et al. Effects of intensive glucose lowering in type 2 diabetes. *N Engl J Med*. 2008;358(24):2545-2559.
18. Gill JM, Cooper AR. Physical activity and prevention of type 2 diabetes mellitus. *Sports Med*. 2008;38(10): 807-824.
19. Granger A, Abdullah I, Huebner F, et al. Histone deacetylase inhibition reduces myocardial ischemia-reperfusion injury in mice. *FASEB J*. 2008;22(10):3549-3560.
20. Greiwe JS, Holloszy JO, Semenkovich CF. Exercise induces lipoprotein lipase and GLUT-4 protein in muscle independent of adrenergic-receptor signaling. *J Appl Physiol*. 2000;89(1):176-181.
21. Groop L, Ling C, Orhu-Melander M. Genetic epidemiology of type 2 diabetes. In: Ekoe J, Williams R, Zimmet P, eds. *The Epidemiology of Diabetes Mellitus*. New York: Wiley; 2008:95-110.
22. Haberland M, Montgomery RL, Olson EN. The many roles of histone deacetylases in development and physiology: implications for disease and therapy. *Nat Rev Genet*. 2009;10(1):32-42.
23. Hales CN, Barker DJ. Type 2 (non-insulin-dependent) diabetes mellitus: the thrifty phenotype hypothesis. *Diabetologia*. 1992;35(7):595-601.
24. Heijmans BT, Tobi EW, Stein AD, et al. Persistent epigenetic differences associated with prenatal exposure to famine in humans. *Proc Natl Acad Sci USA*. 2008;105(44): 17046-17049.
25. Hofmann MA, Schiekofer S, Kanitz M, et al. Insufficient glycemic control increases nuclear factor-kappa B binding activity in peripheral blood mononuclear cells isolated from patients with type 1 diabetes. *Diab Care*. 1998;21(8): 1310-1316.
26. Jiang MH, Fei J, Lan MS, et al. Hypermethylation of hepatic Gck promoter in ageing rats contributes to diabetogenic potential. *Diabetologia*. 2008;51(8):1525-1533.
27. Jin CJ, Park HK, Cho YM, et al. S-adenosyl-L-methionine increases skeletal muscle mitochondrial DNA density and whole body insulin sensitivity in OLETF rats. *J Nutr*. 2007;137(2):339-344.
28. Kelley DE, He J, Menshikova EV, Ritov VB. Dysfunction of mitochondria in human skeletal muscle in type 2 diabetes. *Diabetes*. 2002;51(10):2944-2950.
29. Kraniou Y, Cameron-Smith D, Misso M, Collier G, Hargreaves M. Effects of exercise on GLUT-4 and glycogenin gene expression in human skeletal muscle. *J Appl Physiol*. 2000;88(2):794-796.
30. Kuroda A, Rauch TA, Todorov I, et al. Insulin gene expression is regulated by DNA methylation. *PLoS ONE*. 2009;4(9):e6953.
31. Li Y, Reddy MA, Miao F, et al. Role of the histone H3 lysine 4 methyltransferase, SET7/9, in the regulation of NF-kappaB-dependent inflammatory genes. Relevance to diabetes and inflammation. *J Biol Chem*. 2008;283(39): 26771-26781.
32. Lillycrop KA, Slater-Jefferies JL, Hanson MA, Godfrey KM, Jackson AA, Burdge GC. Induction of altered epigenetic regulation of the hepatic glucocorticoid receptor in the offspring of rats fed a protein-restricted diet during pregnancy suggests that reduced DNA methyltransferase-1 expression is involved in impaired DNA methylation and changes in histone modifications. *Br J Nutr*. 2007;97(6): 1064-1073.
33. Lillycrop KA, Phillips ES, Torrens C, Hanson MA, Jackson AA, Burdge GC. Feeding pregnant rats a protein-restricted diet persistently alters the methylation of specific cytosines in the hepatic PPAR alpha promoter of the offspring. *Br J Nutr*. 2008;100(2):278-282.
34. Lin J, Handschin C, Spiegelman BM. Metabolic control through the PGC-1 family of transcription coactivators. *Cell Metab*. 2005;1(6):361-370.
35. Lin SJ, Guarente L. Nicotinamide adenine dinucleotide, a metabolic regulator of transcription, longevity and disease. *Curr Opin Cell Biol*. 2003;15(2):241-246.
36. Lindsay RS, Bennett PH. Type 2 diabetes, the thrifty phenotype - an overview. *Br Med Bull*. 2001;60:21-32.
37. Ling C, Poulsen P, Simonsson S, et al. Genetic and epigenetic factors are associated with expression of respiratory

chain component NDUFB6 in human skeletal muscle. *J Clin Invest.* 2007;117(11):3427-3435.
38. Ling C, Del Guerra S, Lupi R, et al. Epigenetic regulation of PPARGC1A in human type 2 diabetic islets and effect on insulin secretion. *Diabetologia.* 2008;51(4):615-622.
39. Liu Y, Dentin R, Chen D, et al. A fasting inducible switch modulates gluconeogenesis via activator/coactivator exchange. *Nature.* 2008;456(7219):269-273.
40. Lyssenko V, Nagorny CL, Erdos MR, et al. Common variant in MTNR1B associated with increased risk of type 2 diabetes and impaired early insulin secretion. *Nat Genet.* 2009;41(1):82-88.
41. Mack CP. An epigenetic clue to diabetic vascular disease. *Circ Res.* 2008;103(6):568-570.
42. McCarthy MI, Zeggini E. Genome-wide association studies in type 2 diabetes. *Curr Diab Rep.* 2009;9(2):164-171.
43. McGee SL, Howlett KF, Starkie RL, Cameron-Smith D, Kemp BE, Hargreaves M. Exercise increases nuclear AMPK alpha2 in human skeletal muscle. *Diabetes.* 2003;52(4):926-928.
44. McGee SL, Hargreaves M. Exercise and myocyte enhancer factor 2 regulation in human skeletal muscle. *Diabetes.* 2004;53(5):1208-1214.
45. McGee SL, Hargreaves M. Exercise and skeletal muscle glucose transporter 4 expression: molecular mechanisms. *Clin Exp Pharmacol Physiol.* 2006;33(4):395-399.
46. McKinsey TA, Zhang CL, Olson EN. Activation of the myocyte enhancer factor-2 transcription factor by calcium/calmodulin-dependent protein kinase-stimulated binding of 14-3-3 to histone deacetylase 5. *Proc Natl Acad Sci USA.* 2000;97(26):14400-14405.
47. Melzner I, Scott V, Dorsch K, et al. Leptin gene expression in human preadipocytes is switched on by maturation-induced demethylation of distinct CpGs in its proximal promoter. *J Biol Chem.* 2002;277(47):45420-45427.
48. Miao F, Gonzalo IG, Lanting L, Natarajan R. In vivo chromatin remodeling events leading to inflammatory gene transcription under diabetic conditions. *J Biol Chem.* 2004;279(17):18091-18097.
49. Miao F, Smith DD, Zhang L, Min A, Feng W, Natarajan R. Lymphocytes from patients with type 1 diabetes display a distinct profile of chromatin histone H3 lysine 9 dimethylation: an epigenetic study in diabetes. *Diabetes.* 2008;57(12):3189-3198.
50. Milagro FI, Campion J, Garcia-Diaz DF, Goyenechea E, Paternain L, Martinez JA. High fat diet-induced obesity modifies the methylation pattern of leptin promoter in rats. *J Physiol Biochem.* 2009;65(1):1-9.
51. Milicevic Z, Raz I, Beattie SD, et al. Natural history of cardiovascular disease in patients with diabetes: role of hyperglycemia. *Diab Care.* 2008;31(Suppl 2):S155-S160.
52. Mootha VK, Lindgren CM, Eriksson KF, et al. PGC-1alpha-responsive genes involved in oxidative phosphorylation are coordinately downregulated in human diabetes. *Nat Genet.* 2003;34(3):267-273.
53. Mutskov V, Raaka BM, Felsenfeld G, Gershengorn MC. The human insulin gene displays transcriptionally active epigenetic marks in islet-derived mesenchymal precursor cells in the absence of insulin expression. *Stem Cells.* 2007;25(12):3223-3233.
54. Neufer PD, Carey JO, Dohm GL. Transcriptional regulation of the gene for glucose transporter GLUT4 in skeletal muscle. Effects of diabetes and fasting. *J Biol Chem.* 1993;268(19):13824-13829.
55. Neufer PD, Dohm GL. Exercise induces a transient increase in transcription of the GLUT-4 gene in skeletal muscle. *Am J Physiol.* 1993;265(6 Pt 1):C1597-C1603.
56. Newman B, Selby JV, King MC, Slemenda C, Fabsitz R, Friedman GD. Concordance for type 2 (non-insulin-dependent) diabetes mellitus in male twins. *Diabetologia.* 1987;30(10):763-768.
57. Noer A, Sorensen AL, Boquest AC, Collas P. Stable CpG hypomethylation of adipogenic promoters in freshly isolated, cultured, and differentiated mesenchymal stem cells from adipose tissue. *Mol Biol Cell.* 2006;17(8):3543-3556.
58. Pandorf CE, Haddad F, Wright C, Bodell PW, Baldwin KM. Differential epigenetic modifications of histones at the myosin heavy chain genes in fast and slow skeletal muscle fibers and in response to muscle unloading. *Am J Physiol Cell Physiol.* 2009;297(1):C6-C16.
59. Park JH, Stoffers DA, Nicholls RD, Simmons RA. Development of type 2 diabetes following intrauterine growth retardation in rats is associated with progressive epigenetic silencing of Pdx1. *J Clin Invest.* 2008;118(6):2316-2324.
60. Patel A, MacMahon S, Chalmers J, et al. Intensive blood glucose control and vascular outcomes in patients with type 2 diabetes. *N Engl J Med.* 2008;358(24):2560-2572.
61. Patti ME, Butte AJ, Crunkhorn S, et al. Coordinated reduction of genes of oxidative metabolism in humans with insulin resistance and diabetes: potential role of PGC1 and NRF1. *Proc Natl Acad Sci USA.* 2003;100(14):8466-8471.
62. Petersen KF, Befroy D, Dufour S, et al. Mitochondrial dysfunction in the elderly: possible role in insulin resistance. *Science.* 2003;300(5622):1140-1142.
63. Pogribny IP, Tryndyak VP, Bagnyukova TV, et al. Hepatic epigenetic phenotype predetermines individual susceptibility to hepatic steatosis in mice fed a lipogenic methyl-deficient diet. *J Hepatol.* 2009;51(1):176-186.
64. Poirier LA, Brown AT, Fink LM, et al. Blood S-adenosylmethionine concentrations and lymphocyte methylenetetrahydrofolate reductase activity in diabetes mellitus and diabetic nephropathy. *Metabolism.* 2001;50(9):1014-1018.
65. Ravelli AC, van der Meulen JH, Michels RP, et al. Glucose tolerance in adults after prenatal exposure to famine. *Lancet.* 1998;351(9097):173-177.
66. Ravelli AC, van Der Meulen JH, Osmond C, Barker DJ, Bleker OP. Obesity at the age of 50 y in men and women exposed to famine prenatally. *Am J Clin Nutr.* 1999;70(5):811-816.
67. Ravelli GP, Stein ZA, Susser MW. Obesity in young men after famine exposure in utero and early infancy. *N Engl J Med.* 1976;295(7):349-353.
68. Reddy MA, Villeneuve LM, Wang M, Lanting L, Natarajan R. Role of the lysine-specific demethylase 1 in the proinflammatory phenotype of vascular smooth muscle cells of diabetic mice. *Circ Res.* 2008;103(6):615-623.
69. Rodgers JT, Lerin C, Haas W, Gygi SP, Spiegelman BM, Puigserver P. Nutrient control of glucose homeostasis through a complex of PGC-1alpha and SIRT1. *Nature.* 2005;434(7029):113-118.
70. Ronn T, Poulsen P, Hansson O, et al. Age influences DNA methylation and gene expression of COX7A1 in human skeletal muscle. *Diabetologia.* 2008;51(7):1159-1168.

71. Ronn T, Wen J, Yang Z, et al. A common variant in MTNR1B, encoding melatonin receptor 1B, is associated with type 2 diabetes and fasting plasma glucose in Han Chinese individuals. *Diabetologia.* 2009;52:830-833.
72. Roy S, Sala R, Cagliero E, Lorenzi M. Overexpression of fibronectin induced by diabetes or high glucose: phenomenon with a memory. *Proc Natl Acad Sci USA.* 1990;87(1):404-408.
73. Saxena R, Voight BF, Lyssenko V, et al. Genome-wide association analysis identifies loci for type 2 diabetes and triglyceride levels. *Science.* 2007;316(5829):1331-1336.
74. Scott LJ, Mohlke KL, Bonnycastle LL, et al. A genome-wide association study of type 2 diabetes in Finns detects multiple susceptibility variants. *Science.* 2007;316(5829):1341-1345.
75. Shanmugam N, Reddy MA, Guha M, Natarajan R. High glucose-induced expression of proinflammatory cytokine and chemokine genes in monocytic cells. *Diabetes.* 2003;52(5):1256-1264.
76. Sinclair KD, Allegrucci C, Singh R, et al. DNA methylation, insulin resistance, and blood pressure in offspring determined by maternal periconceptional B vitamin and methionine status. *Proc Natl Acad Sci USA.* 2007;104(49):19351-19356.
77. Smith JA, Kohn TA, Chetty AK, Ojuka EO. CaMK activation during exercise is required for histone hyperacetylation and MEF2A binding at the MEF2 site on the Glut4 gene. *Am J Physiol Endocrinol Metab.* 2008;295(3):E698-E704.
78. Sparks LM, Xie H, Koza RA, et al. A high-fat diet coordinately downregulates genes required for mitochondrial oxidative phosphorylation in skeletal muscle. *Diabetes.* 2005;54(7):1926-1933.
79. Stein AD, Kahn HS, Rundle A, Zybert PA, van der Pal-de Bruin K, Lumey LH. Anthropometric measures in middle age after exposure to famine during gestation: evidence from the Dutch famine. *Am J Clin Nutr.* 2007;85(3):869-876.
80. Stoger R. In vivo methylation patterns of the leptin promoter in human and mouse. *Epigenetics.* 2006;1(4):155-162.
81. Suwa M, Nakano H, Kumagai S. Effects of chronic AICAR treatment on fiber composition, enzyme activity, UCP3, and PGC-1 in rat muscles. *J Appl Physiol.* 2003;95(3):960-968.
82. Suwa M, Egashira T, Nakano H, Sasaki H, Kumagai S. Metformin increases the PGC-1alpha protein and oxidative enzyme activities possibly via AMPK phosphorylation in skeletal muscle in vivo. *J Appl Physiol.* 2006;101(6):1685-1692.
83. Suwa M, Nakano H, Radak Z, Kumagai S. Endurance exercise increases the SIRT1 and peroxisome proliferator-activated receptor gamma coactivator-1alpha protein expressions in rat skeletal muscle. *Metabolism.* 2008;57(7):986-998.
84. Szyf M. Epigenetics, DNA methylation, and chromatin modifying drugs. *Annu Rev Pharmacol Toxicol.* 2009;49:243-263.
85. Tateishi K, Okada Y, Kallin EM, Zhang Y. Role of Jhdm2a in regulating metabolic gene expression and obesity resistance. *Nature.* 2009;458(7239):757-761.
86. Tikoo K, Tripathi DN, Kabra DG, Sharma V, Gaikwad AB. Intermittent fasting prevents the progression of type I diabetic nephropathy in rats and changes the expression of Sir2 and p53. *FEBS Lett.* 2007;581(5):1071-1078.
87. Tobi EW, Lumey LH, Talens RP, et al. DNA methylation differences after exposure to prenatal famine are common and timing- and sex-specific. *Hum Mol Genet.* 2009;18(21):4046-4053.
88. Turner RC, Holman RR. Lessons from UK prospective diabetes study. *Diab Res Clin Pract.* 1995;28(Suppl):S151-S157.
89. Villeneuve LM, Reddy MA, Lanting LL, Wang M, Meng L, Natarajan R. Epigenetic histone H3 lysine 9 methylation in metabolic memory and inflammatory phenotype of vascular smooth muscle cells in diabetes. *Proc Natl Acad Sci USA.* 2008;105(26):9047-9052.
90. Wellen KE, Hatzivassiliou G, Sachdeva UM, Bui TV, Cross JR, Thompson CB. ATP-citrate lyase links cellular metabolism to histone acetylation. *Science.* 2009;324(5930):1076-1080.
91. Wild S, Roglic G, Green A, Sicree R, King H. Global prevalence of diabetes: estimates for the year 2000 and projections for 2030. *Diab Care.* 2004;27(5):1047-1053.
92. Wong RH, Chang I, Hudak CS, Hyun S, Kwan HY, Sul HS. A role of DNA-PK for the metabolic gene regulation in response to insulin. *Cell.* 2009;136(6):1056-1072.
93. Wren JD, Garner HR. Data-mining analysis suggests an epigenetic pathogenesis for type 2 diabetes. *J Biomed Biotechnol.* 2005;2005(2):104-112.
94. Yang H, Yang T, Baur JA, et al. Nutrient-sensitive mitochondrial NAD+ levels dictate cell survival. *Cell.* 2007;130(6):1095-1107.
95. Yokomori N, Tawata M, Onaya T. DNA demethylation during the differentiation of 3T3-L1 cells affects the expression of the mouse GLUT4 gene. *Diabetes.* 1999;48(4):685-690.
96. Yokomori N, Tawata M, Onaya T. DNA demethylation modulates mouse leptin promoter activity during the differentiation of 3T3-L1 cells. *Diabetologia.* 2002;45(1):140-148.
97. Zeggini E, Weedon MN, Lindgren CM, et al. Replication of genome-wide association signals in UK samples reveals risk loci for type 2 diabetes. *Science.* 2007;316(5829):1336-1341.
98. Zhang BB, Zhou G, Li C. AMPK: an emerging drug target for diabetes and the metabolic syndrome. *Cell Metab.* 2009;9(5):407-416.

10 Epigenetic Regulation of Asthma and Allergic Diseases

Andrew L. Durham and Ian M. Adcock

10.1 Introduction

When the human body encounters a foreign antigen, the immune response uses a variety of mechanisms to remove that antigen. These mechanisms include, among others, localized inflammation (characterized by rubor, calor, dolor, tumor, and loss of function),[27] immune cell recruitment, and antibody production.[70] Inappropriate immune responses lead to diseases, including autoimmunity and hypersensitivity.[70] There are four classes of hypersensitive responses: the IgE mediated (type I), IgG or IgM mediated (type II), immune complex mediated (type III), and cell mediated (type IV).[47,70]

The IgE-mediated type I hypersensitivity reaction is responsible for allergies (e.g., seafood, peanut, and bee sting venom), anaphylactic shock,[41] and asthma. Type 1 hypersensitivity is induced by specific antigens called allergens resulting in B cell maturation, antibody induction, and the production of memory B cells.[5] The difference between type 1 hypersensitivity and the normal humoral immune response is the production of IgE (normally produced in response to parasitic infection), as opposed to IgM and IgG, in response to the activation of allergen-specific type 2 T helper (T_H2) cells.[5,51] IgE binds with high affinity to the Fc receptors on granulocytes, which become sensitized (Fig. 10.1a).[34,91] Subsequent exposure of the allergen to sensitized cells results in the cross-linking of the IgE on the cell's surface.[91] This triggers the release of inflammatory mediators and, in the case of asthma bronchoconstrictor agents, such as histamines, leukotrienes (LTs), and prostaglandins (PGs) into the surrounding tissues[91,98] (Fig. 10.1b).

The release of these powerful mediators can produce localized or systemic responses.[9] Systemic responses, when severe, can lead to life-threatening anaphylactic shock, for example, in response to peanut allergy and bee stings.[41,111,120] The localized responses (atopy) are generally less severe and their symptoms vary depending on the organs affected; from diarrhea and vomiting due to food allergy, runny noses and eyes in allergic rhinitis (hay fever), to more severe skin allergies (atopic dermatitis). The response can even be life threatening when affecting vital organs such as the lungs in asthma, the clinical features of which can include intermittent wheezing, dyspnea, cough, and chest tightness.[40,62,71,110]

Of the allergic reactions, asthma is the most important due to its life-threatening symptoms and high prevalence.[70] The World Health Organization (WHO) estimates that 300 million people suffer from asthma and that 255,000 people died of the disease in 2005.[89] By 2025 the number of asthma sufferers may increase by 100 million.[89] Asthma attacks are triggered by allergens, such as pollen, dust mites, and fumes (allergic asthma), and in about 10% of cases by other factors, such as exercise or cold, which are independent of the presence of allergen (intrinsic asthma).[59] All patients with asthma have a specific pattern of inflammation in the airways, characterized by degranulated mast cells, an infiltration of eosinophils, and an increased number of activated T_H2 cells.[62] An asthma attack is triggered by allergens, cold air, exercise, or environmental pollutants that cause mast cell degranulation in the lower respiratory tract. The release of histamine, LTs (particularly LTC_4), and PGs from mast cells results in the contraction of bronchial smooth muscle and

A.L. Durham (✉)
Department of Airways Disease, National Heart and Lung Institute, Imperial College London, Dovehouse Street, London SW3 6LY, UK
e-mail: a.durham@imperial.ac.uk

H.I. Roach et al. (eds.), *Epigenetic Aspects of Chronic Diseases*,
DOI: 10.1007/978-1-84882-644-1_10,

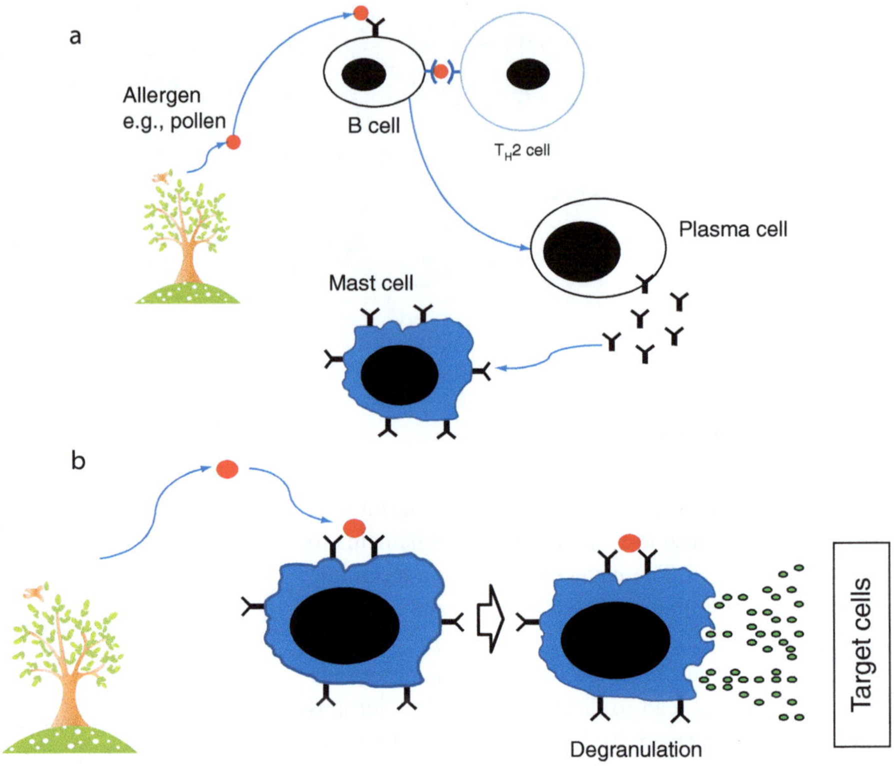

Fig. 10.1 (**a**) Allergic sensitization of mast cells. An allergen (such as tree pollen) is recognized by B cells, and presented to T_H2 cells. This causes the B cell to mature into a plasma cell and to produce IgE antibodies. The IgE antibodies are bound by mast cells and basophils, resulting in sensitization of these cells. (**b**) Mast cell degranulation. Repeated exposure to the allergen results in the IgE cross-linking on the surface of the mast cells. Cross-linking results in degranulation of the granulocytes cells, releasing histamines, leukotrines, prostaglandins, etc., all of which act as effector molecules on the surrounding tissues, smooth muscle (resulting in contraction), blood vessels (vasodilatation), mucus-producing cells (hypersecretion), and eosinophils

bronchoconstriction.[21,98] Edema, mucus hypersecretion, and inflammation also contribute to the severity of the airway obstruction.[39,104] Current asthma treatments include the use of controller drugs (e.g., glucocorticoids) that prevent airway inflammation and of relievers such as beta agonists that help the airway smooth muscle to relax after contraction.[14]

10.2 Development of Allergy

The increasing levels of asthma prevalence worldwide has been especially noticeable in the developed world,[11,24,117] as in the USA, where asthma prevalence has increased from 3.6% in 1980 to 10% in 2002.[46] The cause of this increase is unclear and may be multifactorial, including changes in diet, lifestyle (the hygiene hypothesis), and better diagnosis.[86] Genetics, however, plays an important role in the asthma development, even though genome-wide analysis (GWAS) has not delineated a single key asthma gene.[126] A child is more likely to suffer from asthma or other allergies if one or both parents have the disease. Asthma in these children is often associated with atopic dermatitis.[85] Asthma does not follow a Mendelian pattern of inheritance.[72] Alleles inherited from the mother may give rise to a type of asthma distinct from that inherited from the father, possibly due to imprinting.[127,130] In general, people are thought to inherit a tendency to allergy, rather than a specific allergy.

10.2.1 Maternal Inheritance of Risk of Asthma

Maternal inheritance may be more important for the development of asthma than paternal inheritance.[35] In children under 5 years of age, the risk associated with maternal asthma is more than three times greater than the risk associated with paternal asthma.[79] Childhood atopy is also strongly linked with maternal asthma and with hay fever in either parent.[108] Studies of the mechanisms of maternal inheritance have revealed several candidate genes. For example, polymorphisms in Fcε(epsilon)RI-β(beta) (the β-chain of the high-affinity receptor for IgE) have shown stronger association with positive skin prick tests and greater allergen-specific IgE levels when inherited from the mother.[35,40,107,119] Another example of maternal inheritance is *HLA-G*, a novel human leukocyte antigen (HLA) gene[78] that is expressed predominantly at the maternal–neonatal interface and is involved in immunomodulation, down-regulating NK and T cells.[78] *HLA-G* has been linked to asthma[95] and the *HLA-G* allele is overexpressed in children with bronchial hyperresponsiveness if the mother is also affected by bronchial hyperresponsiveness.[95] The mechanism of maternal inheritance in asthma is not known, but may be due to allelic imprinting or other epigenetic mechanisms.

10.3 Epigenetic Mechanisms

Gene regulation has been thought as solely due to DNA of the gene and its promoter regions. It is now clear, however, that factors other than changes in the genetic code determine cell function.[3] Moreover, polymorphisms in the genetic code cannot account for all variations in the expression of specific genes. Differences may be due to epigenetic modifications.

Epigenetic modifications can alter the structure of DNA, such as DNA methylation, or alter the structure of chromatin through alterations to scaffolding proteins, such as histones.[74] DNA methylation and chromatin modifications are not only critical for generating cellular diversity during development, but also for maintaining distinct gene expression profiles.[2] DNA methylation is a reversible modification of DNA structure, adding a methyl group to a cytosine, often part of a CpG island or cluster. Methylation results in gene silencing.[74] Cytosine methylation occurs also at sites distinct from CpG islands (CpG shores).[60] As DNA methylation is affected by diet, stress, and other environmental factors – including heavy metals, pesticides, diesel exhaust, and tobacco smoke – it is one mechanism to explain how dietary and environmental risk factors contribute to allergy development.[16]

The other main aspect of epigenetics is alteration to the structure of chromatin, specifically the nucleosome, a sequence of DNA that encompasses 146 base pairs (bp) and is wrapped around an octomer of core histone proteins (2 molecules each of H2A, H2B, H3, and H4).[3,74] Nucleosomes are compacted together to form 30 nM chromatin fibers which, in turn, are compacted in the chromosome structure.[74] The local chromatin structure plays an important role in controlling gene expression.[74] In general, acetylation of histones H3 and H4, along with histone H3 lysine (K)4 methylation, leads to the formation of tags that allow subsequent recruitment of other transcriptional complexes, loosening the local nucleosomal structure and allowing gene transcription to occur.[125] These changes occur predominantly at the transcriptional start site,[105,124] are closely associated with transcription factor DNA binding and gene expression,[53] and appear invariant across many cell types, in contrast to the greater variability seen in histone modifications in enhancer regions.[53] Less-frequently transcribed or silenced genes have repressive marks and remain more tightly packaged and therefore less accessible to the transcriptional mechanisms of the cell.[74,125] In addition, the activating H3K4 methylation mark is associated with poised transcription, whereby RNA polymerase II is located at the transcriptional start site, but does not have the correct acetylation marks to allow transcription to occur. Additional stimuli may be needed to induce a rapid response.[124]

The changes in DNA methylation and histone modifications are maintained through many subsequent cell divisions, leading to specific gene expression patterns that are determined by the epigenetic profile.[74] The variable histories of asthma patients (i.e., development, incidence, and remission) may, therefore, be the result of epigenetic changes due to environmental exposures.[92] Prenatal development, early childhood, and adolescence are the times when a person is most susceptible to the environmental exposures that trigger asthma and other allergies.[48,83]

10.4 Epigenetic Imprinting

During gametogenesis, the existing pattern of DNA methylation is removed by a genome-wide demethylation event and a methylation pattern determined by the parent's sex is imposed.[74] For the male parent the methylation pattern specific for sperm is established in the spermatocyte, with additional changes occurring after fertilization. For the female the maternal pattern is established during the meiotic stages of oogenesis. In both sexes the genes are predominantly methylated and silenced.[74]

Subsequent to fertilization, the CpG sites in the promoters of genes are demethylated in cells where the gene is to be expressed. The specific patterns of methylation result in the phenomenon of imprinting where specific alleles inherited from the parents are expressed or silenced, irrespective of the sequence of the DNA.[74] Whether the mechanism of imprinting explains the parent of origin effect in the inheritance of asthma is not known.

10.5 Epigenetic Regulation of the Immune Response

The role of DNA methylation in the immune response is especially important in the regulation of the major histocompatibility (MHC) genes that are involved in antigen presentation.[122] The region encoding the MHC on chromosome 6 has been implicated in asthma and atopy.[93] As MHC has such a wide-ranging role in the innate and adaptive immune response pattern it has been difficult to identify candidate genes. Two genes that have been implicated are *HLA-DRB1* and the maternally associated *HLA-G* genes.[93,95]

DNA methylation often acts in concert with histone modifications to regulate gene function by altering chromatin structure, stability, and accessibility to the transcriptional machinery.[25] Histone modification is a dynamic process with numerous moieties, including acetyl, methyl, and phospho groups that are added or removed from specific lysine, arginine, or serine residues by distinct enzymes. One of the key modifications associated with inflammatory gene induction and cell proliferation is histone acetylation.[3] However, the histone code hypothesis proposes that different combinations of histone modifications may bring about different outcomes, as far as chromatin-regulated functions are concerned.[3]

Transcriptional co-activators such as CBP, SRC-1, TIF2, GRIP-1, and p300/CBP-associated factor (PCAF) have intrinsic histone acetyltransferase (HAT) activity. An increase in inflammatory/immune gene transcription is associated with an increase in histone acetylation at transcriptional start sites.[3,124] As detailed in Chap. 4, changes in the acetylation status of specific lysine residues in the N-terminal tails of histones 3 and 4 form a molecular tag for the recruitment of ATP-dependent chromatin remodeling enzymes such as Brg1.[17] This allows local chromatin unwinding and the recruitment of other transcription factors, the basal transcriptional complex and RNA polymerase II. This mechanism is common to all genes, including those involved in differentiation, proliferation, DNA repair, and activation of cells.[3] In particular, pro-inflammatory transcription factors such as AP-1 and NF-κB require recruitment of HAT-containing complexes in order to induce inflammatory and immunomodulatory genes.[3]

With the aid of ChIP-seq (chromatin immunoprecipitation linked to high throughput sequencing) HDACs have been shown to be also associated with actively transcribed genes.[124] The relative level of HATs and HDACs at specific promoters allows active gene expression or can keep the gene silenced until subsequent transcription. It is therefore evident that the model of epigenetic action has become more complex.

In contrast to the activation effect of acetylation, the removal of the acetyl group from the nucleosome, often in conjunction with a change in DNA methylation status, is associated with gene repression.[3] HDACs interact with corepressor molecules, which aid HDACs in gene repression and may provide specificity by selecting which genes are regulated by individual HDACs. The balance between HAT and HDAC activity is a key component of gene expression, including inflammatory gene expression. Evidence for a key role of HDACs in inflammatory gene expression is provided by the fact that HDAC inhibitors such as trichostatin A (TSA) enhance NF-κB-driven inflammatory gene transcription in vitro in different cell types, in response to many distinct stimuli.[3] The development of selective HAT inhibitors such as anacardic acid provides further evidence for the key role played by HAT activity in inflammatory gene expression in primary human airway cells.[128]

10.6 Histone Acetylation and Inflammatory Gene Expression

Changes in histone acetylation induce pro-inflammatory genes in human lung epithelial and smooth muscle cells.[3] In human airway smooth muscle (HASM) cells TNFα-stimulated eotaxin release is associated with NF-κB binding and histone H4 acetylation of the *eotaxin* gene promoter region.[96] In contrast, IFNγ markedly inhibits TNFα-induced expression of the NF-κB-sensitive genes *IL-6*, *IL-8*, and *eotaxin*.[69] IFNγ decreases TNFα-induced p65-associated HAT activity and increases total nuclear HDAC activity. Importantly, the HDAC inhibitor TSA prevents the inhibitory effect of IFNγ on TNFα-induced gene expression. Therefore, inhibition of TNFα-induced cytokine and chemokine expression is mediated through HDAC activity and/or expression in HASM cells.[69]

10.7 Nonhistone Substrates of HDACS

The view that HDAC enzymes act as gene repressors by targeting acetylated histones cannot be totally correct because as many genes are suppressed by non-selective HDAC inhibitors as are enhanced.[3] An expanding number of nonhistone proteins are substrates for HDACs, including transcription factors, cytoplasmic signaling, and structural proteins.[3] Indeed, from a phylogenetic viewpoint, HDACs existed prior to histones.[129] This means that the stability, localization, protein dimerization, and protein–protein interaction of these acetylated nonhistone proteins can be altered. The HATs involved in acetylating these nonhistone proteins are often known, but specific protein deacetylases that regulate the reverse process are not.

10.8 HATS and HDACS in Asthma and Allergic Disease

Allergic diseases are characterized by a T_H2-dominant immune response, with HDACs maintaining preestablished T_H1-like and T_H2-like immunity in human T cells.[112] Phytohemagglutinin (PHA) activation selectively stimulates antigen-driven CD45RO (+) memory T cells, eliciting recall cytokine responses.[112] The HDAC inhibitor TSA provokes total cell hyperacetylation and leads to an increase in T_H2-associated IL-13 and IL-5 cytokine expression and to a reduction in T_H1-associated IFNγ and IP10-associated recall responses. In addition, IL-2 and IL-10 production is also reduced. TSA treatment shifts the $T_H1{:}T_H2$ ratio three- to eight-fold, skewing the recall responses more toward a T_H2-like phenotype, independent of the stimulus and highlighting the role of HDACs in the development and maintenance of $T_H1{:}T_H2$ immunity.[112]

Differential expression and activity of HATs and HDACs favors gene induction in bronchial biopsies of asthmatic patients.[61] HAT/HDAC ratios are higher in peripheral blood cells of patients with more severe disease than in controls. This is true in adults[54] and children.[113] In asthmatic children, increasing bronchial hyperresponsiveness enhances this difference.[113]

Overall, inflammatory stimuli including environmental agents, viral infection, and allergen exposure can lead to enhanced inflammatory gene expression, with T_H2 skewed due to changes in histone acetylation. Enhanced histone acetylation is associated with inflammation; reduced acetylation is associated with decreased inflammation.

10.9 Environmental Factors Altering the Epigenetic Profile

Epigenetic changes do not alter the underlying genetic code of the person, but affect a cell's transcriptional program in response to environmental challenge. The changes are reversible throughout life.[19] Environmental exposure in utero may drive organ-specific gene programming that determines subsequent physiological changes in adult life.[10] Alteration of this programming due to environment stress, including diet, may lead to greater risk of disease in later life.[52] In asthma and other allergic diseases, the alteration of the epigenetic profile is important, particularly in utero, when it may be associated with predisposition to childhood wheeze and subsequent disease (Fig. 10.2).

10.9.1 Effect of Diet on Epigenetics and Allergy

Asthma is becoming more prevalent in the developing world such as India and is associated with the adoption of Western lifestyles. Parts of India have shown a 50% increase in the incidence of asthma, although the more traditional communities have been the least affected.[123] Diet plays an important role in the development and maintenance of epigenetic profiles. For example, a study of 300 Swedish families demonstrated that the grandchildren of males who enjoyed a "surplus of food" during a national period of famine had significantly higher levels of diabetes and heart disease.[67]

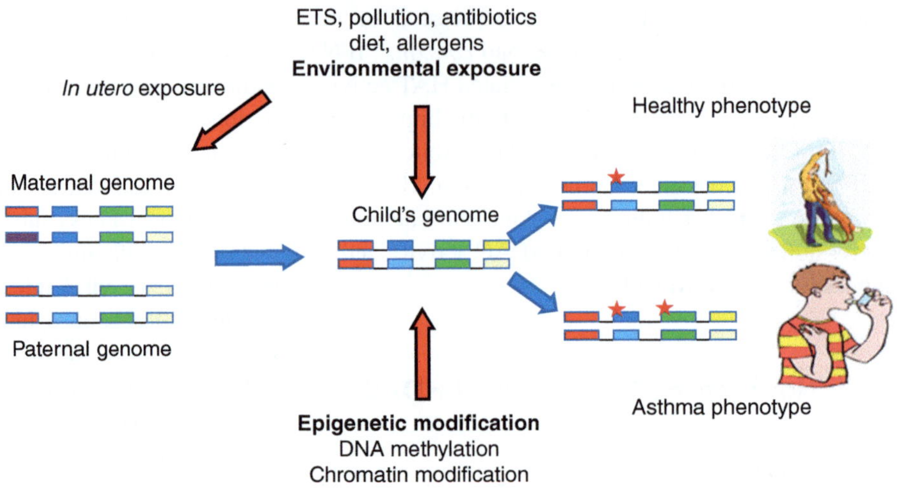

Fig. 10.2 Factors effecting epigenetics. Environmental exposures can alter the epigenetic profile of an individual resulting in a different phenotype without alteration of the genetic profile. The epigenetic changes can be altered by both external environmental factors such as Environmental tobacco smoke (ETS), and by internal factors, such as genetic imprinting. Repression and expression of different genes lead to altered phenotypes, in this case either asthmatic or non-asthmatic (Adapted from Miller and Ho[92])

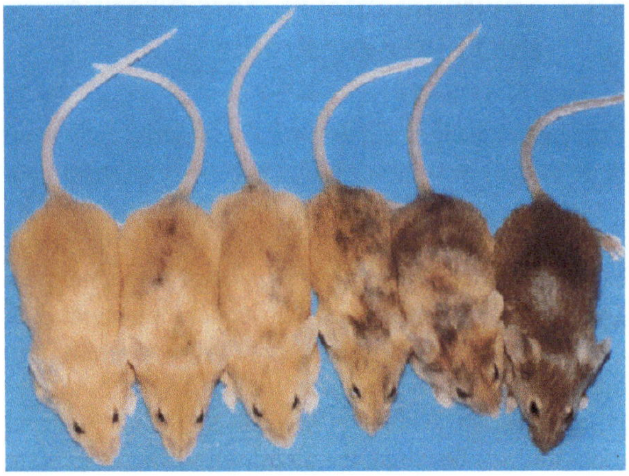

Fig. 10.3 Agouti mice. Genetically identical mice show different coat patterns due to the different levels of agouti gene expression caused by a change in the maternal diet (Reprinted Morgan et al.[94] With permission from Macmillan Publishers Ltd, copyright 1999)

To maintain normal DNA methylation patterns, several essential nutrients are required from the diet, including a source of methyl groups (e.g., methionine or choline) and folate. Folic acid (vitamin B_9) affects the heritable phenotype of the transgenic agouti mouse, named after its mottled yellow coat color.[94] Offspring fed a folic acid supplemented diet had normal coat color and adequate health, because of increased methylation of the CpG island upstream of the *agouti* gene, which determines coat color[36] (Fig. 10.3).

The same pattern maternal diet altering the methylation status and expression of genes has also been shown with allergy in mice.[57] Hollingsworth et al. demonstrated that increasing the levels of methyl donors in the diet of female mice increased the levels of allergic airway disease in their offspring.[57] The change in diet altered methylation at specific CpG motifs and thus altered the expression of key genes that regulate the development of an adaptive immune response, increasing T_H2 immunity, lung eosinophilia, and airway remodeling and enhancing the heritable risk of developing allergic airway disease. The maternal dietary intake of methyl donors enhanced the severity of allergic airway disease in offspring of the mice and this trait is inherited transgenerationally. However, the risk/benefits of folic acid in the diet in man remain controversial. Increasing the folic acid intake may have a protective effect, counteracting the effects of air pollution and DNA methylation.[8]

10.9.2 Postnatal Risk

In addition to changes in the epigenetic profile of an individual, environmental factors play a role in the development of allergy. Early exposure to an allergen can modify the risk of subsequent allergy, although the mechanisms involved remain elusive. In some instances, early exposure is detrimental, such as exposure to house dust mite allergen which may predispose to childhood asthma.[118] By contrast, evidence from Israel indicates that early exposure to allergens improves immune tolerance of the allergen.

Israel has a low level of peanut allergy and peanuts are included in baby food.[42] However, a direct link to early exposure and epigenetic regulation has not been demonstrated.

10.9.3 Effect of Smoking on Epigenetics and Asthma

There is a large body of evidence that prenatal exposure to environmental tobacco smoke (ETS) is associated with impaired respiratory function and an increase in risk of transient wheeze or asthma.[76,82,103] Smoking in the last trimester of pregnancy is correlated with asthma in the offspring by one year of age[108] and may be associated with changes in global and gene-specific DNA methylation patterns.[23] Furthermore, ETS has also been linked to the development of adult asthma.[101] Smoking is thought to alter DNA methylation through a process of oxidative stress[45] which can cause lesions in the DNA that interfere with the binding of DNA methyltransferases, resulting in hypomethylation.[45]

Interaction between genes and tobacco smoke leads to asthma. Thus short nucleotide polymorphisms (SNP) in the *CD14* gene combined with ETS significantly alter levels of IgE production.[30] In addition, children who lack the glutathione S-transferase M1 (GSTM1) enzyme (involved in the detoxification of tobacco smoke and reactive oxygen species [ROS]) are more susceptible to the effects of environmental tobacco smoke exposure in utero[49,68]; this is also true with respect to glutathione S-transferase T1 (GSTT1).[68] Children with the common GSTM1 null genotype who were exposed in utero to tobacco smoke displayed altered methylation status of the DNA repetitive element LINE1. Variants in detoxification genes may therefore modulate the effects of in utero exposure due to the action of epigenetic factors. Other findings, however, have not shown a direct effect of smoking on global DNA methylation, even though the association between global methylation patterns in fathers and their offspring was lost if the offspring smoked.[55]

These alterations to the epigenetic profile are heritable and may even cross to the F2 generation in both animal models[57] and in man.[23] Indeed, a child whose grandmother smoked has double the chance of developing asthma than one whose grandmother did not.[77] This risk is further increased if the pregnant mother smokes.[77]

10.9.4 Effect of Air Pollution on Epigenetics of Asthma

Exposure to air pollution has long been associated with asthma and other lung diseases.[50,90] Air pollution, as from traffic or industry, is due to both gaseous (e.g., gasoline) and particulate matter, including the particulate matter of diesel exhaust (DEP). Given of the small size of the latter, it reaches the depths of the airways. The effects of traffic pollution on public health have been extensively studied. The ambient level of black carbon particles, used as a tracer for traffic pollution, has been consistently associated with a variety of adverse health outcomes.[9] The role played by pollutants in making asthma worse is established, but the role air pollution plays in causing asthma is less well defined.[81,99]

Pollutants such as ozone have been shown to damage the airways directly[33] and have been linked to epigenetic changes in the lung. For example, air pollution, like smoking, induces oxidative stress that leads to DNA lesions and hypomethylation.[45,121] Furthermore, pollution associated with oxidative stress decreases methylation across the genome.[9]

Benzene, toluene, xylene, and other volatile organic compounds are associated with asthma.[7] Childhood exposure to benzene is associated with an eightfold increase in asthma risk.[7] In addition, exposure to benzene has been associated with altered DNA methylation.[18,106]

In utero exposure to particulate pollutants such as polycyclic aromatic hydrocarbons has been linked to asthma status.[100] Particulate matter alters DNA methylation in vitro[114] and in animal models in vivo.[13] Pollutants such as DEP can increase levels of ROS, which in turn leads to an increase pro-inflammatory cytokines, through redox sensitive transcription factors such as NF-κB and MAP kinase pathways.[115] Exposure to particulate matter from the steel industry has been associated with demethylation of nitric oxide synthetase (NOS)2 promoter DNA and increased *NOS2* gene transcription.[116]

DEP induces pulmonary inflammation and exacerbates asthma in vivo and is a potent inducer of inflammatory responses in human airway epithelial cells.[26] DEP induces the expression of cyclooxygenase-2 (COX-2) in BEAS-2B cells at both the transcriptional and protein levels. The induction of *COX-2* gene expression is associated with p300-mediated induction of

Fig. 10.4 Mechanism of GC action. (**a**) The glucocorticoid steroid (*GC*) enters the cytoplasm, where (**b**) it binds to the cytosolic glucocorticoid receptor (*GR*) complex (including chaperone proteins HSP90 and HSP51). The GC/GR complex translocates into the nucleus (**c**), where it either binds the glucocorticoid responsive element (*GRE*), resulting in the upregulation of anti-inflammatory gene expression (**d**), or acts to prevent pro-inflammatory transcription factors such as NF-κB from downregulating pro-inflammatory gene expression (**e**)

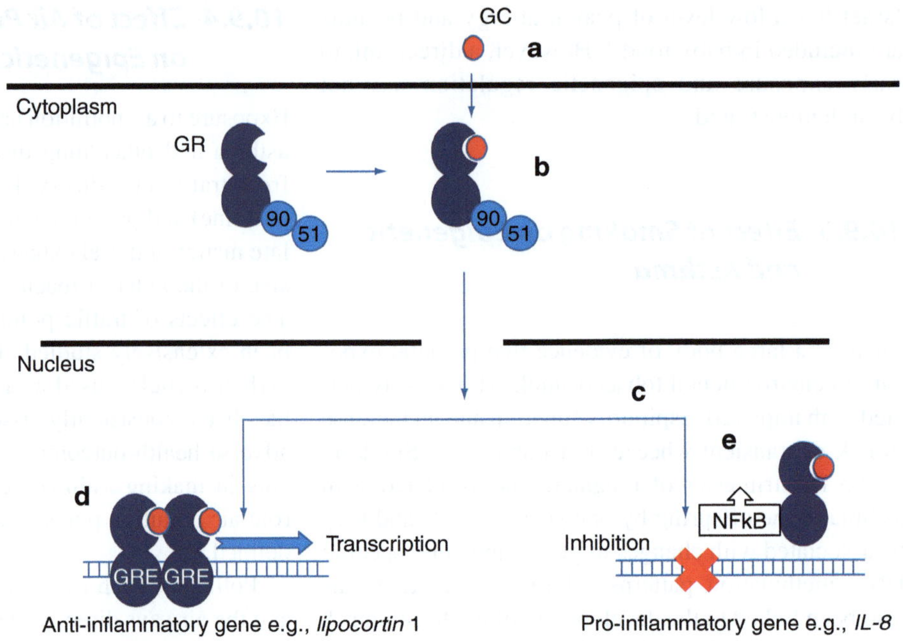

histone H4 acetylation at the native *COX-2* promoter start site. DEP has been shown to promote selective degradation of HDAC1. In this study selective knockdown of HDAC1 using siRNA along with specific overexpression of HDAC1 confirmed the role of HDAC1 in regulating DEP-induced *COX-2* transcription.[26]

Changes in CpG methylation in mice have been linked to T_H2 polarization. In mice exposed to a combination of DEP and the fungus *Aspergillus fumigatus* the promoter was hypomethylated, the IFNγ promoter was hypermethylated and IgE production was increased.[80] Air pollution thus appears to alter the epigenetic state of the genome and may play an important role in the regulation of asthma-associated genes.

10.10 The Role of Epigenetics in the Treatment of Asthma: HDACS and Glucocorticoid Function

Glucocorticoids (GC), either alone or in concert with other drugs, are very effective in suppressing inflammatory gene expression in most chronic inflammatory diseases. GCs function by binding to the cytosolic glucocorticoid receptor (GR). The resulting complex translocates to the nucleus where it binds to glucocorticoid responsive elements (GRE) that upregulate anti-inflammatory genes (transactivation) or interact with transcription factors that repress pro-inflammatory genes (Fig. 10.4).

Nonselective HDAC inhibitors in vitro prevent GR repression of inflammatory genes. This confirms a role for HDACs in GR function.[3] In lung epithelial cells, HDAC2 is important in GR-mediated *GM-CSF* and *IL-8* gene repression.[63] HDAC2 does not affect GR nuclear translocation, DNA binding, or transactivation.[63] Upon binding at aa492-495 GR is rapidly acetylated. Only the deacetylated form of GR can associate with p65. Recruitment of HDAC2 by acetylated GR leads to deacetylation of GR, to interaction with the p65 -NF-κB-activated complex. This in turn brings about reversal of local histone acetylation and suppresses inflammatory gene expression. It is this mechanism that explains how GR distinguishes between recruitment of co-activator and co-repressor proteins and why it can either transactivate or repress gene transcription.[63]

Other HDACs have also been implicated in GR actions. Thus, dexamethasone-induced repression of T_H2 cytokines required GR to attract HDAC1 to GATA3 located on the promoters of T_H2 genes.[66] HDAC3 can also modulate GR functions: HDAC3 complexes with silencing mediator for retinoid and thyroid receptors (SMRT) and nuclear receptor corepressor (N-CoR) for full activity. When recruited to

GR, transactivation is repressed due to the action of athanogene 1M (Bag-1M) associated with the chaperone protein Bcl-2.[58] Heat shock protein 90 (Hsp90) and its accessory co-chaperones ensure the correct structure of client proteins, including GR.[22] The function of Hsp90 is regulated by reversible acetylation under the control of HDAC6. Inactivation of HDAC6 leads to Hsp90 hyperacetylation, its dissociation from an essential co-chaperone, p23, and a loss of chaperone activity.[1] Changes in Hsp90 acetylation alter the dynamic GR.hsp90 heterocomplex assembly/disassembly, causing reduction in GR ligand binding, nuclear translocation, and a 100-fold reduction in glucocorticoid responsiveness.[1,22] SIRT1 (a member of the sirtuin family of deacetylases) is a major repressor of the expression of dexamethasone-induced uncoupling protein-3 (UCP3). It does this through deacetylation of the *ucp3* promoter, indirectly impairing the association of p300 with GR.[6]

The anti-inflammatory actions of glucocorticoids depend on the presence or activation of a number of HDACs, each of which targets different aspects of glucocorticoid action. Although suppressing inflam mation in most asthmatics, glucocorticoids are ineffective in 5–10% of cases.[56] Clearly steroid-insensitive asthmatics (SIA) account for higher costs (economic, morbidity, and mortality) than GC-responsive patients.[56]

An alteration in HDAC/HAT activities may, therefore, affect glucocorticoid responsiveness in asthmatic patients.[3] The suppression of LPS-induced cytokine release (MCP-1, MIP-1α, RANTES, TNFα, IL-1β, IL-8, IFNγ, IL-6, IL-10, and GM-CSF) by dexamethasone from PBMCs of SIA patients is less than in steroid-sensitive asthmatics (SSA).[54] HDAC and HAT activities are also lower in patients with SIA than in patients with SSA, with the reduction in HDAC activity varying with steroid insensitivity. On the other hand, reduction in HAT activity is a function of corticosteroid use rather than of asthma severity.[54] Findings are similar in children where severity of disease is measured by bronchial hyperresponsiveness.[113] However, a reduction in HDAC2 expression is not seen in all SIA patients, reflecting perhaps the heterogeneity of the severe asthma phenotype.[15]

In asthma and other chronic inflammatory diseases CD4+ and CD8+ T cells often do not respond to glucocorticosteroids in the same fashion. As a result CD8+ cells remain, the systemic GC treatment notwithstanding.[31,43,73] The mechanism of GC action is the same in both cell types, but the effects of GC on histone acetylation differ,[75] with the difference attributed to deficient HAT activation by ATF2 in the CD8+ cells, which in turn results in reduced gene silencing.[75]

HDAC2 expression and activity are decreased in smokers, in asthmatics who smoke,[3] in patients with COPD[3,11,65] or cystic fibrosis,[12] all conditions that are insensitive to the anti-inflammatory effects of glucocorticoids.[3] In addition, total HDAC activity in alveolar macrophages falls as the repressive effect of dexamethasone on cytokine production goes up in both smokers and nonsmokers (3). Overexpression of HDAC2, but not of HDAC1, in primary macrophages from COPD patients restores the efficacy of dexamethasone in suppressing LPS-induced GM-CSF release to levels in cells from healthy controls.[65] Furthermore, knockdown by RNAi of HDAC2 in the macrophages of sputum from healthy nonsmokers reduces the inhibitory effect of dexamethasone.[65]

In support of a major role for HDAC2 in GR function is the observation[17] that GR-mediated repression of proopiomelanocortin (POMC) requires the ATPase-dependent chromatin remodeling enzyme Brg1 and the recruitment of HDAC2, so that the POMC promoter can be deacetylated. Interestingly, 50% of glucocorticoid-resistant human and dog corticotroph adenomas, a hallmark of Cushing's disease, lack Brg1 and/or HDAC2 expression.[17]

10.11 Epigenetic Regulation of Tolerance

Bronchodilators and glucocorticoids are effective treatments for asthma, but only treat the symptoms and do not provide a cure. One approach to treat the underlying mechanism of allergy is allergen immunotherapy, that is, administration of gradually increasing quantities of an allergen. This ameliorates the symptoms that arise upon exposure to the allergen.[20] Allergen immunotherapy was first used to treat allergic rhinitis almost 100 years ago[97] but due to the high risk of anaphylactic shock has not been successful in humans. Recently, however, allergen was successfully used to reduce peanut allergy in children to the point where eating ten peanuts led to only mild symptoms.[32]

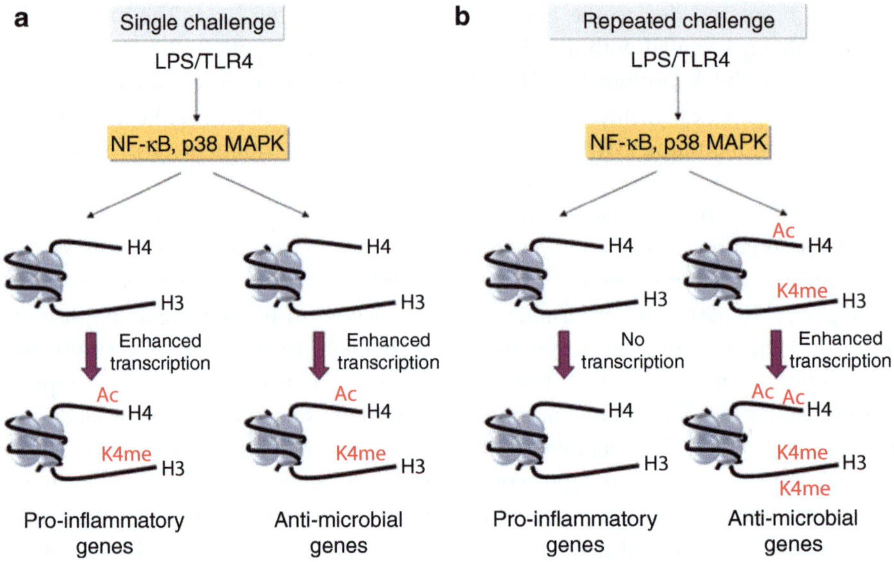

Fig. 10.5 Tolerance. Histone acetylation and methylation are involved in the induction of tolerance in macrophages following a repeated pathogen challenge. (**a**) Bacterial lipopolysaccharide (*LPS*) acting through its receptor, TLR4, induces a signaling cascade that involves NF-κB and p38 MAPK pathways. These in turn induce expression of both antimicrobial and pro-inflammatory genes following tagging of their promoter regions with epigenetic marks (histone H4 acetylation, H4 Ac and histone H3 lysine 4 tri-methylation, K4 Me3). (**b**) However, on restimulation with LPS only antimicrobial genes are activated. Antimicrobial gene promoters retain the epigenetic marks that prepare cells to respond to subsequent LPS stimulation. Anti-inflammatory gene promoters lose their histone marks after initial stimulation, thereby preventing expression on restimulation. This process prevents excessive inflammation, which can be deadly, while retaining the antimicrobial actions of macrophages (Adapted from Foster et al.[44])

In vitro studies with murine macrophages have shown that their LPS tolerance can be raised by repeated exposure to LPS without affecting their sensitivity to antimicrobial drugs, thereby preventing systemic inflammation and septic shock[44] (Fig. 10.5). Tolerant (pro-inflammatory) and non-tolerant (antimicrobial) genes have distinct patterns in their promoters of histone acetylation and methylation to the initial LPS challenge, with this pattern maintained with repeated LPS-stimulation. The increased levels of histone H4 acetylation and H3 K4 tri-methylation of the antibacterial genes prime the genes to respond to LPS stimulation with higher levels of gene expression. In contrast, inflammatory and immunomodulatory genes that are not induced in response to LPS do not have these epigenetic marks, cannot recruit RNA polymerase II, and fail to respond to trichostatin A.[44] Similar processes may occur in vivo, where prolonged exposure to inhaled allergen leads to persistent tolerance in murine models of asthma[44] by an as yet unknown mechanism. Histone acetylation and methylation are crucial for the development of tolerance in macrophages and for CpG methylation of T regulatory cell development and function.[4]

10.12 Effect of HAT/HDAC Modulators In Vivo

Selective HAT inhibitors have been developed (e.g., Cyclopentylidene-(4-(4′-chlorophenyl)thiazol-2-yl) hydrazone[28,84]) that suppress gene expression but also induce others, again reflecting acetylation of histones and nonhistone proteins. For example, TSA represses IL-1β and LPS plus IFNγ-induced expression of NOS2 in murine mesangial and macrophage-like cells, but increases LPS-stimulated NOS2 expression in murine N9 and primary rat microglial cells.[1] Respiratory syncytial virus (RSV)-induced IL-8 expression attenuates after 24 h in A549 cells, even though viral replication continues because HDAC1 has been recruited to the IL-8 promoter. If TSA is repressed, the effect is lost. This indicates that HDAC inhibitors enhance viral inflammation.[1] The differential effects of nonselective HDAC inhibitors on distinct gene targets is highlighted by the fact that TSA enhanced LPS-stimulated IL-8, but repressed IL-12 p40 expression in BEAS-2B cells.[1]

In contrast, in an ovalbumin (OVA)-challenged murine model of asthma, a prophylactic dose of TSA suppressed OVA-induced airway inflammation and airway hyperresponsiveness.[29] TSA caused a ~30–40% reduction in BAL eosinophils and lymphocytes and a 24% reduction in infiltrating mucosal inflammatory cells. BAL IL-4 and IL-5 levels were also significantly reduced in this model, with no adverse effects. This report suggests that HDAC inhibitors with greater bioavailability and longer-lasting effects have the potential of anti-inflammatory agents in the treatment of allergic airways disease. Whether TSA targets a histone or nonhistone protein is not known.[29] The anti-inflammatory effects of TSA may reflect a general effect on T-cell function/death seen with nonselective HDAC inhibitors, but treatment with nonselective HDAC inhibitors is also detrimental in some animal models of inflammation.[1]

As HDAC2 expression and activity are reduced in asthma, in smokers, in smoking asthmatics, and in patients with COPD, elevation of HDAC2 activity by low doses of theophylline may be of benefit in these diseases, particularly when combined with corticosteroids.[3,38] The combination is also effective in conditions of oxidative stress.[38] It is blocked by TSA, involving changes in inflammatory gene promoter histone acetylation status.[88] In vivo studies in smoking asthmatics and patients with COPD indicate that low-dose theophylline enhances the anti-inflammatory effects of steroids by increasing HDAC activity and reducing CXCL8 and TNFα in sputum during COPD exacerbations.[37] This combination also improved lung function and symptoms in smokers with asthma.[109] Whether the beneficial effect of combined steroid/theophylline treatment relates to effects on HDAC activity is still under study.[62]

10.12.1 Clinical Implications of HAT/ HDAC Modulation

In inflammation, specific upregulation of HDAC2 is likely to restore glucocorticoid responsiveness. Drugs such as theophylline[64] and selective PI3K inhibitors,[87] which increase HDAC2 activity suppressed by oxidative stress, may together restore steroid sensitivity in severe asthma, smoking asthmatics, and COPD patients. Thus it may be desirable to target other HDAC enzymes. Alternatively, with the aid of HAT inhibitors it may become possible to repress inflammatory and immunomodulatory genes.

Currently, nonselective HDAC inhibitors such as voronistat are approved for the treatment of cutaneous T-cell lymphoma, but these drugs also have efficacy in inflammatory diseases and in the treatment of cancer, possibly because they preferentially induce T-cell apoptosis. However, the nonselective nature of these agents may limit their clinical effectiveness and increase the potential to cause metastasis,[102] just as nonselective kinase inhibitors induce cell death, whereas selective inhibitors can be beneficial.[72] More data are needed to determine whether selective HDAC inhibitors act on eosinophils, T cells, or macrophages; this would make them useful for treating allergic diseases. Whether the chronic use of these powerful drugs is advisable in nonfatal allergic diseases also needs to be evaluated.

10.13 Summary

Asthma is a chronic inflammatory disease of the airways in which genetics do not adequately explain the heritability and susceptibility to disease. Recent evidence suggests that epigenetic changes may underlie these processes; however, the true role of epigenetics in asthma and allergy still requires much research. It is hoped that the combination of animal and human studies will further understanding of epigenetics as an underlying cause of asthma and other allergies. The role played in these diseases by airborne pollutants such as diesel particles and tobacco smoke, and the effect of parental and grandparental diet, along with genetic predisposition needs to be elucidated.

Glucocorticoids which are currently used to treat the inflammatory component of asthma can modulate the epigenetic environment by recruiting corepressor proteins such as HDAC2 to the sites of inflammatory gene transcription. As our understanding of epigenetics evolves, new tools and compounds are likely to become available, including more selective HDAC and HAT inhibitors/activators and drugs that selectively target histone methyltransferases. These in turn will aid our understanding of how specific histone modifications regulate inflammation and the immune response in cell and animal models of disease. New therapeutic agents may be developed particularly for situations in which current therapies remain suboptimal. The potential of epigenetics to help us understand and subsequently treat what is a theoretically preventable environmental disease is immense.

Acknowledgments We thank the Biotechnology and Biological Sciences Research Council (BBSRC), the Medical Research Council (UK), the Wellcome Trust, GlaxoSmithKline, and Pfizer for supporting current research in our laboratories.

References

1. Adcock IM. Histone deacetylase inhibitors as novel anti-inflammatory agents. *Curr Opin Investig Drugs*. 2006;7:966-973.
2. Adcock IM, Brown CR, Kwon O, Barnes PJ. Oxidative stress induces NF-κB DNA binding and inducible NOS mRNA in human epithelial cells. *Biochem Biophys Res Commun*. 1994;199:1518-1524.
3. Adcock IM, Ford PA, Ito K, Barnes PJ. Epigenetics and airways disease. *Respir Res*. 2006;7:21.
4. Adcock IM, Tsaprouni L, Bhavsar P, Ito K. Epigenetic regulation of airway inflammation. *Curr Opin Immunol*. 2007;19:694-700.
5. Akdis M. Healthy immune response to allergens: T regulatory cells and more. *Curr Opin Immunol*. 2006;18:738-744.
6. Amat R, Solanes G, Giralt M, Villarroya F. SIRT1 is involved in glucocorticoid-mediated control of uncoupling protein-3 gene transcription. *J Biol Chem*. 2007;282:34066-34076.
7. Archivist. Some risk factors for asthma. *Arch Dis Child Ed Pract*. 2004;89:ep78.
8. Baccarelli A, Cassano PA, Litonjua A, et al. Cardiac autonomic dysfunction: effects from particulate air pollution and protection by dietary methyl nutrients and metabolic polymorphisms. *Circulation*. 2008;117:1802-1809.
9. Baccarelli A, Wright RO, Bollati V, et al. Rapid DNA methylation changes after exposure to traffic particles. *Am J Respir Crit Care Med*. 2009;179:572-578.
10. Barker DJP, Hales CN, Fall CHD, Osmond C, Phipps K, Clark PMS. Type 2 (non-insulin-dependent) diabetes mellitus, hypertension and hyperlipidaemia (syndrome X): relation to reduced fetal growth. *Diabetologia*. 1993;36:62-67.
11. Barnes PJ, Adcock IM, Ito K. Histone acetylation and deacetylation: importance in inflammatory lung diseases. *Eur Respir J*. 2005;25:552-563.
12. Bartling TR, Drumm ML. Loss of CFTR results in reduction of histone deacetylase 2 in airway epithelial cells. *Am J Physiol Lung Cell Mol Physiol*. 2009;297:L35-L43.
13. Belinsky SA, Snow SS, Nikula KJ, Finch GL, Tellez CS, Palmisano WA. Aberrant CpG island methylation of the p16INK4a and estrogen receptor genes in rat lung tumors induced by particulate carcinogens. *Carcinogenesis*. 2002;23:335-339.
14. van den Berge M, ten Hacken NH, Kerstjens HAM, Postma DS. Management of asthma with ICS and LABAs: different treatment strategies. *Clin Med Ther*. 2009;2009:77.
15. Bergeron C, Fukakusa M, Olivenstein R, et al. Increased glucocorticoid receptor-β expression, but not decreased histone deacetylase 2, in severe asthma. *J Allergy Clin Immunol*. 2006;117:703-705.
16. Bhavsar P, Ahmad T, Adcock IM. The role of histone deacetylases in asthma and allergic diseases. *J Allergy Clin Immunol*. 2008;121:580-584.
17. Bilodeau S, Vallette-Kasic S, Gauthier Y, et al. Role of Brg1 and HDAC2 in GR trans-repression of the pituitary POMC gene and misexpression in Cushing disease. *Genes Dev*. 2006;20:2871-2886.
18. Bollati V, Baccarelli A, Hou L, et al. Changes in DNA methylation patterns in subjects exposed to low-dose benzene. *Cancer Res*. 2007;67:876-880.
19. Bollati V, Schwartz J, Wright R, et al. Decline in genomic DNA methylation through aging in a cohort of elderly subjects. *Mech Ageing Dev*. 2009;130:234-239.
20. Bousquet J, Lockey R, Malling HJ. Allergen immunotherapy: therapeutic vaccines for allergic diseases A WHO position paper. *Mech Ageing Dev*. 1998;02:558-562.
21. Bousquet J, Jeffery PK, Busse WW, Johnson M, Vignola AM. Asthma. From bronchoconstriction to airways inflammation and remodeling. *Am J Respir Crit Care Med*. 2000;161:1720-1745.
22. Boyault C, Sadoul K, Pabion M, Khochbin S. HDAC6, at the crossroads between cytoskeleton and cell signaling by acetylation and ubiquitination. *Oncogene*. 2007;26:5468-5476.
23. Breton CV, Byun HM, Wenten M, Pan F, Yang A, Gilliland FD. Prenatal tobacco smoke exposure affects global and gene-specific DNA methylation. *Am J Respir Crit Care Med*. 2009;180:462-467.
24. Bruce IN, Harland RW, McBride NA, MacMahon J. Trends in the prevalence of asthma and dyspnoea in first year university students, 1972-89. *QJM*. 1993;86:425-430.
25. Callinan PA, Feinberg AP. The emerging science of epigenomics. *Hum Mol Genet*. 2006;15:R95-R101.
26. Cao D, Bromberg PA, Samet JM. COX-2 expression induced by diesel particles involves chromatin modification and degradation of HDAC1. *Am J Respir Cell Mol Biol*. 2007;37:232-239.
27. Chandrasoma P, Taylor CR. Part A. General pathology, Section II. The host response to injury, Chapter 3. The acute inflammatory response. In: *Concise Pathology*. 3(3). New York, NY: McGraw-Hill; 1998.
28. Chimenti F, Bizzarri B, Maccioni E, et al. A novel histone acetyltransferase inhibitor modulating Gcn5 network: cyclopentylidene-(4-(4′-chlorophenyl)thiazol-2-yl)hydrazone. *J Med Chem*. 2008;52:530-536.
29. Choi J-H, Oh S-W, Kang M-S, Kwon HJ, Oh G-T, Kim D-Y. Trichostatin A attenuates airway inflammation in mouse asthma model. *Clin Exp Allergy*. 2005;35:89-96.
30. Choudhry S, Avila PC, Nazario S, et al. CD14 tobacco gene-environment interaction modifies asthma severity and immunoglobulin E levels in Latinos with asthma. *Am J Respir Crit Care Med*. 2005;172:173-182.
31. Chrousos GP, Kino T. Intracellular glucocorticoid signaling: a formerly simple system turns stochastic. *Sci Signal*. 2005;2005:e48.
32. Clark AT, Islam S, King Y, Deighton J, Anagnostou K, Ewan PW. Successful oral tolerance induction in severe peanut allergy. *Allergy*. 2009;64(8):1218-1220.

33. Committee of the Environmental and Occupational Health Assembly of the American Thoracic Society. Health effects of outdoor air pollution. *Am J Respir Crit Care Med.* 1996;153:3-50.
34. Cook JPD, Henry AJ, McDonnell JM, Owens RJ, Sutton BJ, Gould HJ. Identification of contact residues in the IgE binding site of human FcepsilonRIalpha. *Biochemistry.* 1997;36: 15579.
35. Cookson WO, Young RP, Sandford AJ, et al. Maternal inheritance of atopic IgE responsiveness on chromosome 11q. *Lancet.* 1992;340:381-384.
36. Cooney CA, Dave AA, Wolff GL. Maternal methyl supplements in mice affect epigenetic variation and DNA methylation of offspring. *J Nutr.* 2002;132:2393S-2400S.
37. Cosio BG, Iglesias A, Rios A, et al. Low-dose theophylline enhances the anti-inflammatory effects of steroids during exacerbations of COPD. *Thorax.* 2009;64:424-429.
38. Cosio BG, Tsaprouni L, Ito K, Jazrawi E, Adcock IM, Barnes PJ. Theophylline restores histone deacetylase activity and steroid responses in COPD macrophages. *J Exp Med.* 2004;200:689-695.
39. Dale HH, Richards AN. The vasodilator action of histamine and of some other substances. *J Physiol.* 1918;52:110-165.
40. Dold S, Wjst M, von Mutius E, Reitmeir P, Stiepel E. Genetic risk for asthma, allergic rhinitis, and atopic dermatitis. *Arch Dis Child.* 1992;67:1018-1022.
41. Dombrowicz D, Brini AT, Flamnd V, Hicks E, Snouwaert JKN. Anaphylaxis mediated through a humanized high affinity IgE receptor. *J Immunol.* 1996;157:1645.
42. Du Toit G, Katz Y, Sasieni P, et al. Early consumption of peanuts in infancy is associated with a low prevalence of peanut allergy. *J Allergy Clin Immunol.* 2008;122:984-991.
43. Duma D, Jewell CM, Cidlowski JA. Multiple glucocorticoid receptor isoforms and mechanisms of post-translational modification. *J Steroid Biochem Mol Biol.* 2006;102:11-21.
44. Foster SL, Hargreaves DC, Medzhitov R. Gene-specific control of inflammation by TLR-induced chromatin modifications. *Nature.* 2007;447:972-978.
45. Franco R, Schoneveld O, Georgakilas AG, Panayiotidis MI. Oxidative stress, DNA methylation and carcinogenesis. *Cancer Lett.* 2008;266:6-11.
46. Fuhlbrigge AL, Adams RJ, Guilbert TW, et al. The burden of asthma in the United States: level and distribution are dependent on interpretation of the National Asthma Education and Prevention Program Guidelines. *Am J Respir Crit Care Med.* 2002;166:1044-1049.
47. Gell PGH, Coombs RRA. *Clinical Aspects of Immunology.* 1st ed. Oxford, England: Blackwell; 1963.
48. Gilliland FD, Berhane K, Li YF, Rappaport EB, Peters JM. Effects of early onset asthma and in utero exposure to maternal smoking on childhood lung function. *Am J Respir Crit Care Med.* 2003;167:917-924.
49. Gilliland FD, Li YF, Dubeau L, et al. Effects of glutathione *S*-transferase M1, maternal smoking during pregnancy, and environmental tobacco smoke on asthma and wheezing in children. *Am J Respir Crit Care Med.* 2002;166:457-463.
50. Gilmour MI, Jaakkola MS, London SJ, Nel AE, Rogers CA. How exposure to environmental tobacco smoke, outdoor air pollutants, and increased pollen burdens influences the incidence of asthma. *Environ Health Perspect.* 2006;114: 627-633.
51. Gould HJ, Sutton BJ, Beavil AJ, et al. The biology of IGE and the basis of allergic disease. *Annu Rev Immunol.* 2003;21:579-628.
52. Hales CN, Desai M, Ozannes SE. The thrifty phenotype hypothesis: how does it look after 5 years? *Diabet Med.* 2009;14:189-195.
53. Heintzman ND, Hon GC, Hawkins RD, et al. Histone modifications at human enhancers reflect global cell-type-specific gene expression. *Nature.* 2009;459:108-112.
54. Hew M, Bhavsar P, Torrego A, et al. Relative corticosteroid insensitivity of peripheral blood mononuclear cells in severe asthma. *Am J Respir Crit Care Med.* 2006;174:134-141.
55. Hillemacher T, Frieling H, Moskau S, et al. Global DNA methylation is influenced by smoking behaviour. *Eur Neuropsychopharmacol.* 2008;8:295-298.
56. Holgate ST, Polosa R. The mechanisms, diagnosis, and management of severe asthma in adults. *Lancet.* 2006;368: 780-793.
57. Hollingsworth JW, Maruoka S, Boon K, et al. In utero supplementation with methyl donors enhances allergic airway disease in mice. *J Clin Invest.* 2009;118(10):3462-3469.
58. Hong W, Baniahmad A, Li J, Chang C, Gao W, Liu Y. Bag-1M inhibits the transactivation of the glucocorticoid receptor via recruitment of corepressors. *FEBS Lett.* 2009;583:2451-2456.
59. Humbert M, Menz G, Ying S, et al. The immunopathology of extrinsic (atopic) and intrinsic (non-atopic) asthma: more similarities than differences. *Immunol Today.* 1999;20: 528-533.
60. Irizarry RA, Ladd-Acosta C, Wen B, et al. The human colon cancer methylome shows similar hypo- and hypermethylation at conserved tissue-specific CpG island shores. *Nat Genet.* 2009;41:178-186.
61. Ito K, Caramori G, Lim S, et al. Expression and activity of histone deacetylases in human asthmatic airways. *Am J Respir Crit Care Med.* 2002;166:392-396.
62. Ito K, Chung KF, Adcock IM. Update on glucocorticoid action and resistance. *J Allergy Clin Immunol.* 2006;117: 522-543.
63. Ito K, Ito M, Elliott WM, et al. Decreased histone deacetylase activity in chronic obstructive pulmonary disease. *N Engl J Med.* 2005;352:1967-1976.
64. Ito K, Lim S, Caramori G, et al. A molecular mechanism of action of theophylline: induction of histone deacetylase activity to decrease inflammatory gene expression. *Proc Natl Acad Sci USA.* 2002;99:8921-8926.
65. Ito K, Yamamura S, Essilfie-Quaye S, et al. Histone deacetylase 2-mediated deacetylation of the glucocorticoid receptor enables NF-κB suppression. *J Exp Med.* 2006;203:7-13.
66. Jee YK, Gilmour J, Kelly A, et al. Repression of interleukin-5 transcription by the glucocorticoid receptor targets GATA3 signaling and involves histone deacetylase recruitment. *J Biol Chem.* 2005;280:23243-23250.
67. Kaati G, Bygren LO, Edvinsson S. Cardiovascular and diabetes mortality determined by nutrition during parents' and grandparents' slow growth period. *Eur J Hum Genet.* 2002;10:682-688.
68. Kabesch M, Hoefler C, Carr D, Leupold W, Weiland SK, von Mutius E. Glutathione S transferase deficiency and passive smoking increase childhood asthma. *Thorax.* 2004;59:569-573.

69. Keslacy S, Tliba O, Baidouri H, Amrani Y. Inhibition of tumor necrosis factor-α-inducible inflammatory genes by interferon-γ is associated with altered nuclear factor-κB transactivation and enhanced histone deacetylase activity. *Mol Pharmacol*. 2007;71:609-618.
70. Kindt TJ, Goldsby RA, Osborne BA. *Kuby Immunology*. New York, NY: Freeman and Company; 2007.
71. KleinJan A, Vinke JG, Severinjnen LWFM, Fokkens WJ. Local production and detection of (specific) IgE in nasal B-cells and plasma cells of allergic rhinitis patients. *Eur Respir J*. 2000;15:491.
72. Knight ZA, Shokat KM. Features of selective kinase inhibitors. *Chem Biol*. 2005;12:621-637.
73. Leung DYM, Bloom JW. Update on glucocorticoid action and resistance. *J Allergy Clin Immunol*. 2003;111:3-22.
74. Lewin B. *Genes IX*. Sudbury, MA: Jones and Bartlett Publishers; 2008.
75. Li Lb, Leung DYM, Strand MJ, Goleva E. ATF2 impairs glucocorticoid receptor mediated transactivation in human CD8+ T cells. *Blood*. 2007;110:1570-1577.
76. Li Y-F, Gilliland F, Berhane K, et al. Effects of in utero and environmental tobacco smoke exposure on lung function in boys and girls with and without asthma. *Am J Respir Crit Care Med*. 2000;162:2097-2104.
77. Li Y-F, Langholz B, Salam MT, Gilliland FD. Maternal and grandmaternal smoking patterns are associated with early childhood asthma. *Chest*. 2005;127:1232-1241.
78. Lin H, Mosmann TR, Guilbert L, Tuntipopipat S, Wegmann TG. Synthesis of T helper 2-type cytokines at the maternal-fetal interface. *J Immunol*. 1993;151:4562-4573.
79. Litonjua AA, Carey VJ, Burge HA, Weiss ST, Gold DR. Parental history and the risk for childhood asthma. Does mother confer more risk than father? *Am J Respir Crit Care Med*. 1998;158:176-181.
80. Liu J, Ballaney M, Al-alem U, et al. Combined inhaled diesel exhaust particles and allergen exposure alter methylation of T helper genes and IgE production in vivo. *Toxicol Sci*. 2008;102:76-81.
81. London SJ. Gene-air pollution interactions in asthma. *Proc Am Thorac Soc*. 2007;4:217-220.
82. Magnusson LL, Olesen AB, Wennborg H, Olsen J. Wheezing, asthma, hayfever, and atopic eczema in childhood following exposure to tobacco smoke in fetal life. *Clin Exp Allergy*. 2005;35:1550-1556.
83. Mandhane PJ, Greene JM, Cowan JO, Taylor DR, Sears MR. Sex differences in factors associated with childhood- and adolescent-onset wheeze. *Am J Respir Crit Care Med*. 2005;172:45-54.
84. Mantelingu K, Reddy BA, Swaminathan V, et al. Specific inhibition of p300-HAT alters global gene expression and represses HIV replication. *Chem Biol*. 2007;14: 645-657.
85. Martinez FD. Complexities of the genetics of asthma. *Am J Respir Crit Care Med*. 1997;156:S117-S122.
86. Martinez FD, Wright AL, Taussig LM, et al. Asthma and wheezing in the first six years of life. *N Engl J Med*. 1995;332:133-138.
87. Marwick JA, Caramori G, Stevenson CS, et al. Inhibition of PI3Kδ restores glucocorticoid function in smoking-induced airway inflammation in mice. *Am J Respir Crit Care Med*. 2009;179:542-548.
88. Marwick JA, Wallis G, Meja K, et al. Oxidative stress modulates theophylline effects on steroid responsiveness. *Biochem Biophys Res Commun*. 2008;377:797-802.
89. Masoli M, Fabian D, Holt S, Beasley R. The global burden of asthma: executive summary of the GINA Dissemination Committee Report. *Allergy*. 2004;59:469-478.
90. McCreanor J, Cullinan P, Nieuwenhuijsen MJ, et al. Respiratory effects of exposure to diesel traffic in persons with asthma. *N Engl J Med*. 2007;357:2348-2358.
91. Metcalfe DD, Baram D, Mekori YA. Mast cells. *Physiol Rev*. 1997;77:1033.
92. Miller RL, Ho SM. Environmental epigenetics and asthma: current concepts and call for studies. *Am J Respir Crit Care Med*. 2008;177:567-573.
93. Moffatt MF, Schou C, Faux JA, et al. Association between quantitative traits underlying asthma and the HLA-DRB1 locus in a family-based population sample. *Eur J Hum Genet*. 2001;9:341-346.
94. Morgan HD, Sutherland HGE, Martin DIK, Whitelaw E. Epigenetic inheritance at the agouti locus in the mouse. *Nat Genet*. 1999;23:314-318.
95. Nicolae D, Cox NJ, Lester LA, et al. Fine mapping and positional candidate studies identify HLA-G as an asthma susceptibility gene on chromosome 6p21. *Am J Hum Genet*. 2005;76:349-357.
96. Nie M, Knox AJ, Pang L. β_2-Adrenoceptor agonists, like glucocorticoids, repress eotaxin gene transcription by selective inhibition of histone H4 acetylation. *J Immunol*. 2005;175:478-486.
97. Noon L. Prophylactic inoculation against Hay fever. *Lancet*. 1911;177:1572-1573.
98. Parsons ME, Ganellin CR. Histamine and its receptors. *Br J Pharmacol*. 2009;147:S127-S135.
99. Patel MM, Miller RL. Air pollution and childhood asthma: recent advances and future directions. *Curr Opin Pediatr*. 2009;21:235-242.
100. Perera F, Wy T, Herbstman J, et al. Relation of DNA methylation of 5′-CpG island of ACSL3 to transplacental exposure to airborne polycyclic aromatic hydrocarbons and childhood asthma. *PLoS ONE*. 2009;4:e4488.
101. Piipari R, Jaakkola JJK, Jaakkola N, Jaakkola MS. Smoking and asthma in adults. *Eur Respir J*. 2004;24:734-739.
102. Pulukuri SMK, Gorantla B, Rao JS. Inhibition of histone deacetylase activity promotes invasion of human cancer cells through activation of urokinase plasminogen activator. *J Biol Chem*. 2007;282:35594-35603.
103. Raherison C, Pønard-Morand C, Moreau D, et al. In utero and childhood exposure to parental tobacco smoke, and allergies in schoolchildren. *Respir Med*. 2007;101: 107-117.
104. Rogers DF. Airway mucus hypersecretion in asthma: an undervalued pathology? *Curr Opin Pharmacol*. 2004;4: 241-250.
105. Roh TY, Cuddapah S, Cui K, Zhao K. The genomic landscape of histone modifications in human T cells. *Proc Natl Acad Sci USA*. 2006;103:15782-15787.
106. Ross CM. Epigenetics, traffic and firewood. *Schizophr Res*. 2009;109:193.
107. Ruiz RGG, Kemeny DM, Price JF. Higher risk of infantile atopic dermatitis from maternal atopy than from paternal atopy. *Clin Exp Allergy*. 2009;22:762-766.

108. Sears MR, Holdaway MD, Flannery EM, Herbison GP, Silva PA. Parental and neonatal risk factors for atopy, airway hyper-responsiveness, and asthma. *Arch Dis Child.* 1996;75:392-398.
109. Sears MR, Ottosson A, Radner F, Suissa S. Long-acting {beta}-agonists: a review of formoterol safety data from asthma clinical trials. *Eur Respir J.* 2009;33:21-32.
110. Smurthwaite L, Durham SR. Local IgE production in allergic rhinitis and asthma. *Curr Allergy Asthma Rep.* 2002;2:231.
111. Strait RT, Morris SC, Yang M, Qu X-W, Finkelman FD. Pathways of anaphylaxis in the mouse. *J Allergy Clin Immunol.* 2002;109:658.
112. Su RC, Becker AB, Kozyrskyj AL, HayGlass KT. Epigenetic regulation of established human type 1 versus type 2 cytokine responses. *J Allergy Clin Immunol.* 2008; 121:57-63.
113. Su RC, Becker AB, Kozyrskyj AL, HayGlass KT. Altered epigenetic regulation and increasing severity of bronchial hyperresponsiveness in atopic asthmatic children. *J Allergy Clin Immunol.* 2009;124:1116-1118.
114. Takiguchi M, Achanzar WE, Qu W, Li G, Waalkes MP. Effects of cadmium on DNA-(Cytosine-5) methyltransferase activity and DNA methylation status during cadmium-induced cellular transformation. *Exp Cell Res.* 2003;286:355-365.
115. Takizawa H. Diesel exhaust particles and their effect on induced cytokine expression in human bronchial epithelial cells. (Miscellaneous). *Curr Opin Allergy Clin Immunol.* 2004;4:355-359.
116. Tarantini L, Bonzini M, Apostoli P, et al. Effects of particulate matter on genomic DNA methylation content and iNOS promoter methylation. *Environ Health Perspect.* 2009;117:217-222.
117. Tirimanna PR, van Schayck CP, den Otter JJ, et al. Prevalence of asthma and COPD in general practice in 1992: has it changed since 1977. *Br J Gen Pract.* 1996;46:277-281.
118. Torrent M, Sunyer J, Garcia R, et al. Early-life allergen exposure and atopy, asthma, and wheeze up to 6 years of age. *Am J Respir Crit Care Med.* 2007;176:446-453.
119. Traherne JA, Hill MR, Hysi P, et al. LD mapping of maternally and non-maternally derived alleles and atopy in FcεRI-β. *Hum Mol Genet.* 2003;12:2577-2585.
120. Ujike A, Ishikawa Y, Ono M, Yuasa T, Yoshino T. Modulation of immunoglobulin (Ig)E-mediated systemic anaphylaxis by low-affiniy Fc receptors for IgG. *J Exp Med.* 1999;189:1573.
121. Valinluck V, Tsai HH, Rogstad DK, Burdzy A, Bird A, Sowers LC. Oxidative damage to methyl-CpG sequences inhibits the binding of the methyl-CpG binding domain (MBD) of methyl-CpG binding protein 2 (MeCP2). *Nucleic Acids Res.* 2004;32:4100-4108.
122. van den Elsen PJ, Holling TM, van der Stoep N, Boss JM. DNA methylation and expression of major histocompatibility complex class I and class II transactivator genes in human developmental tumor cells and in T cell malignancies. *Clin Immunol.* 2003;109:46-52.
123. Vedanthan PK, Mahesh PA, Vedanthan R, Holla AD, Liu AH. Effect of animal contact and microbial exposures on the prevalence of atopy and asthma in urban vs rural children in India. *Ann Allergy Asthma Immunol.* 2006;96: 571-578.
124. Wang Z, Zang C, Cui K, et al. Genome-wide mapping of HATs and HDACs reveals distinct functions in active and inactive genes. *Cell.* 2009;138:1019-1031.
125. Wang Z, Zang C, Rosenfeld JA, et al. Combinatorial patterns of histone acetylations and methylations in the human genome. *Nat Genet.* 2008;40:897-903.
126. Weiss ST, Raby BA, Rogers A. Asthma genetics and genomics 2009. *Curr Opin Genet Dev.* 2009;19:279-282.
127. Wills-Karp M, Ewart SL. Time to draw breath: asthma-susceptibility genes are identified. *Nat Rev Genet.* 2004;5:376-387.
128. Wort SJ, Ito M, Chou PC, et al. Synergistic induction of endothelin-1 by tumor necrosis factor α and interferon γ is due to enhanced NF-κB binding and histone acetylation at specific κB sites. *J Biol Chem.* 2009;284:24297-24305.
129. Xu WS, Parmiginai RB, Marks PA. Histone deacetylase inhibitors: molecular mechanisms of action. *Oncogene.* 2007;26:5541-5552.
130. Zhang Y, Nicholas I, Anderson GG, et al. Positional cloning of a quantitative trait locus on chromosome 13q14 that influences immunoglobulin E levels and asthma. *Nat Genet.* 2003;34:181-186.

Epigenetics in Psychiatry

11

Hamid Mostafavi-Abdolmaleky, Stephen J. Glatt, and Ming T. Tsuang

11.1 Introduction

Diversity, perhaps the most prominent aspect of life, is consistent with the Darwinian theory of evolution, the result of random mutation. Yet the diversity of cells and tissues during the development of a single animal (even though the genetic make-up of all cells of an organism is constant) obviously cannot be attributed to random mutations. While the Lamarckian concept of species development is not considered as a valid explanation of evolution, it may offer clues about the origin of diversity in the development of many cells that make up a single organism.

The zygocytes of multicellular organisms generate diverse cells/tissues with the same genetic materials. Waddington (1939) first proposed epigenetic modulation of gene expression as the underlying mechanism of cell differentiation and of the corresponding diverse structural and functional identities of those cells.[114] However, the molecular mechanism of epigenetic regulations remained obscure and unclear for decades. DNA methylation was among the earliest discovered epigenetic mechanisms mediating gene–gene and gene–environment interactions, followed by RNA editing, histone modifications, and RNA interference.

As a general rule, the organic bases of DNA (genetic codes) are fixed and can transfer long-term genetic memory. However, epigenetic marks such as DNA methylation and histone acetylation are flexible and dynamic, and adapt to a variety of environmental conditions granting flexibility and adaptability to the organism.[17,87] The details of the processes that govern the enzymatic fine-tuning of epigenetic marks are not well known. Among epigenetic mechanisms, microRNAs (miRNAs) appear to be involved in homeostatic regulation at the cellular level, preventing excess protein synthesis,[75] and overconsumption of nucleic acids or RNA production at toxic levels. Histone acetylation and methylation are likely mechanisms that enable a rapid response to the signals of other cells or to the environment, with the responses mediated by regulatory proteins, elements, hormones, and transmitters. The cellular DNA methylation status seems to be more stable and provides cell-specific epigenetic memory. This memory in turn is inherited or acquired during the life of the cell and then transmitted to daughter cells. Once acquired, the methylation status governs the establishment of corresponding histone codes.[17,71]

Environmental insults such as exposure to chemicals, infections, and malnutrition can change epigenetic memory and bring about sporadic and, eventually, inherited developmental diseases. The homeostasis of the cell, its ecological conditions, and even factors related to the social milieu[68,104] can modulate epigenetic memory consistent with a given micro/macro environment, but these molecular changes may be inappropriate for other conditions. It seems reasonable, therefore, to infer that evolution may be driven by epigenetic mechanisms. It seems likely that the continued growth in our knowledge and understanding of epigenetics will add to the Darwinian theory of evolution. Rather than a passive structure for the acquisition of random mutations, the genome will come to be

M.T. Tsuang (✉)
Center for Behavioral Genomics, Department of Psychiatry, Institute of Genomic Medicine, University of California, San Diego, 9500 Gilman Drive, La Jolla, CA 92039, USA and
Veterans Affairs San Diego Healthcare System, 3350 La Jolla Village Drive San Diego, CA 92161, USA and
Harvard Institute of Psychiatric Epidemiology and Genetics, Harvard Departments of Epidemiology and Psychiatry, 25 Shattuck Street, Boston, MA 02115, USA
e-mail: mtsuang@ucsd.edu

H.I. Roach et al. (eds.), *Epigenetic Aspects of Chronic Diseases*,
DOI: 10.1007/978-1-84882-644-1_11, © Springer-Verlag London Limited 2011

viewed as interacting with the environment so as to produce specific proteins in the amount needed to deal with the ever-changing environment, including the familial and social milieu.

The factors that interfere with these adaptive mechanisms need to be scrutinized for their potential impact on human development and on neurodevelopmental diseases such as schizophrenia (SCZ), bipolar disorder (BD), and other mood disorders. In fact, the brain is the fertile ground for transforming the environmental and social cues into a model in the neuronal network. This structural and functional modeling is dynamic and subject to continuous remodeling, based on the incoming environmental signals that have been translated or transduced to biological language. In turn, this information is transmitted to the epigenetic machinery which may then bring about appropriate adaptive modifications. Any interfering factor that disturbs the epigenetic machinery and codes may derail adaptive neurodevelopmental events that are required for neuronal remodeling. SCZ is a neurodevelopmental disease that affects cognition, social judgment, and interpersonal relationships, and is associated with interhemispheric brain dysfunctions that may well be due to epigenetic aberrations, perhaps together with genetic mutations, but not to mutations alone.

Extensive genetic studies, including the most recent genome-wide association scans in thousands of cases and controls[84] have found neither a single gene nor a group of genes with major effects to be responsible for the pathogenesis of major mental diseases such as SCZ and BD. Instead, they found a large number of common single nucleotide polymorphisms with small effects involved in the genesis of these psychiatric diseases, as was suggested by population and familial genetic analysis many decades ago.[107]

Advanced imaging techniques during the last two decades have made it possible to identify psychiatric endophenotypes such as atrophy of the cingulate gyrus and dorsolateral frontal cortex or gliosis of the corpus callosum[10,69,83,112,113] mostly in the left brain hemisphere (see Table 11.1). However, the underlying mechanisms responsible for the genesis of these endophenotyes have not been resolved at the molecular level. Analyses of the brain transcriptome using high-throughput techniques such as microarray have provided strong evidence that hundreds of genes are

Table 11.1 Supporting evidence for the loss of brain laterality/dominance in SCZ

Method/sample	Objectives	Findings	Ref.
Postmortem (~40 SCZ & 30 Cont.)	Ventricular size	Enlargement of the left ventricle: SCZ is an "anomaly of development of cerebral asymmetry"	30
MRI (high resolution 3D); (SCZ & Cont. 20M/20F)	Ventricular size	Enlargement of the left ventricle (male)	121
fMRI (high-functioning); SCZ & Cont.)	Cortical activation during verbal fluency	Left Broca's area in Cont. bilateral in SCZ "reduced language lateralization of the frontal cortex in SCZ"	117
Postmortem (10 SCZ/10 Cont.)	Asymmetry of pyramidal cell density in DLFC	Greater density in the left in Cont. Loss or reversal of brain asymmetry in SCZ (layer 3)	31
Auditory evoked magnetic field	Left-Hemispheric activation to right-ear stimulation	Absence of contralateral dominance in response to auditory stimuli in SCZ; "failure to establish clear left-hemispheric dominance of the phonological loop"	85
fMRI (15 inpatients acutely ill SCZ)	Cortical activation in verbal & spatial working memory task	Absence of prefrontal lateralization comparing verbal and spatial working memory in SCZ	115
fMRI (24 SCZ)	Cerebral activation during (1) Auditory verbal Hallucination (2) silently generated words	(1) Activation of right homolog of Broca's area, but not Broca's area; (2) activation in Broca's & Wernicke's areas & their right-sided homologs	98
Gamma synchrony	Neuronal synchrony in first episode SCZ and 2.5 years later	Progressive loss of the gray matter in SCZ, more in left & progressive disruption to the laterality of early gamma synchrony	118
Mapping of scalp electrical activity	Mean Response Times in linguistic tasks	Increase in mean response times & loss of left frontal activity during phonological task in SCZ	8

differentially expressed in the affected brain regions of patients with major mental diseases, but not in most other tissues.[21,42,52] Genetic polymorphisms alone cannot be responsible for specific dysregulation of gene expression in a given brain region. Rather, several lines of evidence suggest that epigenetic aberrations play a significant role in tissue-specific dysregulation of affected genes, which then result in disease phenotypes, as reviewed below.

11.2 Epigenetic Aberrations Reported in Mental Diseases

Most psychiatric mood disorders such as a major depressive disorder or BD are episodic and may have a spontaneous remission that sometimes can last many years. Even SCZ and obsessive/compulsive disorder have a fluctuating course. This suggests that genetic mutations alone cannot be responsible for the presentation of the disease phenotype, because spontaneous remission and fluctuation do not occur in purely genetic diseases. On the other hand, dysregulation of epigenetic machinery due to macro- or microenvironmental exposures/factors can result in episodes of remission or disease. For instance, neuronal activation can alter DNA methylation of the brain-derived neurotrophic factor (BDNF) promoter at CRE binding sites that induce gene expression.[63] Expression of the human glutamate transporter gene is also regulated by promoter DNA methylation, and increase in cAMP level (known to be influenced by environmental and nutritional factors such as tea and coffee intake) is associated with an increase in gene expression linked to changes in promoter DNA methylation.[124] Early life experience has been shown to alter promoter DNA methylation of steroid receptors in mouse hippocampus, but can be adjusted later in life, depending on environmental influences.[67,116] These observations indicate that epigenetic memory like genetic codes can be affected by environmental factors, but, unlike genetic mutations, epigenetic changes are reversible. This provides the organism an opportunity to recover epigenetic aberrations and their corresponding phenotypes. However, severe and/or continuous environmental impacts may lead to the loss of epigenetic memory and thus to a stable disease phenotype.[65,108]

11.2.1 Aberrant DNA Methylation in Psychiatric Diseases

In mammals, DNA methylation is a simple reaction catalyzed by the addition of methyl groups to cytosines that are followed by guanine. It now seems that ~25% of the DNA methylation reactions are made on cytosines that are followed by other bases, at least in embryonic cells.[59] Several enzymes such as, DNMT1 (DNA methyltransferase-1, involved in the maintenance of DNA methylation), DNMT3a, and DNMT3b (de novo methyltransferases) mediate the DNA methylation reaction. Methyl groups are taken from SAM (s-adenosyl methionine, the universal methyl donor), which is made from the processing of the essential amino acid, methionine. Some vitamins including folic acid and vitamin B12 are involved in remethylation of demethylated SAM.[13,36,56] Shortage or surplus of any of these contributors may result in DNA methylation changes. Methylated cytosines are binding targets of several types of methylated DNA-binding domain (MBD) proteins (e.g., MBD1, MBD3, MBD4, MeCP2 and Kaiso-binding sequence) that block the access of transcription factors to the promoter DNA.[18,35]

The first observations suggestive of epigenetic aberrations in mental diseases were linked to fragile X disease and Rett syndrome.[5,6,33,62,103] DNA methylation abnormalities in psychiatric diseases include promoter DNA hypermethylation of reelin[2,43] and SOX10[51] associated with the reduced expression of these genes in SCZ, promoter hypomethylation of membrane-bound catechol-*O*-methyltransferase (MB-COMT) together with increased expression in SCZ and BD,[1] and global DNA hypomethylation in blood cells from male SCZ patients.[95] On the basis of the cited evidence, none of the DNA methylation changes was due to treatment with antipsychotic drugs. When a genome-wide microarray analysis, which included >12,000 CpG islands and depended on methyl-sensitive restriction enzymes to enrich unmethylated DNA, was carried out in *post-mortem* brains of patients with SCZ and BD, sex-specific methylation changes were uncovered for many genes. These include DNA hypomethylation of ribosomal protein L39 (RPL39) in females with BD, DNA hypermethylation of WD repeat domain 18 (WDR18) in male SCZ patients, and aberrant DNA methylation of GABAergic and glutamatergic genes in SCZ and BD.[70] Expression of glutamate transporter EAAT2 (excitatory amino acid transporter 2), one of the key genes in the clearance of the neurotoxic glutamate, was also influenced by the level of DNA methylation of the promoter.[124] This finding makes EAAT2 a high priority candidate gene for further epigenetic analysis.

Other striking findings include: abnormal transmethylation and trans-sulfuration processes (trans-sulfuration is related to oxidative stress and indirectly affects methylation) and genome-wide DNA hypomethylation in autistic patients[53,54] and their parents,[55] in SCZ patients,[72] and in mothers who are in the third trimester of pregnancy and will give birth to individuals who eventually will develop the disorder.[24]

In mood disorders, female monozygotic twins discordant for BD exhibit differential DNA methylation patterns in blood and buccal cells for chromosome X inactivation.[86] However, a study using methylation-sensitive/insensitive restriction enzymes followed by a radioactive DNA polymerase reaction found no global DNA methylation changes in blood leukocytes of euthymic BD patients compared to matched controls.[23] In bipolar II disorder, DNA hypomethylation of PPIEL (peptidylprolyl isomerase E-like) was associated with increased expression.[58] Other reports of interest include: hyper-expression of DNMT1[88,111] and an increase in SAM content in cortical interneurons of patients with SCZ and psychotic BD.[45]

DNA hypermethylation of the gamma-aminobutyric acid A (GABAA) receptor promoter in the frontal cortex may play a role in suicide.[82] Hypermethylation of ribosomal RNA gene promoter in the hippocampus and DNA hypermethylation of the promoter of the neuron-specific glucocorticoid receptor (NR3C1) have been associated with suicide and a history of childhood abuse, respectively.[66,67] Measurement of allele-specific methylation of the 5-hydroxytryptamine (serotonin) receptor 2A (HTR2A) gene in individuals heterozygous for the T102C polymorphism showed that the methylation of the C allele in the prefrontal cortex of suicide victims did not differ from that of controls. However, methylation of the C allele in the DNA of white blood cells differed significantly in schizophrenic individuals who attempted suicide. This was not the case in bipolar individuals who attempted suicide.[32] Epigenetic analysis of other serotonergic genes in white blood cells also showed a trend for an association between DNA hypermethylation of the promoter of the serotonin transporter (5HTT) and a lifetime history of major depression. Consistent with a higher frequency of major depressive disorder in females, DNA methylation of 5HTT was higher and mRNA expression was lower in females compared to males.[80] A rat model of post-traumatic stress disorder (PTSD) showed that DNA hypermethylation of the gene for Disks Large-Associated Protein is associated with hypoexpression of the gene in the hippocampus.[26]

Smoking decreases promoter DNA methylation of monoamine oxidase A (MAOA) in human lymphoblasts[79] and also changes the inherited global DNA methylation pattern.[47] Animal studies have shown that short-term nicotine use (4 days) decreases DNMT1 mRNA and protein in GABAergic interneurons by the mediation of nicotinic acetylcholine receptors (nAChRs), and increases 67 kDa glutamic acid decarboxylase (GAD67) expression in hippocampus and mouse frontal cortex, particularly in layer 1.[90]

In alcoholics, the dopamine transporter 1 (DAT1) promoter hypermethylation is linked to less alcohol craving.[46] DNA methylation of the NR2B (NMDA receptor subtype 2) promoter is negatively correlated with lifetime alcohol consumption and associated with NR2B gene hyperexpression.[16] Patients suffering from alcohol dependence in the early abstinent phase had shown elevated promoter DNA methylation of the vasopressin gene but no significant changes in gene expression. The atrial natriuretic peptide promoter DNA is hypo-methylated in alcoholic patients and is associated with increased mRNA expression.[46]Alcoholism is also linked to DNA hypermethylation of the alpha synuclein promoter, a gene involved in dopamine transmission[20] and of the HERP (homocysteine-induced endoplasmic reticulum protein) gene promoters, correlated with an elevated level of homocysteine.[19]

In patients with the eating disorders, anorexia nervosa and bulimia nervosa, DNA hypermethylation of the promoter of the DAT1 gene is associated with hyperexpression of the gene. In addition, in anorexia nervosa, DNA hypermethylation of the DRD2 promoter is associated with reduced gene expression in the patient's blood.[37]

In Alzheimer's Disease, the circadian genes, PER1 (period homolog 1) and CRY1 (Cryptochrome1), are hyper-methylated in the blood cells[60] and age-related changes of DNA methylation in SORBS3 and S100A2 genes are accelerated.[96]

11.2.2 Aberrant Histone Modification in Major Mental Diseases

DNA is negatively charged electrically, and wrapped around histone proteins with a positive charge. The nucleosome (the basic unit of chromatin) contains five types of histone proteins, including histone 1 (H1), histone 2A &B (H2A &H2B), histone 3 (H3), and 4 (H4), which together shield 146 base pairs of DNA. The terminal amino acids of histone proteins (H3, in particular) known as histone tails are subject to diverse modifications such as acetylation, methylation, and phosphorylation that change the histone electric charges, unleash the attached DNA, and make it available to transcription factors. The extent of histone modifications, their interaction with DNA methylation, and their impacts on transcription are complex and not well understood. In brief, H3 lysine 9 acetylation (H3K9ac) and H3 lysine 4 methylation (H3K4me) are mediated by histone acetyltransferases (HATs) and histone methyltransferase (HMT), respectively. These histone codes correspond to unmethylated promoter DNA which promotes transcription. The activity of histone deacetylases (HDACs) is associated with gene silencing that can be reversed by HDAC inhibitors such as sodium butyrate and valproate.[57,94,109]

A relatively early report indicated that the neutrophils of patients with SCZ had an altered chromatin ultrastructure.[50] Treatment with pimozide, a selective dopamine antagonist, reduced the arginine-rich histones in these patients.[100] Almost three decades later, microarray analysis showed abnormal expression of histone deacetylase 3 in the temporal cortex of patients with SCZ.[9] This was followed by a report that higher levels of H3-(methyl)arginine 17 are associated with increased expression of four metabolic genes (*CRYM, CYTOC/CYC1*, *MDH*, and *OAT*) in the prefrontal cortex of a subset of patients (8 out of 41) with SCZ.[7] When SCZ and BD patients were treated with valproate, H3 acetylation was increased in the lymphocytes of BD patients, in particular.[92] Histone H3-lysine-4 (H3K4) methylation at GAD1 and other GABAergic gene promoters increase in an age-dependent manner in the human prefrontal cortex. In SCZ patients, the observed hypomethylation of histone H3K4 at GAD1 promoter is associated with reduced expression of GAD1.[49] These findings were attributed to the dysfunction of the mixed-lineage leukemia 1 gene which acts as a histone methyltransferase in GABAergic and other cortical neurons. Complementary animal studies showed that clozapine (a potent atypical antipsychotic drug acting both on DRD2 like and HTR2A receptors), but not haloperidol (a classic antipsychotic drug which mainly blocks DRD2 like receptors), increases H3K4 methylation in mouse cerebral cortex.[49]

An analysis of the microarray collection of the National Brain Databank (Harvard Brain Tissue Resource Center) has shown that the expression of HDAC1 is higher in the frontal cortex of SCZ patients than in controls. In addition, mRNA expression of GAD67, a gene involved in SCZ pathogenesis and regulated by epigenetic mechanisms, is inversely correlated with the expression levels of HDAC1, HDAC3, and HDAC4.[91] This is also true in the hippocampus of SCZ patients.[11]

Chromatin immunoprecipitation analysis of the fetal brain, with antibodies against H3K9Ac (histone H3 acetylated at lysine 9) and H3K4me1 (histone H3 mono-methylated at lysine 4), which mark promoters and enhancers of the genes, has provided supporting evidence for the presence of several putative regulatory elements, of core and proximal promoters in genes related to psychiatric diseases; examples are NRG1, DTNBP1, DISC1, DAO, DAOA, PDE4B, and COMT.[76] More epigenetically relevant findings can be expected as high-throughput analysis continues to be applied to clinical situations.

Treatment with the HDACi valproate leads to a significant dose-dependent increase in Acetylated Histone 3 proteins in lymphocytes from individuals with BD and, to a lesser extent, in persons with SCZ.[92] Interestingly, cultures of lymphocytes from SCZ patients were less responsive to trichostatin A (TSA) than lymphocytes of controls.[39] The baseline levels of dimethylated lysine 9 of histone 3 (H3K9me2), a repressive chromatin mark, were higher in lymphocytes cultured from SCZ patients than in lymphocytes from healthy controls.[40] These investigators also reported that the age of disease onset correlated negatively with levels of H3K9me2 and that valproate and TSA induced an almost fourfold increase in GAD67 mRNA, a twofold increase in total acetylation of H3K9,K14 (acetylated histone H3 at lysine 9 and 14) and a fivefold increase in acetylation of H3K9,K14 attachment to the GAD67 promoter in the lymphocytes. Gavin et al. also report that BD patients had higher baseline levels of H3K9,K14 acetylation compared to SCZ patients, and that in subjects with therapeutic levels of valproate (>65 μg/mL) the expression of GAD67 mRNA in lymphocytes was higher.[41] In mouse studies, MS-275, a potent benzamide derivative histone deacetylase (HDAC) inhibitor, selectively increased GAD67 and RELN acetylhistone 3 (Ac-H3) in the frontal cortex and hippocampus, but not in the striatum.[97] There is also evidence that valproate-mediated chromatin remodeling is facilitated by clozapine or sulpiride (a benzamide used in psychotic patients), but not by haloperidol or olanzapine, an atypical antipsychotic drug.[44]

In patients with major depression, quantitative real-time PCR analysis showed, in comparison with controls, an increase in the expression of HDAC2&5 mRNA in peripheral white blood cells in the depressive phase, but not in the remissive phase. In BD, as compared to controls, HDAC4 mRNA expression increased in the depressive phase, and in HDAC6&8 expressions decreased in both depressive and remissive phases. First-degree relatives of the patients, however, showed no significant changes in gene expression levels. In subsequent animal studies the expression of these histones was not found to be affected by chronic treatment with antidepressants or mood stabilizers.[48]

Based on animal studies, extinction of conditioned fear is linked to histone H4 acetylation of the BDNF gene promoter, leading to an increase in gene expression. Valproic acid, a widely used mood stabilizer in psychiatry, further increases the extent of the histone modifications.[22] Hypo-acetylation of the BDNF III and IV promoters occurs in mice exposed to chronic social defeat. This can be prevented by long-term imipramine treatment, an antidepressant which selectively downregulates histone deacetylase 5.[105] Single immobilization stress in mice leads to a significant decrease in the levels of acetylated histone H3 at the promoters of BDNF exons I, IV, and VI. This in turn brings about, within 2 h, a significant decrease in the BDNF exons I, IV, and in total, mRNA. BDNF protein is reduced in 4 h.[38] In cell culture experiments neuronal activity is associated with a decrease in the expression of DNMT1 and DNMT3a, which leads to an increase in the expression of BDNF exon I and IV mRNA.[93] DNMT inhibitors can also increase activity-dependent de-methylation of BDNF in hippocampal neurons by mediation of NMDA receptors.[73]

Both lithium, by direct inhibition of glycogen synthase kinase-3 (GSK-3), and valproate, by inhibition of HDAC increase promoter activity of BDNF IV.[120] This involves responsive elements that are located upstream, but are independent of the three calcium-responsive elements (CaREs) that cause depolarization-induced BDNF expression.[120] Electroconvulsive therapy alters the level of acetylation of BDNF III/IV, as well as c-fos and CREB (cAMP response element-binding protein) gene promoters, with corresponding increases in gene expression in animal studies.[106] These findings suggest that epigenetic factors are not only involved in the pathogenesis of mental diseases, but that some psychiatric drugs, electroconvulsive therapy, even behavioral therapy may act via epigenetic mechanisms. However, much work is required to identify the details, time-point, and causes of epigenetic alterations in affected tissues and to classify the subtype of diseases based on genetic/epigenetic aberrations and translate this knowledge to clinical practice. Although current epigenetic therapeutics are nonspecific at this time, targeted epigenetic modifications using other methods such as pallet implantation or transcranial magnetic stimulating are in the horizon.

11.2.3 Dysregulation of miRNA in Mental Diseases

MicroRNAs (miRNAs) are small noncoding RNAs, 21–25-nucleotides in length that are key regulators of messenger RNA (mRNA). By binding to target mRNAs, miRNAs inhibit the translation of mRNA to protein or induce the degradation of RNA.[74] Almost 1,000 miRNAs have been identified in mammals to date. Interestingly, each miRNA can target hundreds of genes and each gene could be regulated by several miRNAs. The first step in miRNA biogenesis takes place in the nucleus and is mediated by the microprocessor complex that contains the RNase III-like enzyme Drosha and its cofactor DGCR8 (a double-stranded RNA binding protein, DiGeorge syndrome critical region gene 8), generating precursor miRNA (pre-miRNA) from primary miRNA (pri-miRNA). Next, mature miRNAs are generated from pre-miRNAs by the Dicer/TRBP complex in the cytoplasm.[101] The association of a complex of NF90-NF45 (nuclear factors 90&45) with pri-miRNAs prevents the access of the microprocessor complex to the pri-miRNAs, thereby reducing the production of mature miRNA.[89]

Long-term potentiation and long-term depression involved in synaptic plasticity regulate the expression of hippocampal miRNAs.[75] Microarray analyses of human postmortem brain[77] have shown that of 260 miRNAs in the frontal lobe of patients affected by SCZ and schizoaffective disorder, reduced expression was present in 16 miRNAs. Quantitative real-time PCR analysis confirmed significantly reduced expression of four miRNAs, miRNA-24, miRNA-26b, miRNA-30e, and miRNA-92. In addition, hsa-miRNA-106b was hyperexpressed in the patients compared with controls. To rule out the effect of antipsychotic drugs, rats were treated with haloperidol. Three miRNAs (miRNA-199a, miRNA-128a, and miRNA-128b) were hyperexpressed in the treated rats, none of which was however differentially expressed in the prefrontal cortex of the patients. Further experiments led to the conclusion that miRNA changes are likely due to altered regulation of miRNA biogenesis, which then is the potential cause of SCZ pathogenesis.[77]

Beveridge et al. analyzed the superior temporal gyrus and the dorsolateral prefrontal cortex and report an increase in cortical microRNA biogenesis as the result of elevated primary microRNA processing that leads to the increase of DGCR8 (a RNA binding protein, and a component of microprocessor) in SCZ.[14] Among highly expressed miRNAs in superior temporal gyrus and dorsolateral prefrontal cortex were miRNA-15a, miRNA-15b, miRNA-195, and miRNA-107, which are thought to target *BDNF*, *NRG1*, *RELN*, *DRD1*, *HTR4*, *GABR1*, *GRIN1*, *GRM7*, *CHRM1*, and *ATXN2*, all of which are known to be involved in SCZ pathogenesis. The same group of researchers also reported upregulation of miRNA 181b associated with decreased expression of VSNL1 (calcium sensor gene visinin-like 1) and GRIA2 (inotropic glutamate receptor AMPA 2) in postmortem gray matter of the temporal cortex in patients with SCZ.[15] Among other putative target genes are glutamate receptors (GRM5, GRM7, GRIK2, and GRID), GABRA1, HTR1B, and HTR2C. Further studies[15] in cell culture verified that synthetic miRNA-181b suppresses the expression of VSNL1 and GRIN2, reported to be involved in SCZ pathogenesis.[12,110]

miRNA-346 and glutamate receptor ionotropic delta 1 (GRID1) are both at a lower level of expression in postmortem brain of SCZ patients as compared to normal controls. Although the miRNA-346 sequence is located in intron 2 of GRID1, the expression of miRNA-346 and GRID1 were less correlated in SCZ than in bipolar patients or normal controls. Zhu et al. suggest that miRNA-346 targets known SCZ "susceptibility genes more frequently than expected relative to other genes in the genome." [123]

Very rare mutations in the microRNA of X-chromosomes of male SCZ patients have been reported.[34] Lithium and valproate act on miRNA expression (e.g., let-7b, let-7c, miRNA-128a, miRNA-24a, miRNA-30c, miRNA-34a, miRNA-221, and miRNA-144) in vivo as in vitro.[25,122] Glutamate receptor, metabotropic 7 (GRM7) is thought to be acted on by miRNA-34a. Lithium or VPA may decrease the level of miRNA-34a and increase the level of GRM7. Because treatment with the miRNA-34a precursor lowers GRM7 level, and a miRNA-34a inhibitor increases GRM7 levels, "endogenous miRNA-34a (may) regulate(s) GRM7 level and contribute(s) to the effects of lithium and VPA on GRM7."[122]

It is thus apparent, as indicated above, that miRNAs are involved in the pathogenesis of major mental diseases and the effects of known psychiatric drugs may be mediated by miRNA modulation. Given every miRNA regulates hundreds of genes and each gene can be regulated by several miRNAs, much work remains to be done before miRNA-based drugs can be developed to prevent and treat psychiatric diseases.

11.3 Epigenetic Aberrations as Potential Causes of the Paternal Age Effects, Sexual Differences, and the Loss of Brain Laterality in Mental Diseases

Among unexplained dilemmas in mental diseases are the observed relationships with paternal age, living in urban areas, and gender differences. Epigenetic research in neuroscience now provides convincing evidence that, in addition to the possibility of de novo mutations, the impact of paternal age on SCZ pathogenesis could be due to epigenetic aberrations that may accumulate in the sperm DNA in older men.[61,78] There has been some speculation, based on limited evidence, that the social milieu influences and is influenced by the epigenome that affects the prevalence and outcome of psychiatric diseases.[68,104] Additionally, epigenetic may play a role in "sex differences in the brain," sexual differentiation and sex behavior, and sexual orientation.[64]

Another interesting domain is the potential role of epigenetics in functional laterality of the brain hemispheres as a key determinant of the human mind and cognitive state. The disturbance in interhemispheric synchrony underlying the left brain logical dominance may account for a wide range of symptoms, from psychosis to cognitive dysfunction in SCZ and BD.[118] A progressive loss of the gray matter, especially in the left brain, leading to the loss of brain hemispheric normal functional asymmetry is one the most consistent endophenotypes identified by brain imaging studies in SCZ (Table 11.1).

For more than two decades it has been argued that genetic mutations are involved in the loss of brain laterality and may induce SCZ pathogenesis. However, because all brain cells have the same genetic make-up, and the genome of the left and right brain is the same, it seems unlikely that a mutation would result in a unilateral brain defect. In fact, the assumption that genetic mutations responsible for the development of language may also be involved in the emergence of brain laterality contradicts the fundamental principle of genetics that genetic materials are distributed equally in developing cells. Epigenetic dysregulation, rather than genetic mutations, have now been postulated to constitute the underlying mechanisms of SCZ pathogenesis and disruption of human brain lateralization.[27–29] Indeed, epigenetic modifications, the primary leading mechanism of cell and tissue differentiation, may be a more plausible mechanism of the normal functional laterality (asymmetry) of human brain hemispheres disrupted in SCZ.

Abnormal brain asymmetry frequently reported in major mental diseases such as SCZ (Table 11.1) may be a consequence of the loss of inherited brain laterality due to the loss of epigenetic memory. This inference derives support from the studies that describe brain laterality of promoter DNA methylation of DRD2[81] and MB-COMT in normal individuals, which is lost in SCZ and BD.[1] Several environmental factors such as choline, methionine, and folic acid deficiency and/or oxidative stress are linked to the failure of the DNA methylation machinery and of the loss of the epigenetic memory (for more description, see Abdolmaleky et al.[4]) SCZ has also been linked to maternal nutritional deficiency such as exposure to famine during pregnancy.[99,102,119]

11.4 Dilemmas of the Brain Epigenetic Analysis and Methodological Challenges

As epigenetic modulation is a leading mechanism of cell and tissue differentiation, epigenome patterns are cell- and tissue-specific. Even with the use of current high-throughput techniques, the human epigenome project cannot be accomplished in the short term because human tissues at all developmental periods need to be epigenotyped to uncover underlying mechanisms of diseases. In fact, the human epigenome project has to be expanded in two dimensions, spatial (tissue and site specific) and temporal (age- and generation-specific). The brain, the most complicated structure of the living organism, includes very large numbers of nuclei and pathways subject to epigenotyping. At least one-third of human genes are expressed in the brain in a dynamic manner. Moreover, the epigenetic code of a specific group of cells in a pathway is also subject to geographic and environmental variations.

11.5 Conclusion

Epigenetic abnormalities that affect dopaminergic and serotoninergic (e.g., DRD2, DAT1, MB-COMT, MAOA, HTR2A, and 5HTT) as well as GABAergic and glutaminergic genes (e.g., GAD1, GAD67, and RELN) and BDNF are involved in the pathogenesis of major mental diseases. However, underlying etiologies of these aberrations are still not understood sufficiently to lead to prevention or therapeutic use. As a consequence of the key role played by epigenetics, especially in human development, interference with or loss of the epigenetic memory may lead to disease phenotype. Factors involved include nutrition, oxidative stress, inflammation and autoimmune diseases, genetic traits governing the functionality of epigenetic machinery, retroviral infections, maternal stress, paternal age, contaminants, and toxins.[3,4,61] Furthermore, inasmuch as epigenetic alteration of even one gene in a pathway can lead to epigenetic dysregulation of other genes in the pathway, pathway analysis is key to revealing the nature of epigenetic aberrations in psychiatric diseases. It is likely that further epigenetic analyses of the affected tissues will provide valuable information to tease out the etiopathology of major mental diseases in the future. This knowledge may also lead to the discovery of novel preventive, diagnostic, and therapeutic strategies to deal with these devastating diseases that affect so many millions of people worldwide.

References

1. Abdolmaleky HM, Cheng KH, Faraone SV, et al. Hypomethylation of MB-COMT promoter is a major risk factor for schizophrenia and bipolar disorder. *Hum Mol Genet.* 2006;15(21):3132-3145.
2. Abdolmaleky HM, Cheng KH, Russo A, et al. Hypermethylation of the reelin (RELN) promoter in the brain of schizophrenic patients: a preliminary report. *Am J Med Genet B Neuropsychiatr Genet.* 2005;134(1):60-66.
3. Abdolmaleky HM, Smith CL, Faraone SV, et al. Methylomics in psychiatry: modulation of gene-environment interactions may be through DNA methylation. *Am J Med Genet B Neuropsychiatr Genet.* 2004;127B(1):51-59.
4. Abdolmaleky HM, Zhou JR, Thiagalingam S, Smith CL. Epigenetic and pharmacoepigenomic studies of major psychoses and potentials for therapeutics. *Pharmacogenomics.* 2008;9(12):1809-1823. Review.
5. Akbarian S. The neurobiology of Rett syndrome. *Neuroscientist.* 2003;9(1):57-63.
6. Akbarian S, Jiang Y, Laforet G. The molecular pathology of rett syndrome: synopsis and update. *Neuromolecular Med.* 2006;8(4):485-494.
7. Akbarian S, Ruehl MG, Bliven E, et al. Chromatin alterations associated with down-regulated metabolic gene expression in the prefrontal cortex of subjects with schizophrenia. *Arch Gen Psychiatry.* 2005;62(8):829-840.
8. Angrilli A, Spironelli C, Elbert T, Crow TJ, Marano G, Stegagno L. Schizophrenia as failure of left hemispheric dominance for the phonological component of language. *PLoS One.* 2009;4(2):e4507.
9. Aston C, Jiang L, Sokolov BP. Microarray analysis of postmortem temporal cortex from patients with schizophrenia. *J Neurosci Res.* 2004;77(6):858-866.
10. Bangalore SS, Goradia DD, Nutche J, Diwadkar VA, Prasad KM, Keshavan MS. Untreated illness duration correlates with gray matter loss in first-episode psychoses. *Neuroreport.* 2009;20(7):729-734.
11. Benes FM, Lim B, Matzilevich D, Walsh JP, Subburaju S, Minns M. Regulation of the GABA cell phenotype in hippocampus of schizophrenics and bipolars. *Proc Natl Acad Sci USA.* 2007;104(24):10164-10169.
12. Bernstein HG, Braunewell KH, Spilker C, et al. Hippocampal expression of the calcium sensor protein visinin-like protein-1 in schizophrenia. *Neuroreport.* 2002;13(4):393-396.
13. Bestor TH. The DNA methyltransferases of mammals. *Hum Mol Genet.* 2000;9:2395-2402.
14. Beveridge NJ, Gardiner E, Carroll AP, Tooney PA, Cairns MJ. Schizophrenia is associated with an increase in cortical microRNA biogenesis. Mol Psychiatry. 2010;5(12): 1176-1189.
15. Beveridge NJ, Tooney PA, Carroll AP, et al. Dysregulation of miRNA 181b in the temporal cortex in schizophrenia. *Hum Mol Genet.* 2008;17(8):1156-1168.
16. Biermann T, Reulbach U, Lenz B, et al. N-methyl-D-aspartate 2b receptor subtype (NR2B) promoter methylation in patients during alcohol withdrawal. *J Neural Transm.* 2009;116(5):615-622.
17. Bird A. DNA methylation patterns and epigenetic memory. *Genes Dev.* 2002;16:6-21.
18. Bird A. The methyl-CpG-binding protein MeCP2 and neurological disease. *Biochem Soc Trans.* 2008;36(Pt 4):575-583.
19. Bleich S, Lenz B, Ziegenbein M, et al. Epigenetic DNA hypermethylation of the HERP gene promoter induces downregulation of its mRNA expression in patients with alcohol dependence. *Alcohol Clin Exp Res.* 2006;30:587-591.
20. Bonsch D, Lenz B, Kornhuber J, Bleich S. DNA hypermethylation of the alpha synuclein promoter in patients with alcoholism. *Neuroreport.* 2005;16:167-170.
21. Bowden NA, Scott RJ, Tooney PA. Altered gene expression in the superior temporal gyrus in schizophrenia. *BMC Genomics.* 2008;9:199.
22. Bredy TW, Wu H, Crego C, Zellhoefer J, Sun YE, Barad M. Histone modifications around individual BDNF gene promoters in prefrontal cortex are associated with extinction of conditioned fear. *Learn Mem.* 2007;14(4):268-276.
23. Bromberg A, Bersudsky Y, Levine J, Agam G. Global leukocyte DNA methylation is not altered in euthymic bipolar patients. *J Affect Disord.* 2009;118(1–3):234-239.
24. Brown AS, Bottiglieri T, Schaefer CA, et al. Elevated prenatal homocysteine levels as a risk factor for schizophrenia. *Arch Gen Psychiatry.* 2007;64(1):31-39.
25. Chen H, Wang N, Burmeister M, McInnis MG. MicroRNA expression changes in lymphoblastoid cell lines in response to lithium treatment. *Int J Neuropsychopharmacol.* 2009; 2:1-7.
26. Chertkow-Deutsher Y, Cohen H, Klein E, Ben-Shachar D. DNA methylation in vulnerability to post-traumatic stress in rats: evidence for the role of the post-synaptic density protein Dlgap2. Int J Neuropsychopharmacol. 2010;13(3): 347-359.
27. Crow TJ. How and why genetic linkage has not solved the problem of psychosis: review and hypothesis. *Am J Psychiatry.* 2007;164(1):13-21. Review.
28. Crow TJ. Craddock & Owen vs Kraepelin 85 years late, mesmerised by "polygenes". *Schizophr Res.* 2008;103(1–3): 156-160.
29. Crow TJ. A theory of the origin of cerebral asymmetry: epigenetic variation superimposed on a fixed right-shift. *Laterality.* 2009;13:1-15.
30. Crow TJ, Ball J, Bloom SR, et al. Schizophrenia as an anomaly of development of cerebral asymmetry. A postmortem study and a proposal concerning the genetic basis of the disease. *Arch Gen Psychiatry.* 1989;46(12):1145-1150.
31. Cullen TJ, Walker MA, Eastwood SL, Esiri MM, Harrison PJ, Crow TJ. Anomalies of asymmetry of pyramidal cell density and structure in dorsolateral prefrontal cortex in schizophrenia. *Br J Psychiatry.* 2006;188:26-31.

32. De Luca V, Viggiano E, Dhoot R, Kennedy JL, Wong AH. Methylation and QTDT analysis of the 5-HT2A receptor 102C allele: analysis of suicidality in major psychosis. *J Psychiatr Res*. 2009;43(5):532-537.
33. de Vries BB, Jansen CC, Duits AA, et al. Variable FMR1 gene methylation of large expansions leads to variable phenotype in three males from one fragile X family. *J Med Genet*. 1996;33(12):1007-1010.
34. Feng J, Sun G, Yan J, et al. Evidence for X-chromosomal schizophrenia associated with microRNA alterations. *PLoS One*. 2009;4(7):e6121.
35. Filion GJ, Zhenilo S, Salozhin S, Yamada D, Prokhortchouk E, Defossez PA. A family of human zinc finger proteins that bind methylated DNA and repress transcription. *Mol Cell Biol*. 2006;26(1):169-181.
36. Fontecave M, Atta M, Mulliez E. S-adenosylmethionine: nothing goes to waste. *Trends Biochem Sci*. 2004;29(5): 243-249.
37. Frieling H, Römer KD, Scholz S, et al. Epigenetic dysregulation of dopaminergic genes in eating disorders. *Int J Eat Disord*. 2010;43(7):577-583.
38. Fuchikami M, Morinobu S, Kurata A, Yamamoto S, Yamawaki S. Single immobilization stress differentially alters the expression profile of transcripts of the brain-derived neurotrophic factor (BDNF) gene and histone acetylation at its promoters in the rat hippocampus. *Int J Neuropsychopharmacol*. 2009;12(1):73-82.
39. Gavin DP, Kartan S, Chase K, Grayson DR, Sharma RP. Reduced baseline acetylated histone 3 levels, and a blunted response to HDAC inhibition in lymphocyte cultures from schizophrenia subjects. *Schizophr Res*. 2008;103:330-332.
40. Gavin DP, Kartan S, Chase K, Jayaraman S, Sharma RP. Histone deacetylase inhibitors and candidate gene expression: an in vivo and in vitro approach to studying chromatin remodeling in a clinical population. *J Psychiatr Res*. 2009;43(9):870-876.
41. Gavin DP, Rosen C, Chase K, Grayson DR, Tun N, Sharma RP. Dimethylated lysine 9 of histone 3 is elevated in schizophrenia and exhibits a divergent response to histone deacetylase inhibitors in lymphocyte cultures. *J Psychiatry Neurosci*. 2009;34(3):232-237.
42. Glatt SJ, Everall IP, Kremen WS, et al. Comparative gene expression analysis of blood and brain provides concurrent validation of SELENBP1 up-regulation in schizophrenia. *Proc Natl Acad Sci USA*. 2005;102(43):15533-15538.
43. Grayson DR, Jia X, Chen Y, et al. Reelin promoter hypermethylation in schizophrenia. *Proc Natl Acad Sci USA*. 2005;102(26):9341-9346.
44. Guidotti A, Dong E, Kundakovic M, Satta R, Grayson DR, Costa E. Characterization of the action of antipsychotic subtypes on valproate-induced chromatin remodeling. *Trends Pharmacol Sci*. 2009;30(2):55-60.
45. Guidotti A, Ruzicka W, Grayson DR, et al. S-adenosyl methionine and DNA methyltransferase-1 mRNA overexpression in psychosis. *Neuroreport*. 2007;18(1):57-60.
46. Hillemacher T, Frieling H, Hartl T, Wilhelm J, Kornhuber J, Bleich S. Promoter specific methylation of the dopamine transporter gene is altered in alcohol dependence and associated with craving. *J Psychiatr Res*. 2009;43(4):388-392.
47. Hillemacher T, Frieling H, Moskau S, et al. Global DNA methylation is influenced by smoking behaviour. *Eur Neuropsychopharmacol*. 2008;18(4):295-298.
48. Hobara T, Uchida S, Otsuki K, et al. Altered gene expression of histone deacetylases in mood disorder patients. *J Psychiatr Res*. 2010;44(5):263-270.
49. Huang HS, Matevossian A, Whittle C, et al. Prefrontal dysfunction in schizophrenia involves mixed-lineage leukemia 1-regulated histone methylation at GABAergic gene promoters. *J Neurosci*. 2007;27(42):11254-11262.
50. Issidorides MR, Stefanis CN, Varsou E, Katsorchis T. Altered chromatin ultrastructure in neutrophils of schizophrenics. *Nature*. 1975;258(5536):612-614.
51. Iwamoto K, Bundo M, Yamada K, et al. DNA methylation status of SOX10 correlates with its downregulation and oligodendrocyte dysfunction in schizophrenia. *J Neurosci*. 2005;25(22):5376-5381.
52. Iwamoto K, Kakiuchi C, Bundo M, Ikeda K, Kato T. Molecular characterization of bipolar disorder by comparing gene expression profiles of postmortem brains of major mental disorders. *Mol Psychiatry*. 2004;9(4):406-416.
53. James SJ, Cutler P, Melnyk S, et al. Metabolic biomarkers of increased oxidative stress and impaired methylation capacity in children with autism. *Am J Clin Nutr*. 2004;80(6): 1611-1617.
54. James SJ, Melnyk S, Jernigan S, et al. Metabolic endophenotype and related genotypes are associated with oxidative stress in children with autism. *Am J Med Genet B Neuropsychiatr Genet*. 2006;141B(8):947-956.
55. James SJ, Melnyk S, Jernigan S, Hubanks A, Rose S, Gaylor DW. Abnormal Transmethylation/transsulfuration Metabolism and DNA Hypomethylation Among Parents of Children with Autism. *J Autism Dev Disord*. 2008;38(10): 1966-1975.
56. Kim GD, Ni J, Kelesoglu N, Roberts RJ, Pradhan S. Co-operation and communication between the human maintenance and de novo DNA (cytosine-5) methyltransferases. *Embo J*. 2002;21:4183-4195.
57. Kouzarides T, Berger SL. Chromatin modifications and their mechanisms of action. In: Allis CD, Jenuwein T, Reinberg D, eds. *Epigenetics*. Cold Spring Harbor, NY: Cold Spring Harbor Laboratory Press; 2007:191-197.
58. Kuratomi G, Iwamoto K, Bundo M, et al. Aberrant DNA methylation associated with bipolar disorder identified from discordant monozygotic twins. *Mol Psychiatry*. 2008;13(4): 429-441.
59. Lister R, Pelizzola M, Dowen RH, et al. Human DNA methylomes at base resolution show widespread epigenomic differences. *Nature*. 2009;462(7271):315-322.
60. Liu HC, Hu CJ, Tang YC, Chang JG. A pilot study for circadian gene disturbance in dementia patients. *Neurosci Lett*. 2008;435(3):229-233.
61. Malaspina D, Perrin M, Kleinhaus KR, Opler M, Harlap S. Growth and schizophrenia: aetiology, epidemiology and epigenetics. *Novartis Found Symp*. 2008;289:196-203. discussion 203–7:238–40. Review.
62. Malmgren H, Steén-Bondeson ML, Gustavson KH, et al. Methylation and mutation patterns in the fragile X syndrome. *Am J Med Genet*. 1992;43(1–2):268-278.
63. Martinowich K, Hattori D, Wu H, et al. DNA methylation-related chromatin remodeling in activity-dependent BDNF gene regulation. *Science*. 2003;302(5646):890-893.
64. McCarthy MM, Auger AP, Bale TL, et al. The epigenetics of sex differences in the brain. *J Neurosci*. 2009;29(41):-12815-12823. Review.

65. McGowan PO, Meaney MJ, Szyf M. Diet and the epigenetic (re)programming of phenotypic differences in behavior. *Brain Res*. 2008;1237:12-24.
66. McGowan PO, Sasaki A, D'Alessio AC, et al. Epigenetic regulation of the glucocorticoid receptor in human brain associates with childhood abuse. *Nat Neurosci*. 2009;12(3): 342-348.
67. McGowan PO, Sasaki A, Huang TC, et al. Promoter-wide hypermethylation of the ribosomal RNA gene promoter in the suicide brain. *PLoS One*. 2008;3(5):e2085.
68. McGowan PO, Szyf M. The epigenetics of social adversity in early life: implications for mental health outcomes. *Neurobiol Dis*. 2010;39(1):66-72.
69. Meda SA, Giuliani NR, Calhoun VD, et al. A large scale (N=400) investigation of gray matter differences in schizophrenia using optimized voxel-based morphometry. *Schizophr Res*. 2008;101(1–3):95-105.
70. Mill J, Tang T, Kaminsky Z, et al. Epigenomic profiling reveals DNA-methylation changes associated with major psychosis. *Am J Hum Genet*. 2008;82(3):696-711.
71. Monk M. Epigenetic programming of differential gene expression in development and evolution. *Dev Genet*. 1995; 17:188-197.
72. Muntjewerff JW, Kahn RS, Blom HJ, den Heijer M. Homocysteine, methylenetetrahydrofolate reductase and risk of schizophrenia: a meta-analysis. *Mol Psychiatry*. 2006;11(2):143-149.
73. Nelson ED, Kavalali ET, Monteggia LM. Activity-dependent suppression of miniature neurotransmission through the regulation of DNA methylation. *J Neurosci*. 2008;28(2): 395-406.
74. Okamura K, Lai EC. Endogenous small interfering RNAs in animals. *Nat Rev Mol Cell Biol*. 2008;9(9):673-678.
75. Park CS, Tang SJ. Regulation of microRNA expression by induction of bidirectional synaptic plasticity. *J Mol Neurosci*. 2009;38(1):50-56.
76. Pedrosa E, Locker J, Lachman HM. Survey of schizophrenia and bipolar disorder candidate genes using chromatin immunoprecipitation and tiled microarrays (ChIP-chip). *J Neurogenet*. 2009;23(3):341-352.
77. Perkins DO, Jeffries CD, Jarskog LF, et al. microRNA expression in the prefrontal cortex of individuals with schizophrenia and schizoaffective disorder. *Genome Biol*. 2007;8(2):R27.
78. Perrin MC, Brown AS, Malaspina D. Aberrant epigenetic regulation could explain the relationship of paternal age to schizophrenia. *Schizophr Bull*. 2007;33(6):1270-1273.
79. Philibert RA, Beach SR, Gunter TD, Brody GH, Madan A, Gerrard M. The effect of smoking on MAOA promoter methylation in DNA prepared from lymphoblasts and whole blood. *Am J Med Genet B Neuropsychiatr Genet*. 2010; 153B(2):619-628.
80. Philibert RA, Sandhu H, Hollenbeck N, Gunter T, Adams W, Madan A. The relationship of 5HTT (SLC6A4) methylation and genotype on mRNA expression and liability to major depression and alcohol dependence in subjects from the Iowa Adoption Studies. *Am J Med Genet B Neuropsychiatr Genet*. 2008;147B(5):543-549.
81. Popendikyte V, Laurinavicius A, Paterson AD, Macciardi F, Kennedy JL, Petronis A. DNA methylation at the putative promoter region of the human dopamine D2 receptor gene. *Neuroreport*. 1999;10(6):1249-1255.
82. Poulter MO, Du L, Weaver IC, et al. GABAA receptor promoter hypermethylation in suicide brain: implications for the involvement of epigenetic processes. *Biol Psychiatry*. 2008;64(8):645-652.
83. Prasad KM, Keshavan MS. Structural cerebral variations as useful endophenotypes in schizophrenia: do they help construct "extended endophenotypes"? *Schizophr Bull*. 2008; 34(4):774-790. Review.
84. Purcell SM, Wray NR, Stone JL, et al. Common polygenic variation contributes to risk of schizophrenia and bipolar disorder. International Schizophrenia Consortium. *Nature*. 2009;460(7256):748-752.
85. Rockstroh B, Clementz BA, Pantev C, Blumenfeld LD, Sterr A, Elbert T. Failure of dominant left-hemispheric activation to right-ear stimulation in schizophrenia. *Neuroreport*. 1998;9(17):3819-3822.
86. Rosa A, Picchioni MM, Kalidindi S, et al. Differential methylation of the X-chromosome is a possible source of discordance for bipolar disorder female monozygotic twins. *Am J Med Genet B Neuropsychiatr Genet*. 2008;147B(4): 459-462.
87. Russo V, Martienssen R, Riggs A. *Epigenetic Mechanisms of Gene Regulation*. Cold Spring Harbor, NY: Cold Spring Harbor Laboratory Press; 1996.
88. Ruzicka WB, Zhubi A, Veldic M, Grayson DR, Costa E, Guidotti A. Selective epigenetic alteration of layer I GABAergic neurons isolated from prefrontal cortex of schizophrenia patients using laser-assisted microdissection. *Mol Psychiatry*. 2007;12(4):385-397.
89. Sakamoto S, Aoki K, Higuchi T, et al. The NF90-NF45 complex functions as a negative regulator in the microRNA processing pathway. *Mol Cell Biol*. 2009;29(13):37543769.
90. Satta R, Maloku E, Zhubi A, et al. Nicotine decreases DNA methyltransferase 1 expression and glutamic acid decarboxylase 67 promoter methylation in GABAergic interneurons. *Proc Natl Acad Sci USA*. 2008;105(42): 16356-16361.
91. Sharma RP, Grayson DR, Gavin DP. Histone deactylase 1 expression is increased in the prefrontal cortex of schizophrenia subjects: analysis of the National Brain Databank microarray collection. *Schizophr Res*. 2008;98(1–3):111-117.
92. Sharma RP, Rosen C, Kartan S, et al. Valproic acid and chromatin remodeling in schizophrenia and bipolar disorder: preliminary results from a clinical population. *Schizophr Res*. 2006;88:227-231.
93. Sharma RP, Tun N, Grayson DR. Depolarization induces downregulation of DNMT1 and DNMT3a in primary cortical cultures. *Epigenetics*. 2008;3(2):74-80.
94. Shilatifard A. Molecular implementation and physiological roles for histone H3 lysine 4 (H3K4) methylation. *Curr Opin Cell Biol*. 2008;20(3):341-348.
95. Shimabukuro M, Sasaki T, Imamura A, et al. Global hypomethylation of peripheral leukocyte DNA in male patients with schizophrenia: a potential link between epigenetics and schizophrenia. *J Psychiatr Res*. 2007;41(12): 1042-1046.
96. Siegmund KD, Connor CM, Campan M, et al. DNA methylation in the human cerebral cortex is dynamically regulated throughout the life span and involves differentiated neurons. *PLoS ONE*. 2007;2(9):e895.
97. Simonini MV, Camargo LM, Dong E, et al. The benzamide MS-275 is a potent, long-lasting brain region-selective

inhibitor of histone deacetylases. *Proc Natl Acad Sci USA*. 2006;103(5):1587-1592.
98. Sommer IE, Diederen KM, Blom JD, et al. Auditory verbal hallucinations predominantly activate the right inferior frontal area. *Brain*. 2008;131(Pt 12):3169-3177.
99. Song S, Wang W, Hu P. Famine, death, and madness: schizophrenia in early adulthood after prenatal exposure to the Chinese Great Leap Forward Famine. *Soc Sci Med*. 2009;68(7):1315-1321. Epub 2009 Feb 14.
100. Stefanis CN, Issidorides MR. Histochemical changes in the blood cells of schizophrenic patients under pimozide treatment. *Biol Psychiatry*. 1976;11(1):53-68.
101. Sun G, Yan J, Noltner K, et al. Sep;SNPs in human miRNA genes affect biogenesis and function. *RNA*. 2009;15(9):1640-1651.
102. Susser E, Neugebauer R, Hoek HW, et al. Schizophrenia after prenatal famine. Further evidence. *Arch Gen Psychiatry*. 1996;53(1):25-31.
103. Sutcliffe JS, Nelson DL, Zhang F, et al. DNA methylation represses FMR-1 transcription in fragile X syndrome. *Hum Mol Genet*. 1992;1(6):397-400.
104. Szyf M. The early life environment and the epigenome. *Biochim Biophys Acta*. 2009;1790(9):878-885. Review.
105. Tsankova NM, Berton O, Renthal W, Kumar A, Neve RL, Nestler EJ. Sustained hippocampal chromatin regulation in a mouse model of depression and antidepressant action. *Nat Neurosci*. 2006;9(4):519-525.
106. Tsankova NM, Kumar A, Nestler EJ. Histone modifications at gene promoter regions in rat hippocampus after acute and chronic electroconvulsive seizures. *J Neurosci*. 2004;24(24):5603-5610.
107. Tsuang MT, Gilbertson MW, Faraone SV. The genetics of schizophrenia. Current knowledge and future directions. *Schizophr Res*. 1991;4(2):157-171. Review.
108. Unterberger A, Szyf M, Nathanielsz PW, Cox LA. Organ and gestational age effects of maternal nutrient restriction on global methylation in fetal baboons. *J Med Primatol*. 2009;38(4):219-227.
109. Vaissière T, Sawan C, Herceg Z. Epigenetic interplay between histone modifications and DNA methylation in gene silencing. *Mutat Res*. 2008;659(1–2):40-48.
110. Vawter MP, Crook JM, Hyde TM, et al. Microarray analysis of gene expression in the prefrontal cortex in schizophrenia: a preliminary study. *Schizophr Res*. 2002;58(1):11-20.
111. Veldic M, Caruncho HJ, Liu WS, et al. DNA-methyltransferase 1 mRNA is selectively overexpressed in telencephalic GABAergic interneurons of schizophrenia brains. *Proc Natl Acad Sci USA*. 2004;101(1):348-353.
112. Venkatasubramanian G, Jayakumar PN, Gangadhar BN, Keshavan MS. Automated MRI parcellation study of regional volume and thickness of prefrontal cortex (PFC) in antipsychotic-naïve schizophrenia. *Acta Psychiatr Scand*. 2008;117(6):420-431.
113. Venkatasubramanian G, Jayakumar PN, Gangadhar BN, Keshavan MS. Neuroanatomical correlates of neurological soft signs in antipsychotic-naive schizophrenia. *Psychiatry Res*. 2008;164(3):215-222.
114. Waddington CH. *Introduction to Modern Genetics*. London: Allen & Unwin; 1939.
115. Walter H, Wunderlich AP, Blankenhorn M, et al. No hypofrontality, but absence of prefrontal lateralization comparing verbal and spatial working memory in schizophrenia. *Schizophr Res*. 2003;61(2–3):175-184.
116. Weaver IC, Cervoni N, Champagne FA, et al. Epigenetic programming by maternal behavior. *Nat Neurosci*. 2004; 8:847-854.
117. Weiss EM, Hofer A, Golaszewski S, et al. Brain activation patterns during a verbal fluency test-a functional MRI study in healthy volunteers and patients with schizophrenia. *Schizophr Res*. 2004;70(2–3):287-291.
118. Williams LM, Whitford TJ, Gordon E, Gomes L, Brown KJ, Harris AW. Neural synchrony in patients with a first episode of schizophrenia: tracking relations with grey matter and symptom profile. *J Psychiatry Neurosci*. 2009; 34(1):21-29.
119. Xu MQ, Sun WS, Liu BX, et al. Prenatal malnutrition and adult schizophrenia: further evidence from the 1959-1961 Chinese famine. *Schizophr Bull*. 2009;35(3):568-576.
120. Yasuda S, Liang MH, Marinova Z, Yahyavi A, Chuang DM. The mood stabilizers lithium and valproate selectively activate the promoter IV of brain-derived neurotrophic factor in neurons. *Mol Psychiatry*. 2009;14(1):51-59.
121. Yotsutsuji T, Saitoh O, Suzuki M, et al. Quantification of lateral ventricular subdivisions in schizophrenia by high-resolution three-dimensional magnetic resonance imaging. *Psychiatry Res*. 2003;122(1):1-12.
122. Zhou R, Yuan P, Wang Y, et al. Evidence for selective microRNAs and their effectors as common long-term targets for the actions of mood stabilizers. *Neuropsychopharmacology*. 2008;34(6):1395-1405.
123. Zhu Y, Kalbfleisch T, Brennan MD, Li Y. A MicroRNA gene is hosted in an intron of a schizophrenia-susceptibility gene. *Schizophr Res*. 2009;109(1–3):86-89.
124. Zschocke J, Allritz C, Engele J, Rein T. DNA methylation dependent silencing of the human glutamate transporter EAAT2 gene in glial cells. *Glia*. 2007;55(7):663-674.

12 Epigenetics and Late-Onset Alzheimer's Disease

Axel Schumacher, Syed Bihaqi, and Nasser H. Zawia

12.1 Introduction

Alzheimer's disease (AD) is the most prevalent neurodegenerative disease, affecting about 20 million people worldwide.[35] It is characterized by progressive loss of memory, declining cognitive function and, ultimately, leads to decreasing physical functions and death. The neuropathological hallmarks of AD are the development of senile plaques and the formation of neurofibrillary tangles, intracellular neuronal lesions deposited in the brain. The neurofibrillary tangles represent bundles of paired helical filaments, which mainly consist of the microtubule-associated protein tau in an abnormally phosphorylated form.[2] The extracellular amyloid plaques mainly consist of the 42-residue long amyloid β-peptide which is proteolytically derived from the much larger amyloid precursor protein (APP).[27] The generation and subsequent aggregation of amyloid beta (Aβ) seems to be at the origin of the disease and is believed to trigger a complex pathological cascade that ultimately causes neuronal dysfunction.[36]

Even though AD is the most prominent form of dementia among the elderly, it is still unclear whether AD is initiated during old age or has its origin earlier in life. Epidemiological studies have shown that people who have a low mental score on intelligence tests during childhood are at higher risk of developing this disease during old age. Most cases of early-onset Alzheimer disease (EOAD) are probably caused by gene mutations that can be passed from parent to the offspring. However, the vast majority of AD cases (~90%) are sporadic, possess no clear genetic association, and account for late-onset AD (LOAD). Current theories about the development of LOAD hinge on the premise that AD arises mainly from heritable causes. Yet, the complex, non-Mendelian disease etiology suggests that an epigenetic component could be involved.[78] Compared to genetic causes, epigenetic factors are probably better suited to explain the observed anomalies in LOAD as aberrant epigenetic patterns that may be acquired during many developmental stages.[78] The epigenome is particularly susceptible to deregulation during early embryonic and neonatal development and puberty. This is also true for old age,[24] which is the most important known risk factor for AD. Multiple lines of evidence indicate that oxidative stress is a contributor to neuronal death in AD. The oxidative damage that occurs to DNA may play a role in both normal aging and neurodegenerative diseases, including AD.[47] Hazardous interactions with DNA by reactive oxygen species (ROS), particularly hydroxyl radicals, can lead to strand breaks, DNA–DNA and DNA–protein cross-linking, sister chromatid exchange and translocation, and formation of at least 20 oxidized base adducts. Modification of DNA bases can lead to mutation and altered protein synthesis. In late-stage AD brain, accumulation of 8 hydroxyguanosine (8-OHdG), an oxidized form of the base guanine, is found in DNA from impacted brain regions of AD.[32] However, whether DNA oxidative damage actually plays a pivotal role in the neurodegenerative cascade or is just an epiphenomenon of the neurodegenerative process is still unclear.

The molecular transition from memory encoding and initial consolidation to progressive long-term memory storage, retrieval, and reconsolidation involves complex layers of local and system-wide epigenetic modifications. These modifications are associated

A. Schumacher (✉)
The Krembil Family Epigenetics Laboratory,
Centre for Addiction and Mental Health,
Toronto, ON, Canada
e-mail: axel_schumacher@camh.net

H.I. Roach et al. (eds.), *Epigenetic Aspects of Chronic Diseases*,
DOI: 10.1007/978-1-84882-644-1_12, © Springer-Verlag London Limited 2011

with transcriptional, posttranscriptional, translational, and posttranslational changes that influence molecular pathways and interactive networks at the intracellular, synaptic, and systemic levels of information processing, integration, transformation, stabilization, and reconfiguration.[53] There is a dynamic interplay between primary DNA and histone modifications, which induce transient and more enduring modifications of the synaptic and neural networks. Epigenetics represents the long sought-after molecular interface that mediates gene–environment interactions during critical periods of the lifecycle. The discipline of environmental epigenomics has begun to identify profiles of environmental stressors that modulate the latency, initiation and progression of specific neurological disorders. Disease biomarkers have been identified and graded molecular responses to emerging therapeutic interventions are now possible.[53]

This chapter summarizes current literature and results from our studies pointing to the possibility of earlier developmental perturbations due to diet, toxicological exposures, or other factors that render the aging brain subject to neurodegenerative damage through epigenetic mechanisms.

12.2 Epigenetics and Brain Disorders

There are many tantalizing clues that suggest neurodegenerative diseases are mediated by aberrant epigenetic mechanisms. The evidence includes the absence of simple Mendelian inheritance patterns, global transcriptional dysregulation, multiple types of pathogenic RNA alterations, aberrant stimulation of developmental and mitogenic signaling pathways, the labile state of differentiation, defects in axon-dendritic transport, the presence of chronic stress, telomere dysfunction, genomic instability, and the importance of environmental factors and multiple transition states associated with disease pathogenesis. Although a role of epigenetics in human disease was originally recognized in oncology,[56] there is now a considerable body of evidence that implicates the disruption of epigenetic mechanisms as a causal basis for human mental illness, in particular fragile X and Rett syndrome.[1,47,71] Fragile X syndrome is a genetic disorder that is caused by instability at the fragile X mental retardation 1 gene (FMR1) gene locus on the long arm of the X chromosome. The (CGG)n triplet repeat in the 5′ untranslated region of FMR1 exhibits considerable instability upon transmission from mothers with premutation alleles (52–200 repeats). As a result, the repeat sequence (>200 copies) in patients is elongated. Hypermethylation of the CGG repeats results in downregulation of the FMR1 gene product FMRP, which plays important roles in learning and memory, development of *axons*, formation of *synapses*, and the wiring and development of neural circuits.[47,71] The phenotype of Fragile X patients is directly linked to the extent of methylation in the triplet repeat.[23]

Schizophrenia (SZ) and bipolar disorder (BD) have also been associated with abnormal epigenetic mechanisms. Both disorders are etiologically related psychiatric disorders, collectively termed major psychosis (MP) affecting approximately 2% of the population. DNA sequence variation and environmental mechanisms do not fully explain many of the epidemiological, clinical and molecular peculiarities associated with MP. As a consequence, a new interpretation of the classical paradigm of "genes plus environment" has emerged in recent years, with the emphasis shifted to epigenetic deregulation as a major etiopathogenic factor.[61,65] There is growing epidemiological and experimental evidence to support a role for epigenetic dysfunction in MP. Indeed, consistent with the epigenetic theory of MP,[65] a number of loci have been found to be epigenetically altered in the brain of SZ and BD patients, compared to controls.[52] It is therefore apparent that dysfunction of the normal epigenetic status of the genome can have marked consequences on normal cognitive function.

12.3 Epigenetics and Alzheimer's Disease

Gene expression in the AD brain has been shown to be altered in a wide variety of reports, including a recent large-scale expression array study of single cell laser-captured entorhinal cortex layer II neurons.[25] Multiple physiologic and molecular pathways are affected, including energy metabolism, inflammation and aberrant cell cycle events. Although individual pathogenic factors such as Aβ peptide and tau phosphorylation are clearly significant factors for disease development, no common principle explaining the consistency, extent, and breadth of the gene expression and of functional and molecular changes in AD has received consensus acceptance.

The development of EOAD seems to be largely genetic, but a different picture is emerging for LOAD, a common, sporadic form of the illness that affects individuals older than 65 years. Accumulating evidence suggests that LOAD not only results from the combined effects of variation in a number of genes and environmental factors, but also from epigenetic abnormalities such as histone modifications or DNA methylation. In comparison to monogenic diseases, LOAD, as many other complex diseases, exhibits numerous non-Mendelian anomalies that suggest an epigenetic component. The anomalies include: (1) Sporadic cases dominate over familial ones; (2) the estimated concordance rate for monozygotic twins is significantly below 100%; (3) gender specific susceptibility and course of disease[41,66]; (4) parent-of-origin effects[9]; (5) late age of onset; (6) brain chromatin abnormalities, including aberrant histone modifications; (7) non-Mendelian inheritance pattern; (8) atypical levels of folate and homocysteine, indicating an abnormal methylation homeostasis in the brain of AD patients; (9) a disturbed control of the epigenetically regulated circadian clock; and (10) monoallelic expression patterns of susceptibility genes.[34]

Aberrant histone regulation is one phenomenon in AD. For example, the cleavage of APP not only generates the Aβ peptide, but also an APP C-terminal peptide (AICD). It is possible that this AICD peptide interacts with Tip60 (a histone acetyltransferase) either directly or through Fe65 (amyloid beta A4 precursor protein-binding family B member 1) interaction.[3] AICD translocates to the nucleus and acts on such genes as neprilysin (NEP), modifying their expression.[11] Furthermore, the overexpression of AICDs in PC12 cells and in rat primary cortical neurons increases acetylation of histones H3K14 and H4K5.[42] Fe65–AICD interaction is necessary for association with chromatin; this also recruits Tip60 to DNA strand breaks.[69] The correct repair of DNA requires Tip60 acetylation of histone 4. This process, which may be altered in AD as an accumulation of phospho-H2AX (an indicator of DNA strand breaks) has been described.[58] Most importantly, neuron-specific overexpression of HDAC2 in mice, but not of HDAC1, decreases dendritic spine density, synapse number, synaptic plasticity, and memory formation.[38] Conversely, HDAC2 deficiency results in an increase in synapse number and in memory facilitation. This fact supports the crucial role of histone acetylation and deacetylation in human diseases associated with memory impairment, as true in AD.

12.4 Methylation Homeostasis in LOAD

DNA methylation is a vital component of the epigenetic machinery that orchestrates changes in multiple genes and helps regulate gene expression in vertebrates. In the human genome, CpG dinucleotides are generally concentrated in regions called CpG islands; these are preferentially located in promoter regions and typically do not contain 5-methylcytosines.[14] However, some physiological processes require DNA methylation of CpG islands. This includes silencing of imprinted genes in situations where only one allele is expressed as result of which expression becomes only paternal or maternal.

Epidemiologic studies have indicated that impaired intrauterine growth and development are associated with a higher risk in the adult offspring of cardiovascular disease, diabetes mellitus, obesity, and osteoporosis (see Chap. 13 for details). Of the many mechanisms by which in utero and early-life conditions affect adult health and diseases, the epigenetic machinery including DNA methylation has been a major one.[12] Epigenetic patterns that were established during the fetal period can be changed in adult life by stochastic or environmental factors such as nutrition. For example, identical twins possess the same genotype and there are no distinguishable epigenetic differences in their early life. The twins show at a later age, however, very different patterns of genomic DNA methylation and histone acetylation. These epigenetic differences may be responsible for differences in gene expression and disease susceptibility[31] and make clear that exposure to a specific environment assumes importance throughout life. In the brain, this environmentally dependent modulatory period may continue into postnatal development. Genome-wide demethylation patterns are also observed shortly after fertilization and followed later by a new wave of methylation.[82] Thus, the process of methylation and demethylation appears to be a programmed event that provides cells with developmental potential and a mechanism that widens the means to regulate the expression of genes and to transmit information beyond that in the genetic code.

Many non-Mendelian characteristics point also to an involvement of epigenetic factors in age-related diseases, but little is known about epigenetic patterns in LOAD. The first indication that epigenetics may play a central role in LOAD came from a study by Wang et al.[78] of DNA methylation patterns in genes with a known role in the etiology of AD. LOAD patients were found to have greater epigenetic distance from the norm in brain tissue than controls, with the distance increasing with age. These findings support a role for epigenetic effects in disease development. Moreover, some genes that play a central role in amyloid-β processing (i.e., PSEN1 and APOE) displayed significant interindividual epigenetic variability. This may contribute to LOAD predisposition. The PSEN1 promoter is known to be regulated by DNA methylation and may be partially silenced in AD.[33] Abnormal PSEN1 methylation patterns have previously been associated with hypomethylation of the promoter. That could induce an overexpression of PSEN1, resulting in an imbalance in amyloid-β production.[73] Intriguingly, PSEN1 and PSEN2 display epigenetic variability already in male germ cells.[29] These patterns may be transmitted through the germline or be re-established post-zygotically. That in turn would contribute to differences in susceptibility to the disease in later life. In the Wang et al.[78] study, the CpG island regions were predominantly unmethylated and even small changes in the methylation levels of these loci are likely to interfere with critical regulatory functions, sufficient to cause the disease. No major DNA methylation changes have been found in the brains of LOAD patients,[8] but the DNA methylation drift that has been established and the unusual methylation patterns often encountered in LOAD, but not in control brains, strengthen and add weight to the role epigenetic deregulation that is likely to play in age-dependent AD. A study of a monozygotic twin pair discordant for AD showed a significantly reduced level of DNA methylation in temporal neocortex neuronal nuclei of the affected co-twin.[49] This finding is consistent with the hypothesis that epigenetic mechanisms influence the effects of life events on LOAD risk.[78,85]

A three-generation family-based study has demonstrated that significant intraindividual changes occur over time.[15] This study also demonstrated familial clustering of decreases or increases in methylation; this indicates that stringent maintenance of global methylation is itself a heritable trait. The very existence of familial clustering may mean that epigenetic stability is directly related to genetic and epigenetic variation, as in genes that regulate DNA methyltransferase activity or 1-carbon metabolism. Intriguingly, *MTHFR* and *DNMT1* genes that give rise to essential components of the 1-carbon metabolism in humans exhibit notable interindividual variation in DNA methylation in LOAD patients,[78] an indication of the age-related instability of these genes. This in turn, could influence methylation homeostasis in cells. As a consequence, even stochastic variations in methylation profiles in some individuals may result in aberrant levels of important methylation metabolism components, such as homocysteine (Hcy) and folate (Fig. 12.1). Indeed, many epidemiological and experimental studies have linked elevated plasma Hcy (called hyperhomocysteinemia) and low serum folate concentrations to age-related conditions, such as Parkinson's disease and AD.[51]

Hyperhomocysteinemia in AD may be due to a point mutation in genes involved in 1-carbon metabolism, or to iron dysregulation and oxidative stress.[26] Elevations in plasma Hcy precede the development of age-dependent dementia, with an inverse linear relationship existing between plasma Hcy concentrations and cognitive performance in older individuals.[64,66] On the basis of these findings, disruption of some epigenetic pathways may precede age-related phenotypes such as formation of amyloid plaques.

Other components of the methylation pathways are also abnormal in AD cases. The level of *S*-adenosymethionine (SAM), required for the methylation of DNA and histones, is markedly decreased in the spinal fluid and brains of AD patients.[18,55] This may be the cause for hypomethylation of protein phosphatase-2A (PP2A) in AD cases. The assembly of functional PP2A heterotrimers is directly dependent on carboxyl methylation by SAM. The reduced methylation of PP2A may also contribute to the increase in phosphorylation of the microtubule-associated protein tau, a major component of neurofibrillary tangles in AD.[75] Tau is modulated by phosphorylation and the ability of tau to interact with microtubules correlates inversely with the degree of phosphorylation. The cellular ratio of SAM:SAH is an inverse function of the concentration of phosphorylated tau (P-tau) in the cerebrospinal fluid (CSF) of

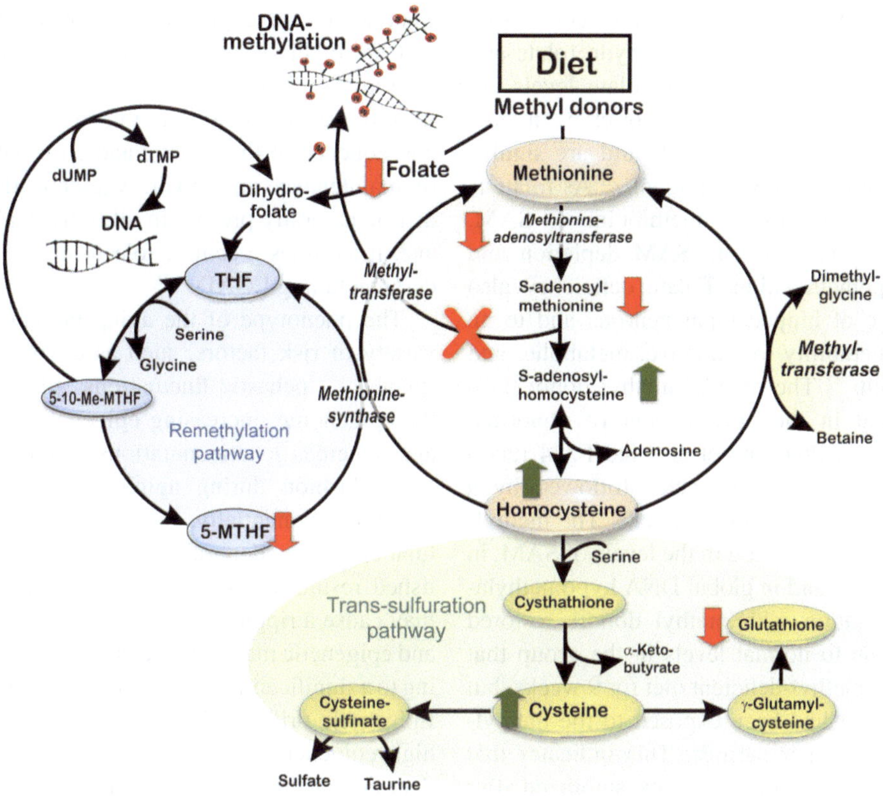

Fig. 12.1 Aberrant one-carbon metabolism in late-onset AD (*LOAD*). Several components of methylation homeostasis, notably *S*-adenosylmethionine (*SAM*) that is required for the methylation of DNA as well as methylation of histones, are frequently found to be abnormal in Alzheimer's (see *arrows*). A reduction of SAM can lead to decreased uptake of folate and vitamin B12, which in turn induces the accumulation of homocysteine (*Hcy*) and *S*-adenosylhomocysteine (SAH). SAH is formed during SAM-dependent methylation reactions and acts as a strong inhibitor of DNA methyltransferases, potentially resulting in DNA hypomethylation. Enhanced hydrolysis of SAH results in increased Hcy levels, which may result in oxidative stress in the cells and impairs DNA repair. In healthy cells, equilibrium Hcy levels are maintained by the remethylation pathway (*blue*) that requires folate and Vitamin B12. Alternately, Hcy may be irreversibly catabolized to cysteine in the trans-sulfuration pathway (*green*) by vitamin B6-dependent enzymes. Clearance of Hcy by methionine synthase maintains a favorable SAM/SAH ratio, an index of cellular methylation potential. Impaired reactions in either pathways, through defects or deficiencies, usually result in the accumulation of Hcy in the affected tissues. These age-dependent metabolic changes may be very important factors that play a paramount role in the genesis of age-related disorders. *THF* tetrahydrofolate, *5-MTHF* 5-methyltetrahydrofolate

patients with neurological disorders, including AD.[59] Aging is related to higher CSF concentrations of Hcy and SAH; it is also related to a decrease in folate levels and a lower SAM: SAH ratio. These age-dependent metabolic changes may play a paramount role in the genesis of LOAD. Total Hcy levels are increased in AD patients, and also in LOAD.[39] Even though the pathological events are very different in EOAD and LOAD, both lead to typical AD phenotypes. The epigenetic role in LOAD derives further support from the fact that relatives of EOAD probands have a significantly higher lifetime risk of developing AD than do relatives of LOAD patients.[45] Elevated plasma Hcy levels are probably involved in the pathogenesis of AD, yet the gene mutations in the Hcy pathway are probably not responsible for the high Hcy levels in many AD cases gene.[87] If genetic factors can be ruled out, environmental factors and epigenetic effects are a plausible explanation for the observed phenotypes. In fact, the main environmental determinants for Hcy levels are sex, age, renal function, and vitamin intake.

Compared to Hcy, the impact of folic acid on epigenetic regulation may be even more significant. Folate is necessary for the generation of additional methyl groups for the one-carbon pathway. As humans cannot synthesize folate, the methyl groups must come from the diet, in the form of 5′-methyltetrahydrofolate and formyltetrahydrofolate. Low serum folate levels are linked to atrophy of the cerebral cortex.[68] In AD, patients' folate levels in the spinal fluid are significantly lower than in healthy individuals.[67] As folate is critical for the conversion of methionine to SAM, folate deficiency may promote SAM depletion and widespread hypomethylation. Folate deficiency also leads to damage of hippocampal neurons and to an increase in vulnerability to oxidative, metabolic, and excitotoxic insults.[43] The time of methyl donor deficiency is critical in the development of abnormal epigenetic patterns. Pogribny et al.[62] fed F344 rats a methyl-deficient diet for 36 weeks, followed by a methyl-rich diet for a total of 54 weeks. The methyl-deficient diet led to a decrease in the levels of SAM, in the SAM:SAH ratio and in global DNA hypomethylation. Supplementation with methyl donors restored DNA methylation to normal levels in the group that had been fed the methyl-deficient diet for 9 weeks, but not in animals that had been exposed to the methyl-deficient diet for longer periods. This indicates that abnormal methylation patterns become stabilized after prolonged folate deficiency.

12.5 Model of Age-Dependent Epigenetic Drift

Epigenetic alterations occur with higher frequency than genetic mutations and may therefore be particularly important in age-related phenotypes.[28]The high frequency of de novo epimutations suggests that epigenetic alterations accumulate in the course of aging. Small epimutations in critical genes may be tolerated, merely reflecting the range of interindividual variance. However, once a critical threshold of epigenetic deregulation is reached, the human brain is likely to malfunction. LOAD could, in this regard, represent a normal part of aging, which would imply that every person has a predisposition to develop AD.[78] In this model, epigenetic effects can accumulate throughout life, from early embryonic stages to old age. Epigenetic effects may even be trans-generational, that is, influenced by epigenetic events in earlier generations. The epigenetic model of LOAD offers a theoretical framework that explains many phenomena of aging that are otherwise not readily explained by classical theories of aging. According to the model, aging results from the accumulation of epigenetic damage, a direct consequence of the genetic and epigenetic limitations on maintenance and repair. Age-dependent epigenetic drift is naturally present in all individuals, but as they age, it assumes a central role in many chronic disorders, including LOAD.[31,78]

The phenotype of the aging human arises from a variety of risk factors, such as environmental, nutritional, or stochastic fluctuations, all of which act on the epigenome, increasing epigenetic variability with age. Whereas genetic mutations accumulate in a nearly linear fashion during aging, epimutations seem to increase exponentially, once a certain threshold of cellular epigenetic deregulation has been reached (unpublished results from our lab). Reaching the threshold may cause a ripple effect that influences other genetic and epigenetic maintenance processes, ultimately leading to a significantly quicker epigenetic drift (Fig. 12.2). Epigenetic drift is likely caused, at least in part, by high epigenetic turnover, as the replication-independent epigenetic maintenance in cells is very dynamic and inherently probabilistic. For example, one of the main genes maintaining a transcriptionally silent state, *Hp1*, which mediates communication between histone and DNA methyltransferases, transiently binds to target chromatin domains. Intriguingly, the turnover time for the entire cellular pool of *Hp1* at a given heterochromatic domain is a matter of seconds.[22]

The rate of epigenetic drift in complex disorders may not only be affected by singular events in the genome, but may also be driven by genome-wide systemic changes. This was demonstrated in a network analysis of postmortem brain samples from schizophrenia cases and controls. Affected individuals had decreased epigenetic modularity (co-regulation) in both brain and germ line. This implies that systemic epigenetic dysfunction may be associated with complex disorders.[54] A systemic effect can be positive or negative in LOAD. It seems reasonable that, in very old healthy individuals, some favorable epigenotypes exist that act as buffers against the deleterious effects of age-related disease genes. The frequency of deleterious epimutations is therefore likely to increase in individuals who have a very long life, their protective epigenotype allowing disease-related genes or epigenetic patterns to accumulate.

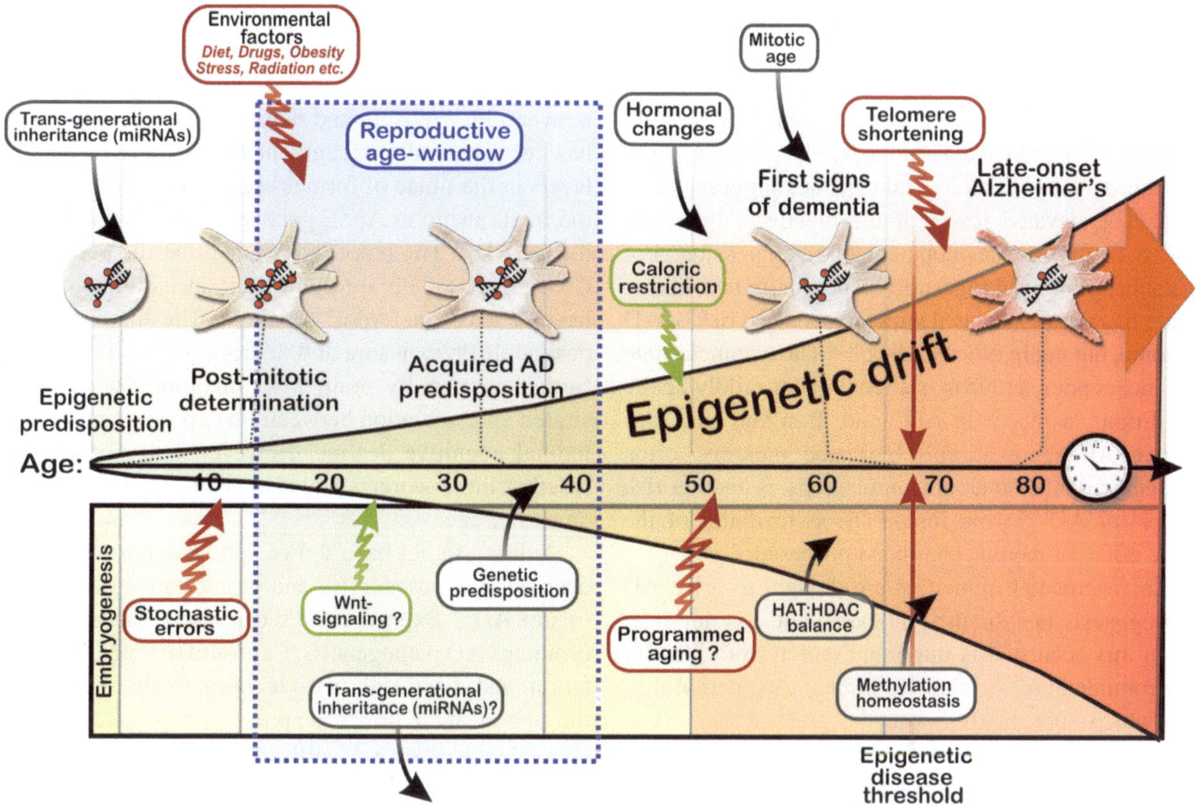

Fig. 12.2 Age-dependent epigenetic drift. Shown here is the example of neuronal epigenetic drift that may lead to age-dependent neurodegenerative disorders, such as late-onset AD (*LOAD*). The relatively high frequency of de novo epimutations suggests that epigenetic alterations accumulate during aging. Small epimutations may be tolerated by the cells; however, once the epigenetic deregulation reaches a critical threshold, the cells no longer function properly. The phenotypic outcome depends on the overall effect of the series of pre- and post-natal impacts on the pre-epimutation. Only some predisposed individuals will reach the "threshold" of epigenetic deregulation that causes the phenotypic changes that meet the diagnostic criteria for LOAD. Epigenetic drift may not only be affected by several internal and external factors – some individuals may already be epigenetically predisposed at birth, due to trans-generational epigenetic effects. Trans-generational epigenetic inheritance results either from incomplete erasure of parental epigenetic marks during phases of epigenetic reprogramming at fertilization, or may be established by small RNA species that pass through the germline. Since all organisms eventually die from different causes, epigenetic patterns beneficial in early life are favored by natural selection over patterns advantageous later in life. Deleterious epigenetic drift occurring after the reproductive phase is relatively neutral to selection, because their bearers have already transmitted their genes (and potentially epigenetic information) to the next generation

12.6 Alzheimer's Disease: Evidence for an Environmental Epigenetic Fingerprint

AD is the most prominent form of dementia among the elderly. In only a minority of patients with AD can the disease be attributed to autosomal familial AD (FAD) mutations in the coding sequences of genes such as APP and PSEN1. This form of AD does not explain the more common sporadic LOAD. Rather, it can be assumed that environmental insults leave an "epigenetic footprint" in mammalian cells, especially in post-mitotic cells, such as neurons. These footprints may cause harm to an organism, even in old age, years after exposure.

Increasing recognition of the neurotoxic potential of many industrial chemicals such as organic solvents raises the question of the occupational and environmental contributions to the etiology of neurodegenerative diseases. Environmental factors that may be AD risk factors include metals, pesticides, solvents, electromagnetic fields, brain injuries, inflammation, educational level, lifestyle, and diet.

Metals have been extensively studied. No direct causal role has as yet been demonstrated for aluminum or other transition metals, that is, zinc, copper, iron, and mercury. Epidemiological evidence suggests, however, that elevated levels of the metals in brain are linked to disease development and progression. For instance, prolonged ingestion of aluminum in drinking water has been associated with an increased risk of AD in some, but not in other studies.[48,52] The chemical state of zinc, copper, and iron is altered under mildly acidic conditions, as occur in AD brain. Iron and zinc ions induce Aβ aggregation.[21] Inorganic mercury, often present in dental amalgam applications, is another risk factor for AD. A role for APOE as mediator of the toxic effect of mercury has been proposed.[57]

Environmental influences are thought to drive AD pathogenesis in sporadic AD cases, but it is not clear when this occurs. It is important therefore to identify environmental triggers and to pinpoint the period during which such factors pose the greatest risk. Twin studies often used to confirm the inheritance pattern of a disease have shown poor concordance in neurodegenerative diseases such as AD. The negative findings of such studies along with the sporadic nature of LOAD suggest a strong role for the environment.

In 1989 David Barker and coworkers demonstrated an inverse relationship between birth weight and the incidence of cardiovascular disease. The Barker hypothesis, also known as the Developmental Origins of Health and Disease (DOHaD) hypothesis, states that many adult diseases have a fetal origin.[4,7,81] A large body of clinical and experimental data has since supported this hypothesis and has shown that diseases of the cardiovascular system, of the hypothalamic-pituitary-adrenal (HPA)-axis, and diabetes can be due to nutritional imbalance during pregnancy.[5,6,85] Diseases such as schizophrenia have also been linked to infection, fetal malnutrition, or hypoxia in the preconceptual, fetal, and infantile phases of life.[13,16,76,79] Lead (Pb) exposure is a major environmental risk that leaves children with cognitive and behavioral deficits that persist into adult life.

Several population-based case-control studies have found that chronic occupational exposure to Pb and other metals is associated with incidence of Parkinson's disease. Other studies have pointed to a relation between high levels of Pb in blood and bone and an increase in the risk of amyotrophic lateral sclerosis (ALS). Pb exposure may therefore play a role in the etiology of the disease. The evidence for a connection between Pb exposure and neurodegenerative diseases has been markedly strengthened by a study of the Pb levels in the tibiae of former organo-lead workers and the relationship to ApoE genotype, a known risk factor for AD.[37] The report concluded that the persistent CNS effects of Pb are more toxic in individuals that have at least one ApoE ε4 allele. The link between past adult Pb exposure and neurodegeneration gained further support by brain MRI imaging that demonstrated an association between Pb exposure and longitudinal cognitive decline. However, it is not known whether these workers had also been exposed to Pb at an earlier age.

Animal studies have shown that Pb exposure during development poses a risk and promotes the pathogenesis of AD.[70] Exposure to Pb can also be a factor that promotes AD pathogenesis. Exposure to Pb during lactation and postnatal periods of animals monitored throughout their life for expression of the APP gene have shown that APP mRNA expression was transiently induced in neonates and overexpressed 20 months after exposure to Pb.[86] Furthermore, the increase in APP gene expression in old age was accompanied by an elevation in APP and its amyloidogenic Aβ (1–42) product. Brains from 23-year-old primates that had been exposed to Pb early in life expressed AD-related genes, APP and beta-site APP cleaving enzyme 1 (BACE1) in old age. Furthermore, developmental exposure to Pb altered the levels, characteristics, and intracellular distribution of Aβ staining and amyloid plaques in the frontal association cortex of these primates.[82] Aβ is known to induce functional disturbances in vivo through its pro-oxidant and neurotoxic properties.[19,60] Given Aβ promotes the formation of reactive oxygen species (ROS), antioxidants can prevent Aβ elicited neurotoxic cascades.[50,77]

Given that oxidative damage is a component of AD pathology, levels of the oxidative DNA marker 8-OHdG have been found to be comparable in the same brain regions of aged primates and rodents that had been exposed to Pb during development.[20] Thus, early exposure to Pb yields a state of increased APP expression with concomitant Aβ deposits, in addition to causing increased oxidative DNA damage in old age.[83] Increases in Aβ could lead to the generation of ROS, promoting the formation of 8-OHdG; alternatively, epigenetic modulation in the methylation pattern of cytosines may interfere with the repair or oxidation potential of adjacent oxidized guanines.[86]

12.7 DNA Methylation and DNA Oxidation in Alzheimer's Disease

One way, in which environmental agents or occupational exposures interfere with DNA methylation is by disrupting enzyme action. In vitro studies have shown that the addition of cadmium (Cd) to hepatic nuclear extracts inhibits DNA-methyltransferase.[46] Sub-chronic Cd exposure inhibits DNA-methyltransferase activity in cultured cells, with chronic exposure enhancing the activity of DNA-methyltransferase.[72] This suggests that the consequences of Cd on DNA methylation may be responsible for its carcinogenic properties.[63] In addition to affecting DNA methylation, environmental agents can also disrupt chromatin restructuring and produce long-term alterations in gene expression. In monkeys, chromatin structure was found altered at environmentally relevant blood Pb concentrations. Moreover, the level of protamine-DNA interactions had decreased, possibly altering sperm chromatin condensation.[30] However, few studies have been conducted on DNA-methylation in the brain, and none has examined whether environmental agents disturb this process.

Oxidative stress contributes to the pathogenesis of disease both through genetic and epigenetic mechanisms. Hydroxyl radicals cause a wide range of DNA lesions, including base modifications, deletions, strand breakage, chromosomal rearrangements, etc. These lesions interfere with DNA functioning as a substrate for the DNMTs, and thus lead to global hypomethylation. More specifically, X-rays,[81] ultraviolet,[10] and γ-rays[40] reduce the methyl-accepting ability of DNA. In addition the presence of 8-OHdG in CpG dinucleotide sequences strongly inhibits methylation of adjacent cytosine residues[80] and interferes with of the cleaving of DNA by restriction nucleases.[74]

When 8-OHdG is substituted for either guanine on the HpaII methylase recognition site (CCGG), DNA methylation of adjacent cytosines and binding to the methyltransferase are inhibited. The extent of inhibition depends on the position of 8-OHdG. Proofreading enzymes may not recognize 8-OHdG, which would then persist as a mutation resulting in G $\rightarrow$ T transversions.[44] Another potentially mutagenic lesion in ROS-induced DNA damage is O^6-methylguanine, a compound that inhibits the binding of DNA methyltransferases and, by blocking methylation of adjacent cytosine molecules, induces hypomethylation. DNA hypomethylation will also result when O^6-methylguanine is spontaneously mispaired with thymine.[84] Finally, single-stranded DNA can signal de novo methylation. If so, the formation of single strand breaks due to oxidative stress contributes to the modification of DNA methylation patterns. Free radicals, derived primarily from molecular oxygen, have long been recognized as risk factors for a variety of human disorders, including neurodegenerative diseases and aging. Oxo-guanosine DNA glycosylase 1 (OGG1) is a major repair enzyme that recognizes and removes 8-OHdG. Oxidative DNA damage and possible deficiencies in OGG1 are a central factor in aging and age-related diseases like AD.[17,32]

Few studies have addressed the epigenetic phenomenon of DNA methylation and DNA oxidative damage simultaneously and little is known on how DNA methylation and DNA oxidation interact. Oxidation of guanine in CpG dinucleotide reduces binding to the methyl group binding domain (MBD). When 5-methylcytosine is oxidized to 5-hydroxymethylcytosine, its affinity to MBD is reduced to the same low level as that of unmethylated cytosine. Methylated CpG also accounts for decreased binding of the transcription factor to the promoter region. Synthesized oligonucleotides that resemble the binding site for the transcription factor Sp1 have been used to mimic the interactions between methylation and oxidation. In some oligonucleotides, the cytosine in CpG has been replaced by methylcytosine and the guanine by an 8-oxo-dG. A third oligonucleotide with methylcytosine and 8-OHdG adjacent to each other has been used in DNA-binding and repair studies. The presence of either 5-methylcytosine or 8-OHdG dramatically suppressed Sp1 DNA binding; however, the combination of both had an effect greater than either alone. Likewise, the repair of 8-OHdG was greatly diminished when 8-OHdG was preceded by 5-methylcytosine. These experiments have shown that methylation greatly impacts gene expression and DNA repair and that oxidized DNA inhibits methylation of an adjacent cytosine.[74,79]

Studies with oxidant-transformed cell lines have also shown unusual changes of methylation patterns of several genes.[85] This suggests that oxidative DNA damage and DNA methylation interact with each other. In turn, this may alter the methylation patterns and transcriptional activity of affected genes. However, oxidation is a dynamic process that occurs whenever there is oxidative stress, typically high during early and late periods of life. Methylation, on the other hand, is still poorly understood and is presumed to occur during early development and from then on is sustained for life. Mammals have DNA methylating enzymes, but, in contrast to plants, have no demethylases. Even though active demethylation has long been thought to occur in mammals, it is not known how this is achieved. Moreover, even if a substance is not a pro-oxidant, it can still bring about oxidative damage through alterations in methylation patterns that impact the repair of adjacent oxidized guanines.

12.8 Conclusion

In summary, findings from many studies suggest that, in addition to genetic determinants and environmental factors, epigenetic effects are important in LOAD. The question then arises whether epigenetic changes precede LOAD and confer a disease risk or whether epigenetic drift is the result. One strong argument in favor of epigenetic changes preceding LOAD is that epigenetic drift increases with aging and also occurs in healthy individuals. It seems likely LOAD predisposition is related to DNA methylation profiles and is influenced by epigenetic drift.

References

1. Akbarian S. The neurobiology of Rett syndrome. *Neuroscientist*. 2003;9:57-63.
2. Alonso AD, Grundke-Iqbal I, Barra HS, Iqbal K. Abnormal phosphorylation of tau and the mechanism of Alzheimer neurofibrillary degeneration: sequestration of microtubule-associated proteins 1 and 2 and the disassembly of microtubules by the abnormal tau. *Proc Natl Acad Sci USA*. 1997;94:298-303.
3. Baek SH, Ohgi KA, Rose DW, Koo EH, Glass CK, Rosenfeld MG. Exchange of N-CoR corepressor and Tip60 coactivator complexes links gene expression by NF-kappaB and beta-amyloid precursor protein. *Cell*. 2002;110:55-67.
4. Barker DJ. Fetal origins of cardiovascular disease. *Ann Med*. 1999;31(Suppl 1):3-6.
5. Barker DJ. The developmental origins of adult disease. *J Am Coll Nutr*. 2004;23:588S-595S.
6. Barker DJ, Eriksson JG, Forsen T, Osmond C. Fetal origins of adult disease: strength of effects and biological basis. *Int J Epidemiol*. 2002;31:1235-1239.
7. Barker DJ, Winter PD, Osmond C, Margetts B, Simmonds SJ. Weight in infancy and death from ischaemic heart disease. *Lancet*. 1989;2:577-580.
8. Barrachina M, Ferrer I. DNA methylation of Alzheimer disease and tauopathy-related genes in postmortem brain. *J Neuropathol Exp Neurol*. 2009;68:880-891.
9. Bassett SS, Avramopoulos D, Perry RT, et al. Further evidence of a maternal parent-of-origin effect on chromosome 10 in late-onset Alzheimer's disease. *Am J Med Genet B Neuropsychiatr Genet*. 2006;141B:537-540.
10. Becker FF, Holton P, Ruchirawat M, Lapeyre JN. Perturbation of maintenance and de novo DNA methylation in vitro by UVB (280–340 nm)-induced pyrimidine photodimers. *Proc Natl Acad Sci USA*. 1985;82:6055-6059.
11. Belyaev ND, Nalivaeva NN, Makova NZ, Turner AJ. Neprilysin gene expression requires binding of the amyloid precursor protein intracellular domain to its promoter: implications for Alzheimer disease. *EMBO Rep*. 2009;10:94-100.
12. Bestor TH. The DNA methyltransferases of mammals. *Hum Mol Genet*. 2000;9:2395-2402.
13. Bilbo SD, Biedenkapp JC, Der-Avakian A, Watkins LR, Rudy JW, Maier SF. Neonatal infection-induced memory impairment after lipopolysaccharide in adulthood is prevented via caspase-1 inhibition. *J Neurosci*. 2005;25:8000-8009.
14. Bird AP. CpG-rich islands and the function of DNA methylation. *Nature*. 1986;321:209-213.
15. Bjornsson HT, Sigurdsson MI, Fallin MD, et al. Intra-individual change over time in DNA methylation with familial clustering. *JAMA*. 2008;299:2877-2883.
16. Boksa P, El-Khodor BF. Birth insult interacts with stress at adulthood to alter dopaminergic function in animal models: possible implications for schizophrenia and other disorders. *Neurosci Biobehav Rev*. 2003;27:91-101.
17. Bolin C, Stedeford T, Cardozo-Pelaez F. Single extraction protocol for the analysis of 8-hydroxy-2′-deoxyguanosine (oxo8dG) and the associated activity of 8-oxoguanine DNA glycosylase. *J Neurosci Methods*. 2004;136:69-76.
18. Bottiglieri T, Godfrey P, Flynn T, Carney MW, Toone BK, Reynolds EH. Cerebrospinal fluid S-adenosylmethionine in depression and dementia: effects of treatment with parenteral and oral S-adenosylmethionine. *J Neurol Neurosurg Psychiatry*. 1990;53:1096-1098.
19. Castellani RJ, Lee HG, Perry G, Smith MA. Antioxidant protection and neurodegenerative disease: the role of amyloid-beta and tau. *Am J Alzheimers Dis Other Demen*. 2006;21:126-130.
20. Cecchi C, Fiorillo C, Sorbi S, et al. Oxidative stress and reduced antioxidant defenses in peripheral cells from familial Alzheimer's patients. *Free Radic Biol Med*. 2002;33:1372-1379.
21. Cherny RA, Legg JT, McLean CA, et al. Aqueous dissolution of Alzheimer's disease Abeta amyloid deposits by biometal depletion. *J Biol Chem*. 1999;274:23223-23228.
22. Cheutin T, McNairn AJ, Jenuwein T, Gilbert DM, Singh PB, Misteli T. Maintenance of stable heterochromatin domains by dynamic HP1 binding. *Science*. 2003;299:721-725.

23. de Vries BB, Jansen CC, Duits AA, et al. Variable FMR1 gene methylation of large expansions leads to variable phenotype in three males from one fragile X family. *J Med Genet*. 1996;33:1007-1010.
24. Dolinoy DC, Das R, Weidman JR, Jirtle RL. Metastable epialleles, imprinting, and the fetal origins of adult diseases. *Pediatr Res*. 2007;61:30R-37R.
25. Dunckley T, Beach TG, Ramsey KE, et al. Gene expression correlates of neurofibrillary tangles in Alzheimer's disease. *Neurobiol Aging*. 2006;27:1359-1371.
26. Dwyer BE, Takeda A, Zhu X, Perry G, Smith MA. Ferric cycle activity and Alzheimer disease. *Curr Neurovasc Res*. 2005;2:261-267.
27. Esch FS, Keim PS, Beattie EC, et al. Cleavage of amyloid beta peptide during constitutive processing of its precursor. *Science*. 1990;248:1122-1124.
28. Feil R. Environmental and nutritional effects on the epigenetic regulation of genes. *Mutat Res*. 2006;600:46-57.
29. Flanagan JM, Popendikyte V, Pozdniakovaite N, et al. Intra- and interindividual epigenetic variation in human germ cells. *Am J Hum Genet*. 2006;79:67-84.
30. Foster WG, McMahon A, Rice DC. Sperm chromatin structure is altered in cynomolgus monkeys with environmentally relevant blood lead levels. *Toxicol Ind Health*. 1996;12: 723-735.
31. Fraga MF, Ballestar E, Paz MF, et al. Epigenetic differences arise during the lifetime of monozygotic twins. *Proc Natl Acad Sci USA*. 2005;102:10604-10609.
32. Franco R, Schoneveld O, Georgakilas AG, Panayiotidis MI. Oxidative stress, DNA methylation and carcinogenesis. *Cancer Lett*. 2008;266:6-11.
33. Fuso A, Seminara L, Cavallaro RA, D'Anselmi F, Scarpa S. S-adenosylmethionine/homocysteine cycle alterations modify DNA methylation status with consequent deregulation of PS1 and BACE and beta-amyloid production. *Mol Cell Neurosci*. 2005;28:195-204.
34. Gimelbrant A, Hutchinson JN, Thompson BR, Chess A. Widespread monoallelic expression on human autosomes. *Science*. 2007;318:1136-1140.
35. Goedert M, Spillantini MG. A century of Alzheimer's disease. *Science*. 2006;314:777-781.
36. Golde TE. Alzheimer disease therapy: can the amyloid cascade be halted? *J Clin Invest*. 2003;111:11-18.
37. Gorell JM, Rybicki BA, Cole JC, Peterson EL. Occupational metal exposures and the risk of Parkinson's disease. *Neuroepidemiology*. 1999;18:303-308.
38. Guan JS, Haggarty SJ, Giacometti E, et al. HDAC2 negatively regulates memory formation and synaptic plasticity. *Nature*. 2009;459:55-60.
39. Guidi I, Galimberti D, Venturelli E, et al. Influence of the Glu298Asp polymorphism of NOS3 on age at onset and homocysteine levels in AD patients. *Neurobiol Aging*. 2005;26:789-794.
40. Kalinich JF, Catravas GN, Snyder SL. The effect of gamma radiation on DNA methylation. *Radiat Res*. 1989;117:185-197.
41. Kaminsky Z, Wang SC, Petronis A. Complex disease, gender and epigenetics. *Ann Med*. 2006;38:530-544.
42. Kim HS, Kim EM, Kim NJ, et al. Inhibition of histone deacetylation enhances the neurotoxicity induced by the C-terminal fragments of amyloid precursor protein. *J Neurosci Res*. 2004;75:117-124.
43. Kruman II, Kumaravel TS, Lohani A, et al. Folic acid deficiency and homocysteine impair DNA repair in hippocampal neurons and sensitize them to amyloid toxicity in experimental models of Alzheimer's disease. *J Neurosci*. 2002;22:1752-1762.
44. Kuchino Y, Mori F, Kasai H, et al. Misreading of DNA templates containing 8-hydroxydeoxyguanosine at the modified base and at adjacent residues. *Nature*. 1987;327:77-79.
45. Lautenschlager NT, Cupples LA, Rao VS, et al. Risk of dementia among relatives of Alzheimer's disease patients in the MIRAGE study: what is in store for the oldest old? *Neurology*. 1996;46:641-650.
46. Lillycrop KA, Phillips ES, Jackson AA, Hanson MA, Burdge GC. Dietary protein restriction of pregnant rats induces and folic acid supplementation prevents epigenetic modification of hepatic gene expression in the offspring. *J Nutr*. 2005;135:1382-1386.
47. Malmgren H, Steen-Bondeson ML, Gustavson KH, et al. Methylation and mutation patterns in the fragile X syndrome. *Am J Med Genet*. 1992;43:268-278.
48. Martyn CN, Coggon DN, Inskip H, Lacey RF, Young WF. Aluminum concentrations in drinking water and risk of Alzheimer's disease. *Epidemiology*. 1997;8:281-286.
49. Mastroeni D, McKee A, Grover A, Rogers J, Coleman PD. Epigenetic differences in cortical neurons from a pair of monozygotic twins discordant for Alzheimer's disease. *PLoS One*. 2009;4:e6617.
50. Mattson MP. Cellular actions of beta-amyloid precursor protein and its soluble and fibrillogenic derivatives. *Physiol Rev*. 1997;77:1081-1132.
51. Mattson MP, Haberman F. Folate and homocysteine metabolism: therapeutic targets in cardiovascular and neurodegenerative disorders. *Curr Med Chem*. 2003;10:1923-1929.
52. McLachlan DR, Bergeron C, Smith JE, Boomer D, Rifat SL. Risk for neuropathologically confirmed Alzheimer's disease and residual aluminum in municipal drinking water employing weighted residential histories. *Neurology*. 1996;46: 401-405.
53. Mehler MF. Epigenetic principles and mechanisms underlying nervous system functions in health and disease. *Prog Neurobiol*. 2008;86:305-341.
54. Mill J, Tang T, Kaminsky Z, et al. Epigenomic profiling reveals DNA-methylation changes associated with major psychosis. *Am J Hum Genet*. 2008;82:696-711.
55. Morrison LD, Smith DD, Kish SJ. Brain S-adenosylmethionine levels are severely decreased in Alzheimer's disease. *J Neurochem*. 1996;67:1328-1331.
56. Mulero-Navarro S, Esteller M. Epigenetic biomarkers for human cancer: the time is now. *Crit Rev Oncol Hematol*. 2008;68:1-11.
57. Mutter J, Naumann J, Sadaghiani C, Schneider R, Walach H. Alzheimer disease: mercury as pathogenetic factor and apolipoprotein E as a moderator. *Neuro Endocrinol Lett*. 2004;25:331-339.
58. Myung NH, Zhu X, Kruman II, et al. Evidence of DNA damage in Alzheimer disease: phosphorylation of histone H2AX in astrocytes. *Age (Dordr)*. 2008;30:209-215.
59. Obeid R, Kasoha M, Knapp JP, et al. Folate and methylation status in relation to phosphorylated tau protein(181P) and beta-amyloid(1-42) in cerebrospinal fluid. *Clin Chem*. 2007;53:1129-1136.

60. Ono K, Hamaguchi T, Naiki H, Yamada M. Anti-amyloidogenic effects of antioxidants: implications for the prevention and therapeutics of Alzheimer's disease. *Biochim Biophys Acta*. 2006;1762:575-586.
61. Petronis A. The origin of schizophrenia: genetic thesis, epigenetic antithesis, and resolving synthesis. *Biol Psychiatry*. 2004;55:965-970.
62. Pogribny IP, Ross SA, Wise C, et al. Irreversible global DNA hypomethylation as a key step in hepatocarcinogenesis induced by dietary methyl deficiency. *Mutat Res*. 2006;593:80-87.
63. Poirier LA, Vlasova TI. The prospective role of abnormal methyl metabolism in cadmium toxicity. *Environ Health Perspect*. 2002;110(Suppl 5):793-795.
64. Quadri P, Fragiacomo C, Pezzati R, et al. Homocysteine, folate, and vitamin B-12 in mild cognitive impairment, Alzheimer disease, and vascular dementia. *Am J Clin Nutr*. 2004;80:114-122.
65. Schumacher A, Petronis A. Epigenetics of complex diseases: from general theory to laboratory experiments. *Curr Top Microbiol Immunol*. 2006;310:81-115.
66. Seeman MV. Psychopathology in women and men: focus on female hormones. *Am J Psychiatry*. 1997;154:1641-1647.
67. Seshadri S, Beiser A, Selhub J, et al. Plasma homocysteine as a risk factor for dementia and Alzheimer's disease. *N Engl J Med*. 2002;346:476-483.
68. Snowdon DA, Tully CL, Smith CD, Riley KP, Markesbery WR. Serum folate and the severity of atrophy of the neocortex in Alzheimer disease: findings from the Nun study. *Am J Clin Nutr*. 2000;71:993-998.
69. Stante M, Minopoli G, Passaro F, Raia M, Vecchio LD, Russo T. Fe65 is required for Tip60-directed histone H4 acetylation at DNA strand breaks. *Proc Natl Acad Sci USA*. 2009;106:5093-5098.
70. Stewart WF, Schwartz BS, Simon D, Kelsey K, Todd AC. ApoE genotype, past adult lead exposure, and neurobehavioral function. *Environ Health Perspect*. 2002;110:501-505.
71. Sutcliffe JS, Nelson DL, Zhang F, et al. DNA methylation represses FMR-1 transcription in fragile X syndrome. *Hum Mol Genet*. 1992;1:397-400.
72. Takiguchi M, Achanzar WE, Qu W, Li G, Waalkes MP. Effects of cadmium on DNA-(Cytosine-5) methyltransferase activity and DNA methylation status during cadmium-induced cellular transformation. *Exp Cell Res*. 2003;286:355-365.
73. Tanzi RE, Bertram L. Twenty years of the Alzheimer's disease amyloid hypothesis: a genetic perspective. *Cell*. 2005;120:545-555.
74. Turk PW, Weitzman SA. Free radical DNA adduct 8-OH-deoxyguanosine affects activity of Hpa II and Msp I restriction endonucleases. *Free Radic Res*. 1995;23:255-258.
75. Vafai SB, Stock JB. Protein phosphatase 2A methylation: a link between elevated plasma homocysteine and Alzheimer's Disease. *FEBS Lett*. 2002;518:1-4.
76. Valdez R, Athens MA, Thompson GH, Bradshaw BS, Stern MP. Birthweight and adult health outcomes in a biethnic population in the USA. *Diabetologia*. 1994;37:624-631.
77. Valinluck V, Tsai HH, Rogstad DK, Burdzy A, Bird A, Sowers LC. Oxidative damage to methyl-CpG sequences inhibits the binding of the methyl-CpG binding domain (MBD) of methyl-CpG binding protein 2 (MeCP2). *Nucleic Acids Res*. 2004;32:4100-4108.
78. Wang SC, Oelze B, Schumacher A. Age-specific epigenetic drift in late-onset Alzheimer's disease. *PLoS One*. 2008;3:e2698.
79. Weitzman SA, Turk PW, Milkowski DH, Kozlowski K. Free radical adducts induce alterations in DNA cytosine methylation. *Proc Natl Acad Sci USA*. 1994;91:1261-1264.
80. Wilson VL, Jones PA, Momparler RL. Inhibition of DNA methylation in L1210 leukemic cells by 5-aza-2′-deoxycytidine as a possible mechanism of chemotherapeutic action. *Cancer Res*. 1983;43:3493-3496.
81. Winneke G, Kramer U, Brockhaus A, et al. Neuropsychological studies in children with elevated tooth-lead concentrations. II. Extended study. *Int Arch Occup Environ Health*. 1983;51:231-252.
82. Wu J, Basha MR, Brock B, et al. Alzheimer's disease (AD)-like pathology in aged monkeys after infantile exposure to environmental metal lead (Pb): evidence for a developmental origin and environmental link for AD. *J Neurosci*. 2008;28:3-9.
83. Wu J, Basha MR, Zawia NH. The environment, epigenetics and amyloidogenesis. *J Mol Neurosci*. 2008;34:1-7.
84. Xiao W, Samson L. In vivo evidence for endogenous DNA alkylation damage as a source of spontaneous mutation in eukaryotic cells. *Proc Natl Acad Sci USA*. 1993;90:2117-2121.
85. Zawia NH, Basha MR. Environmental risk factors and the developmental basis for Alzheimer's disease. *Rev Neurosci*. 2005;16:325-337.
86. Zawia NH, Lahiri DK, Cardozo-Pelaez F. Epigenetics, oxidative stress, and Alzheimer disease. *Free Radic Biol Med*. 2009;46:1241-1249.
87. Zhang YD, Ke XY, Shen W, Liu Y. Relationship of homocysteine and gene polymorphisms of its related metabolic enzymes with Alzheimer's disease. *Chin Med Sci J*. 2005;20:247-251.

13 Epigenetic Mechanisms in the Developmental Origins of Adult Disease

Keith M. Godfrey, Karen A. Lillycrop, Mark A. Hanson, and Graham C. Burdge

13.1 The Developmental Origins of Chronic Disease: Overview of Evidence from Human and Experimental Studies

Non-communicable diseases (NCD), including diabetes, cardiovascular disease, and the metabolic syndrome, account for 60% of all deaths globally.[151] In low to middle income countries, NCD are becoming particularly important as they increase rapidly in countries that undergo socioeconomic improvement. While the increase in NCD is due in part to the adoption of a Western lifestyle, there is growing recognition of the role played by developmental factors. This is in accordance with the fundamental principles of life-course biology, with developmental trajectories established in early life influencing the response to later exposures, such as adult lifestyle (Fig. 13.1). Moreover, the temporal trends in NCD may, in significant part, arise from effects on phenotype established by the interaction between genes and the developmental environment.[56]

A developmental influence on NCD risk is demonstrated by epidemiological cohort studies in which small size at birth and during infancy is associated with a greater risk in later life of chronic disease, including coronary heart disease, hypertension, stroke, type 2 diabetes, and osteoporosis.[55] Early development has also been linked to other common disorders, including affective disorders and cognitive decline, sarcopenia, allergy, and inflammatory conditions.[50] However, the risk of specific cancers, including breast cancer and hepatoblastoma, is increased with higher birth weight.[20] The steep temporal trends in the incidence rates of cardiovascular disease suggest that it is unlikely that associations have arisen only through the pleiotropic effects of genes that regulate both fetal growth and later cardiovascular risk. Moreover, it is now accepted widely that fixed genomic variations, such as single nucleotide polymorphisms, explain only a fraction of the variation in NCD risk in a population.[97]

There has been extensive replication worldwide of the original epidemiological observations linking impaired early development with NCD risk in later

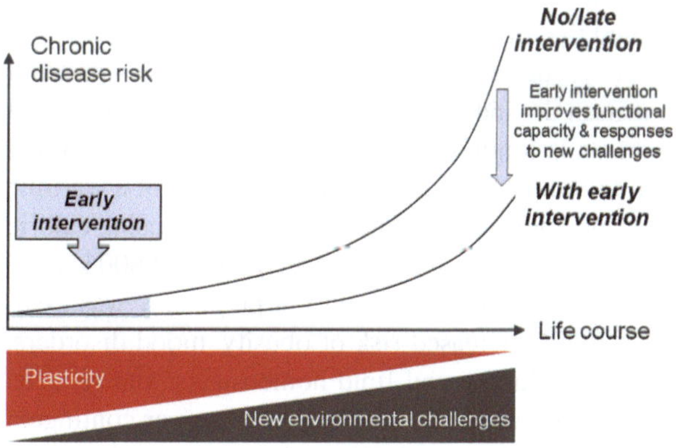

Fig. 13.1 Chronic non-communicable disease differs from the classical medical model, according to which an individual is healthy until contracting disease. Throughout life, risk increases because of declining plasticity and cumulative effects of inadequate responses to new environmental challenges. The greatest increase in risk occurs in adult life, but the trajectory is set much earlier: Epigenetic processes are induced by cues such as the mother's diet and body composition before and during pregnancy, and by nutrition as the child grows and develops. This implies that relatively modest interventions in early life can have a big impact on disease risk in later life. Preventative measures require long-term investment, but are likely to be more effective than screening programs or treatment

K.M. Godfrey (✉)
MRC Epidemiology Resource Centre,
University of Southampton, Southampton General Hospital,
Hampshire, UK
e-mail: kmg@southampton.ac.uk

H.I. Roach et al. (eds.), *Epigenetic Aspects of Chronic Diseases*,
DOI: 10.1007/978-1-84882-644-1_13, © Springer-Verlag London Limited 2011

life[55] and it is now recognized that the associations do not reflect confounding by adult environmental risk factors such as smoking or socioeconomic status.[50] A meta-analysis of 18 studies has shown that the relative risk of adult coronary heart disease was 0.84 for each 1 kg increase in birth weight.[69] This value probably substantially underestimates the developmental influence as there is much experimental evidence that the prenatal environment can induce long-term cardiovascular effects without necessarily affecting birth size.[59] Moreover, profound effects have now been demonstrated if there is a "mismatch" between the early, developmental environment and the subsequent environment in childhood and adult life.[46]

Links between prenatal growth and the later risk of NCD reflect variations in the quality of the intra-uterine environment.[50] Besides the limiting effects of small uterine size, constrained growth may reflect other aspects of the intra-uterine environment such as nutrition, oxygen supply, and hormonal exposure. While both intra-uterine growth restriction and preterm birth appear to have long-term consequences for NCD risk,[65,68] it is important to stress the associations between prenatal development and future risk can be seen across the range of infant size typical for each population. The underlying environmental cues and their mechanisms are still being defined, but it is presumed that nutrition plays a central role.

As yet direct evidence supporting a role for prenatal nutrition in inducing increased future NCD risk in humans is limited. One example is the Dutch Hunger Winter famine which occurred as a result of a wartime blockade during the winter of 1944, decreasing caloric intake from 1,800 kcal to between 400 and 800 kcal.[116] Severe caloric restriction during pregnancy was associated with increased risk of obesity, mood disorders, impaired glucose and lipid homeostasis, and reduced renal function in the offspring in a manner contingent on gestational age at the time of exposure to famine.[116] Further evidence supporting a role for maternal nutrition has come from follow-up studies of people in Motherwell, Scotland, whose mothers were advised to consume 0.45 kg meat and to avoid carbohydrate-rich foods during pregnancy; in this cohort, greater maternal meat and fish intake in pregnancy were associated in the adult offspring with increased blood pressure and accentuated cortisol responses to a stress challenge.[63,118]

The epidemiological observations have resulted in recognition that the quality of the early life environment has major public health implications worldwide (http://www.dohadsoc.org). For example, a World Health Organization Technical Consultation has concluded "The global burden of death, disability, and loss of human capital as a result of impaired fetal development is huge and affects both developed and developing countries."[150] The report advocates a move away from a simple focus on low birth weight to broader considerations of maternal well-being, in the hope of attaining an optimal environment that would maximize the potential for the fetus to achieve a full and healthy life.

In parallel with the epidemiological observations, animal studies have demonstrated the importance of epigenetic changes in mediating effects on adult phenotype and function arising from perturbations of the developmental environment, including maternal diet,[88,140] uterine blood flow,[110] and maternal nursing behavior.[143] The role of epigenetic processes in some forms of cancer is well-established,[83] but evidence has only emerged recently which shows that epigenetic processes also have major implications for understanding variations in human development, reproduction, and degenerative disease. The effects of epigenetic changes during development in animal models mimic aspects of human disease, such as metabolic disease and exaggerated stress responses. As a result, a coherent theory of the role played by epigenetic mechanisms in the early life origins of chronic disease is emerging.

13.2 Developmental Plasticity and Mismatch

Development represents a period of rapid change in the expression of the genome during which environmental cues may induce persistent changes in the phenotype of an organism. The developmental program tends to follow a path in which the characteristics of the wild type or typical phenotype are buffered against genetic and epigenetic changes, termed canalization.[137] However, many organisms respond during development to cues about their likely future environment, and this alters the developmental program and generates altered phenotypes. Such deviation from canalized development allows production of different phenotypes from a single genome more rapidly than could be achieved by mutation. For example, crowding of adult desert locusts (*Schistocerca gregaria*) induces gregarious, diurnal, and migratory offspring, in contrast to the nocturnal, sedentary forms which are produced under low population density[109]; the offspring of *Daphnia* are born with a defensive "helmet" structure if their mother has been exposed to chemicals produced by predators[82]; the duration of day light to which meadow voles (*Microtus pennsylvanicus*) are exposed before conception determines coat thickness in the offspring in anticipation of winter or spring temperatures.[86] Such rapid changes in phenotype may facilitate short-term survival, but may also be genetically assimilated and so produce stable phenotypes on which natural selection can act.[138] As discussed later, increasing evidence suggests that such persistent changes in the expression of the genome involve altered epigenetic regulation of specific genes.

Gluckman and Hanson[5,42,52] and Uller[133] have argued that the developmental environment can produce a range of effects, from overt disruption of development (i.e., teratogenesis), through altered fetal growth, with both its immediate and later consequences, to a range of phenotypes which become manifest only well after birth. This latter class can be induced by maternally mediated cues operating even within the normal range of developmental environments but nonetheless affecting several components of the trajectory of phenotypic development. The responses do not confer any immediate advantage for the fetus but give a Darwinian fitness advantage in later environments, the nature of which is predicted on the basis of the developmental experience. As the phenotype develops, the nature of this advantage may change at different points across the life course. Thus, increased insulin sensitivity may promote adipogenesis, providing nutritional reserves to protect the brain after weaning[80]; earlier puberty enhances fitness in a predicted adverse environment[47,122]; and the development of later insulin resistance confers a degree of "thrift" in a predicted adverse environment, as may reduction in numbers of energy-consuming skeletal and cardiac muscle cells or renal nephrons. This type of response has been termed a predictive adaptive response (PAR),[51] and supportive experimental and clinical studies have now been reported.[27,51,76,121]

According to the PAR model, response accuracy depends on the environment remaining relatively constant throughout the life course. Although environments fluctuate, modeling studies have shown that induced phenotypes persist for several generations and provide an adaptive advantage.[71] Thus, fidelity of predictions made during early life need not be high for PARs to confer a fitness advantage and be selected through evolution. When the anticipated environment is constant over many generations, the predictive trait/response may become fixed, or genetically encoded in a process known as genetic assimilation.[145] This process may include selection of advantageous mutations.

PARs constitute an integrated regulator in early life, establishing a life-course strategy to meet the demands of the predicted later environment.[48] PARs are only adaptive, however, when the post-developmental environment is within the predicted range.[48] If the later environment differs from the predicted range, the individual is "mismatched," having a phenotype that is inappropriate for the actual environment.[51] This can affect a range of traits including abdominal fat deposition, reduced skeletal muscle deposition and sarcopenia, reduced endothelial function, fewer cardiomyocytes, fewer nephrons, earlier puberty (at least in females), alterations in Th_1 to Th_2 cell balance associated with atopic/allergic reactions, reduced DNA repair leading to earlier aging, and a range of effects on behavior, including affective disorders and stress responses which are gender-specific.[45,66] Neither the developmental nor the later environment needs to provide extreme challenges for a mismatch to occur, with the critical determinant being that the phenotype induced by the former is not optimal for responding on a long-term basis to the latter.

Mismatch can result from exposure to an environment that is evolutionarily novel and thus beyond the predictive capacity of the fetus. Indeed, contemporary diets and lifestyles of developed societies constitute such a novel environment for *Homo sapiens*. The risk of NCD is then related to the degree of mismatch rather than to the absolute level of the adult environment per se. This is demonstrated in a number of experimental studies in which pre- and postnatal diets were manipulated.[67,117,136] For example, male sheep exposed to prenatal undernutrition but a normal postnatal diet, or vice versa, had changes in cardiovascular function that were not seen in undernourished animals.[27] In rats exposed to a high-fat diet in utero, endothelial dysfunction was observed in offspring fed a normal post-weaning diet, but not in those fed a high-fat post-weaning diet.[76]

The degree of mismatch can be increased by either poorer environmental conditions during development, or richer condition later or both.[46] Unbalanced maternal diet, altered maternal body composition, or disease can perturb the developmental environment. A rapid increase in energy-dense foods and reduced physical activity levels associated with a Western lifestyle will increase the degree of mismatch through effects on the environment later in life. Such changes are of considerable importance in developing societies that go through rapid socioeconomic transitions. Here we focus on how phenotypic induction through developmental plasticity can produce integrated changes in a range of organs via epigenetic processes, while noting that other, non-genomic mechanisms also operate to alter risk of disease in subsequent generations, e.g., the passage of cultural risk factors such as diet and smoking.

13.3 Epigenetics During Development and Aging

Patterns of DNA methylation are established early in development, and methylation plays a key role in cell differentiation by silencing the expression of specific genes. In mammals, the zygote undergoes rapid demethylation of the male genome within a few hours after fertilization; this is mediated by *Elp3*, a component of the elongator complex.[99,104,106] The female genome is passively demethylated in the course of subsequent mitotic divisions.[87] The overall effect of genome-wide demethylation is to produce pluripotent cells in which all genes are potentially transcriptionally active. Loss of pluripotency leading to cell differentiation and the establishment of adult tissue function depend on changes in the methylation status of individual gene promoters at different time points of development. Between formation and implantation of the blastocyst and gastrulation, the genome of the embryo undergoes de novo methylation, followed by gene-specific demethylation and methylation during cell differentiation.[74] For example, the pluripotency-associated gene Oct-4 is permanently silenced by hypermethylation of its promoter at around E6.5 in mice,[41] while HoxA5 and HoxB5, which are involved in pattern development, are not silenced until early postnatal life.[64] Other genes undergo graded changes in promoter methylation and transcription during development. Δ-Crystallin II and PEPCK promoters are methylated in the early embryo, but undergo progressive demethylation during development and are expressed in the adult.[8,57]

In addition to complete gene silencing by promoter methylation, differential methylation of individual CpG dinucleotides can induce more subtle modulation of transcriptional activity. For example, as a result of methylation of the CpG-rich promoter region, telomerase activity is downregulated in most cells during terminal differentiation in embryogenesis, but is often reactivated in cancer cells. Activation of telomerase in pre-neoplastic cells may be due to regulation shifting from the suppressor WT1 to the activator c-Myc due to changes in the methylation status of specific CpGs in the promoter of the catalytic subunit.[129] In mammals, DNA methylation is induced and maintained by DNA methyltransferases. Deletion or mutation of the genes encoding these enzymes results in embryonic death or severe disruption of development and loss of imprinting.[74] DNA methyltransferase (Dnmt)-1 is responsible for maintaining patterns of CpG dinucleotide methylation through replication cycles. DNA methylation de novo is catalyzed by Dnmt3a and 3b.[105]

The term genomic imprinting describes the monoallelic expression of specific gene loci dependent upon parental origin.[113] The majority of the 53 human genes known to be imprinted are located in CpG-rich domains termed Imprinting Centers, where methylation of CpG dinucleotides represses either the maternal or paternal allele.[29,103] As these methylation patterns are established in the gamete before fertilization, they are excluded from the genome-wide demethylation which occurs after fertilization.[16,84] Impaired imprinting leading to biallelic expression is causally associated with disorders including Angelman, Prader–Willi, and Beckwith–Wiedemann syndromes.[127,132]

Epigenetic marks induced during development largely persist into adulthood. However, aging is associated with tissue-specific epigenetic drift. Senescence is associated with decreasing Dnmt-1 activity and generalized hypomethylation, leading to activation of oncogenes such as c-Myc and c-N-ras.[94] However, hypermethylation of specific tumor-suppressor gene promoters also occurs in aging[115] and is thought to contribute to age-related increases in many malignancies. These findings show that aging is not simply associated with a progressive decline in capacity to maintain methylation of CpG dinucleotides, but also involves selective dysregulation of epigenetic processes. One implication is that the level of methylation induced in early life sets the epigenetic background upon which changes associated with aging operate and so variation in the epigenome induced during development may influence susceptibility to disease in later life.[141]

13.4 Induced Changes in the Epigenetic Regulation of Genes

13.4.1 Nutrition Models of the Early Life Origins of Human Chronic Disease

Maternal nutritional exposures now known to induce long-term effects on the offspring include global dietary restriction, protein restriction (PR), and a high saturated fat diet consumed during specific periods including pre-conception, pregnancy, and lactation. Studies using these diets have demonstrated causal relationships between nutrition in early life and the adult phenotype of the offspring and have begun to define the underlying mechanisms.[3,11]

Global undernutrition during lactation, or for 21 days after weaning, or after puberty reduces cell number in a variety of tissues. This can only be reversed when adequate nutrition is provided in the post-weaning and post-pubertal periods.[147] Reducing nutrient intake in pregnant rats to 30% of ad libitum consumption results in intra-uterine growth retardation, with the offspring developing higher systolic blood pressure, hyperinsulinemia, hyperleptinemia, hyperphagia, reduced locomotion, and obesity.[135,149] Such severe nutrient restriction does not generally occur in developed human societies, but is comparable to what occurred during the Dutch Hunger Winter.[116] A more modest restriction of 15% of the maternal diet during pregnancy impairs cholesterol homeostasis in male guinea pigs[77] and has led to lasting alterations in the hypothalamic-pituitary-adrenal (HPA) axis in sheep.[61]

In rodents, offspring of dams fed a diet with a modest reduction in protein (PR diet) during pregnancy show features similar to cardio-metabolic disease in humans, including graded hypertension, impaired lipid and glucose homeostasis, vascular dysfunction, increased susceptibility to oxidative stress, increased fat deposition, and altered feeding behavior.[7,21,36,130] Exposure at different developmental stages to the PR diet or to diets whose lipid, protein, and energy content are altered has resulted in different phenotypes in the offspring, even though the animal models were superficially similar. For example, in guinea pigs, female offspring born to dams fed a PR diet in the first half of pregnancy have raised mean arterial blood pressure and increased left ventricular wall thickness, but no growth restriction, whereas offspring from dams fed the same PR diet in late gestation are growth restricted, but have no changes in blood pressure or ventricular structure.[12]

13.4.2 Induction of Altered Transcription by Nutrition in Early Life

The induction of changes to the phenotype of the offspring that persist throughout the lifespan implies stable changes to gene transcription, resulting in altered activities of metabolic pathways and homeostatic control processes, and differences in the structure of tissues. The latter may result from changes in stem cell allocation to various lineages, variations in the rate and/or number of mitosis, and the extent of apoptosis. Together these processes provide cellular and molecular explanations for variation between individuals in body structure and functional capacity to respond to environmental challenge.[42] In women, differences in body structure and metabolic capacity induced by the environment experienced before birth may, in turn, have influenced the time of their own sexual maturation[47] and their reproductive success[28] as well as the birth weight of their children.[26] This is one mechanism by which phenotypic changes may be passed to successive generations.

The specificity of induced changes in the transcriptome has been investigated in animal models by microarray analysis. Gheorghe et al.[40] showed that feeding mice a PR diet between gestational days 10.5 and 17.5 altered expression of around 1% of the 22,690 genes in a microarray. There was increased expression of genes involved in the p53 pathway, apoptosis and negative regulation of cell growth and cell metabolism, and of genes related to epigenetic control; genes involved in nucleotide metabolism had lower expression. In pregnant rats fed the PR diet, studies of the livers of the offspring showed that 4,170 genes differed in expression on day 18 by more than twofold compared to those of controls, while in adult offspring, aged 84 days, 2,586 genes differed in expression between the experimental rats and controls.[91] This suggests that in many of the genes whose transcription is altered in response to the maternal diet, the change is transient and therefore unlikely to involve long-term epigenetic change. These genes may be involved in the inductive process but not directly involved in the induced phenotype.

Studies using a candidate gene approach have focused on macronutrient metabolism, HPA axis, and cardiovascular function. They have provided insights into the mechanisms that underlie phenotypes relevant to human disease. Feeding a PR diet to pregnant rats led to an increase in GR and a decrease in 11β-hydroxysteroid dehydrogenase type II expression in the lung, liver, kidney, and brain of the offspring.[13,90] GR expression is increased in the lung, liver, adrenal glands, and kidney of the offspring of sheep fed a restricted diet during pregnancy.[17,54,146] Feeding a PR diet to pregnant and/or lactating rats also upregulates expression in the liver of the offspring of glucokinase,[15] acetyl-CoA carboxylase,[96] PPARα, acyl-CoA oxidase, and carnitine palmitoyl-transferase-1, but not of PPARγ1.[23,24,88,92] In contrast, in adipose tissue, the expression of the adipose-specific isoform of PPARγ (PPARγ2) is reduced.[23] Together these findings show that induction of an altered phenotype involves persistent changes to the expression of a subset of genes that are associated with the altered epigenetic regulation of their promoters or of the promoters of transcription factors that regulate their activity.

While in mammals, the epigenetic processes of developmental plasticity appear to produce graded phenotypic effects, in some other taxa, the effects are dramatic because they induce alternative phenotypes. The honey bee (*Apis mellifera*) provides a good example. Here again the induction of alternative epigenotypes and phenotypes involves nutrition in early life. Queens and worker bees are genetically identical, but their morphology differs, as does their capacity to reproduce, their behavior, and their longevity. These differences arise from the duration of time that larvae are fed Royal Jelly, an as yet poorly defined mixture of proteins, amino acids, vitamins, lipids, and other nutrients; all larvae are fed this for the first 3 days after hatching, but only those destined to become queens are fed Royal Jelly for 3 more days.[95] An epigenetic basis for the difference between queens and workers has recently been established when larvae were injected on day 1 with Dnmt3 siRNA. This produced queens from 72% of the larvae, while injection with control siRNA produced workers from 72% of the larvae.[78] In larvae destined to become queens, lower methylation of specific CpG dinucleotides in the dynactin p62 gene leads to upregulated expression of dynactin p62 in the corpora allata. Thus, one mechanism by which Royal Jelly induces different bee castes is through altered Dnmt3 expression which induces changes in DNA methylation of specific genes.

13.4.3 Human Studies of Maternal Nutrition and Induced Epigenetic Change

As yet little has been published that links maternal nutrition in humans to epigenetic change in the offspring. A study of whole blood genomic DNA has suggested that in adults who were in utero during the Dutch Hunger Winter, the differentially methylated region (DMR) of the imprinted insulin-like growth factor-2 gene (*IGF2* DMR) is hypomethylated, as compared with that region in same-sex siblings.[62] The mean level of methylation of exposed individuals was 52% compared to 49% in unexposed controls, with standard deviations of around 5%. Additional findings from a study of people exposed to the Dutch Winter famine when in utero have provided preliminary evidence of an association of periconceptional famine exposure with altered methylation of the promoter regions of imprinted and non-imprinted genes that are implicated in growth and metabolic disease.[128] Among imprinted genes, *INSIGF* was hypomethylated in exposed individuals and guanine nucleotide–binding protein and *MEG3* were hypermethylated; among non-imprinted genes, interleukin-10, leptin, and ATP-binding cassette A1 were hypermethylated in exposed individuals. However, the maximum difference between exposed and unexposed individuals is only 6%, similar to the analytical error of the technique used.[35] In a recent study measuring five CpGs in the *IGF2* DMR, the whole blood methylation level of children from mothers who took 400-μg folic acid during pregnancy was 49.5%, whereas that of children of mothers who did not take folic acid was 47.4%, further evidence of a small effect of maternal nutrition on epigenetic processes in the fetus.[123]

13.4.4 Effects of Altered Maternal Nutrition During Pregnancy on the Epigenome of the Offspring in Rodents

Three mouse models have been described in which identical alleles are expressed differently as a result of epigenetic modifications; agouti variable yellow (A^{vy}),[34] axin fused ($Axin^{Fu}$),[134] and CDK5 activator-binding protein ($Cabp^{IAP}$).[33] In each case, an intracisternal A particle (IAP) retrotransposon is inserted upstream of the transcription start site which alters the level of transcription by the methylation status of the cryptic promoter. The level of methylation of this promoter can be modified by maternal nutrition and other environmental agents including endocrine disrupters.[31,139,140] For example, feeding pregnant A^{vy} mice diets that contain increased amounts of the methyl donors choline and betaine and of the 1-carbon metabolism cofactors, folic acid, and vitamin B12 changes the phenotype from agouti to pseudoagouti, consistent with increased methylation of the A^{vy}IAP.[148]

In models where maternal nutrition alters cardiovascular function and metabolism in the offspring, feeding a PR diet to rats during pregnancy induces hypomethylation of the PPARα and GR promoters and increased expression of GR and PPARα in the livers of juvenile[88] and adult offspring.[24] Hypomethylation of the GR promoter is associated with an increase in histone modifications that facilitate transcription, i.e., acetylation of histones H3 and H4 and methylation of histone H3 at lysine K4, whereas histone modifications that suppress gene expression are reduced or unchanged.[92] Although functionally consistent, the mechanistic relationship between GR hypomethylation and the associated histone changes is not known. These studies showed for the first time that stable changes to the epigenetic regulation of the expression of transcription factors can be induced in the offspring by modest changes to maternal macronutrient intake during pregnancy. Expression of PPARα and GR, and of their respective target genes, acyl-CoA oxidase and carnitine palmitoyl-transferase-1, and PEPCK was increased in juvenile and adult offspring.[24,88,92] This is consistent with raised plasma β-hydroxybutyrate and glucose concentrations in the fasting offspring.[21] Sequencing analysis of the PPARα promoter showed that four specific CpGs were hypomethylated, and that two CpGs located within transcription factor response elements predicted the level of the transcript.[90] Thus, the effects of the maternal PR diet on the offspring are targeted to specific CpGs.

These findings make it evident that modest dietary protein restriction during pregnancy induces an altered phenotype through epigenetic changes in specific genes. Methylation of the GR and PPARα promoters is also reduced in the heart of the offspring[89] and the PPARα promoter is hypomethylated in the whole umbilical cord.[19] These findings are consistent with increased GR mRNA expression in a range of tissues of the offspring of rats fed a PR diet during pregnancy.[13] However, PPARα methylation does not differ between control and PR offspring in skeletal muscle, spleen, and adipose tissue; in other words, the effects of the maternal diet are tissue specific (K.A. Lillycrop and G.C. Burdge, unpublished data). The GR promoter is hypomethylated in the offspring of mice fed a PR diet during pregnancy,[19] which suggests that the effect of the PR diet is not specific to one species.

The fundamental role that changes in the epigenetic regulation of transcription factor expression play in altering the activity of pathways controlled by their target genes is illustrated by the report[15] that the increase in glucokinase expression in the liver of the PR offspring was not accompanied by changes in the methylation status of the glucokinase promoter. Since GR activity increases glucokinase expression, an increase in glucokinase expression may have been due to the increase in GR activity brought about by hypomethylation of the GR promoter, rather than by a direct effect of prenatal undernutrition on glucokinase promoter methylation.

The process by which environmental cues induce altered epigenetic regulation in the embryo remains unknown. Studies in liver from juvenile offspring have provided some insights into the underlying mechanisms. When pregnant rats are fed a PR diet, Dnmt1 expression is lowered, as is the binding of Dnmt1 to the GR promoter.[92] However, the expression of Dnmt3a, Dnmt3b, and MBD-2, and binding of Dnmt3a to the GR promoter are unaltered.[92] This suggests that hypomethylation of the hepatic GR promoter in the offspring, and perhaps of other genes including PPARα, is induced by a reduction in the capacity to maintain patterns of cytosine methylation during mitosis rather than by a failure of de novo methylation or of active demethylation.[19,92] This interpretation is consistent with lower MeCP2 binding and increased levels of histone modifications that facilitate transcription at the GR promoter. Reduced Dnmt1 activity might be expected to result in global demethylation. However, in vitro studies have shown that the loss of Dnmt1-induced demethylation affects only a subset of genes[73,114] and that therefore Dnmt1 is targeted to specific genes, consistent with selective hypomethylation in the liver of the PR offspring.[88] Dnmt1 activity is also required for progression through mitosis[101] and its expression is substantially reduced in non-proliferating cells.[125] Thus, suppression of Dnmt1 activity in the pre-implantation period could also account for a reduction in cell number during early development in this model.[81]

In contrast to the effects of maternal PR diet on the epigenetic regulation of hepatic genes in the offspring, a 70% reduction of total food intake during pregnancy in rats induced hypermethylation and lower PPARα and GR expression in the liver of 170-day-old offspring.[53] One explanation may lie in the differences in severity of nutritional constraint between these two dietary regimens. If the induction of altered phenotypes is predictive, then it may be anticipated that induced changes in the epigenome represent an effort to match the phenotype to the predicted future environment. The maternal PR diet would therefore represent a moderate nutrient constraint that induces in the offspring an increase in the capacity to use nutrient reserves for energy production. Global undernutrition is a relatively more severe constraint that induces conservation of energy substrates. These interpretations are consistent with the phenotypes induced in the offspring.[21,135]

13.4.5 Neonatal Care and Offspring Stress Responses

Experimental studies have shown that it is not just variations in nutrition that induce altered phenotypes, and there are clear examples of neonatal care inducing long-term epigenetic changes of relevance to developmental influences on later health. Weaver et al.[143] observed that rats who had been licked and groomed more by their mothers in the neonatal period responded less strongly to future stress than offspring who had been nursed more poorly. Better response to stress was due to hypomethylation of a single CpG dinucleotide within the NGF1-A-binding domain in the $GR1_7$ promoter in the hippocampus, which resulted in increased GR mRNA expression.[143] Thus, similar to the effects of maternal diet on the PPARα promoter, maternal nursing behavior led to highly specific changes in the epigenome of the offspring. It therefore appears that an adverse environment in early life induces changes in the offspring that bear on the response to future environmental challenge. A recent study has shown that the methylation status of the equivalent CpG dinucleotide in the GR promoter of the human brain is associated with suicide in individuals who were abused as children.[100] These findings have important implications for understanding how the social environment in early life may induce life-long personality traits and/or facilitate perpetuation of cultural practices.

13.5 Nutritional Interventions to Prevent or Reverse Induced Phenotypes and Epigenetic Changes

13.5.1 Prevention of Phenotypes Induced by Maternal Undernutrition

The identification of epigenetic mechanisms that have led to enhanced disease risk in response to prenatal nutrition raises the possibility of preventing or reversing these processes by nutritional means. Such interventions have been tested in animal models that use nutrients involved in 1-carbon metabolism. So far, these experiments have only reached the "proof of principle" stage, but they have identified possible benefits and pitfalls of relevance to human disease.

Methylation of DNA and histones is closely linked to pathways that supply methyl substrates to their respective methyltransferases. For DNA methylation, methyl groups come primarily from serine by the action of cytoplasmic serine hydroxymethyltransferase which transfers CH_3 to tetrahydrofolate (THF) to form 5,10-methylene THF which in turn is reduced to 5-methyl THF by tetrahydrofolate reductase. This methyl group is used to convert homocysteine to methionine by methionine synthase, with vitamin B_{12} as a cofactor. *S*-Adenosylmethionine is the substrate for Dnmts which compete for CH_3 with phosphatidylethanolamine

N-methyltransferase (PE N-MET) activity. PE N-MET uses three moles of CH_3 to synthesize one mole of phosphatidylcholine. This is the major terminal reaction in 1-carbon metabolism in the liver. Betaine is an alternative substrate for the remethylation of homocysteine, a reaction that is catalyzed by betaine homocysteine methyltransferase. Glycine is used to generate serine as part of the mitochondrial folate cycle, which in turn, is the substrate for cytoplasmic SHMT.

Supplementation of the maternal PR diet with glycine[72] or folic acid[130] prevents hypertension and endothelial dysfunction in the offspring. Supplementation of the maternal PR diet with folic acid also prevents dyslipidemia in the adult offspring.[21] In contrast, supplementation of the control protein diet with folic acid induces impaired endothelial dysfunction and dyslipidemia in the offspring.[21,130] Increasing the folic acid content of the PR diet prevents hypomethylation of the PPARα and GR promoters in the liver of the offspring.[88] However, detailed analysis of the PPARα promoter has shown that even though an increase in maternal folic acid intake prevents hypomethylation of the majority of CpG dinucleotides induced by the PR diet alone, two CpGs become hypermethylated[90] (Fig. 13.2). Thus increasing maternal folic acid intake does not simply prevent the effects of the PR diet, but may induce subtle changes in gene regulation.

13.5.2 Reversal of Phenotypes Induced by Maternal Undernutrition

It is important to consider whether interventions after the neonatal period can reverse the adverse effects of prenatal nutrition. Supplementing the diet of rats with folic acid during their juvenile-pubertal period[22] has shown that, in contrast with when the maternal PR diet is supplemented with folic acid, supplementation during the juvenile-pubertal period results in impaired lipid homeostasis. This includes downregulation of hepatic fatty acid β-oxidation, hepatosteatosis, and an increase in weight gain. Adverse effects are seen irrespective of the maternal diet and are associated with altered methylation of specific genes, including hypomethylation of the insulin receptor in adipose tissue and hypermethylation of PPARα in the liver of the offspring (Fig. 13.3). These findings suggest a period of plasticity between weaning and adulthood that may reflect continued growth and development. This inference is consistent with the view that puberty is one of four periods of increased instability of the epigenome, the other three being prenatal development, neonatal development, and aging.[30] Although folic acid supplementation has deleterious effects in the model described above, the findings do suggest that nutritional interventions before adulthood could reverse adverse effects of prenatal nutrition.

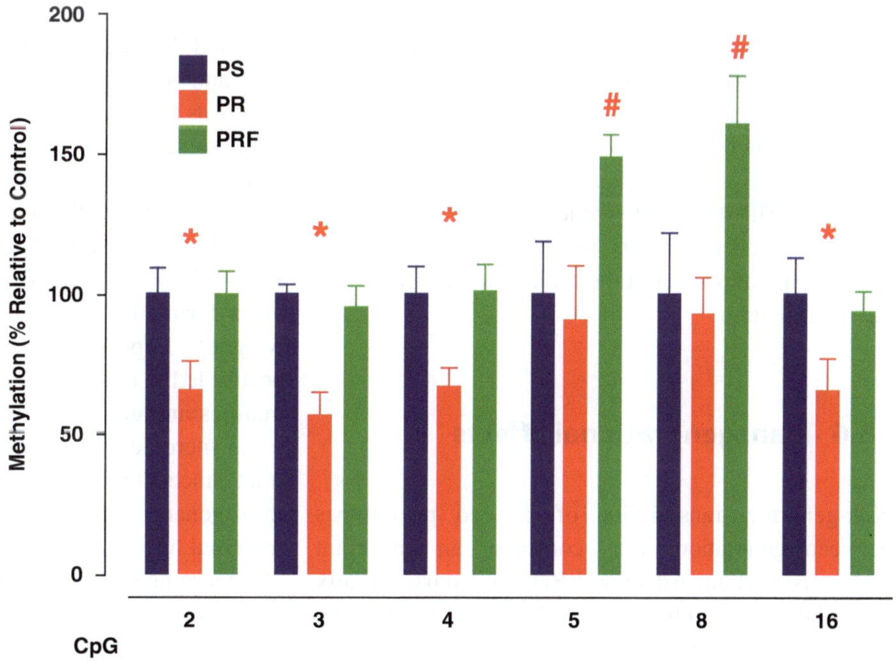

Fig. 13.2 Methylation of individual CpG dinucleotides in the hepatic PPARα promoter. *Four CpGs were hypomethylated in the liver of the offspring of dams fed a protein-restricted (PR) diet during pregnancy compared to the offspring of protein-sufficient (PS) dams. Supplementation of the PR diet with folic acid (PRF) prevented hypomethylation of these four CpGs, but induced hypermethylation of #two CpGs

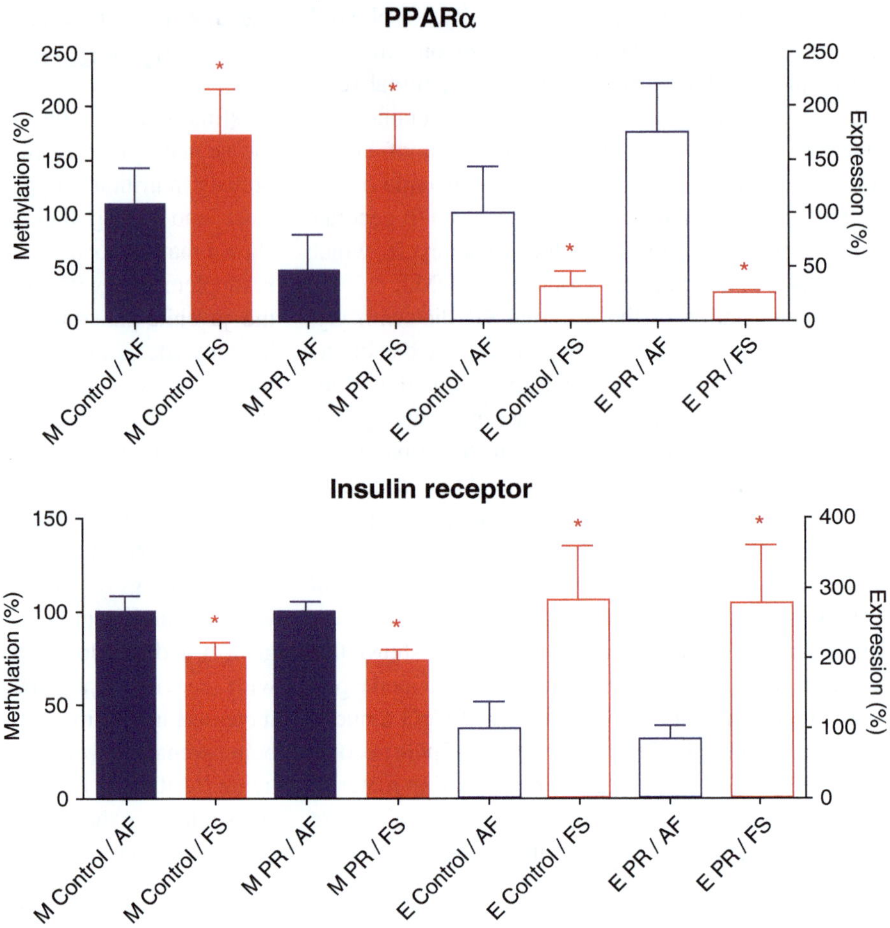

Fig. 13.3 Effect of folic acid supplementation (FS) during the juvenile-pubertal period on promoter methylation (M) and mRNA expression (E) of PPARα in liver and insulin receptor in adipose tissue of adult offspring of dams fed control (protein-sufficient) or protein-restricted (PR) diets during pregnancy. *Values significantly different between FS and folic acid adequate (FA) offspring

The studies of folic acid supplementation during gestation or after weaning have shown that the outcomes of such interventions are influenced by the timing of the intervention and that interactions between folic acid and the background diet cannot be readily predicted. The design of supplementation regimens to reverse epigenetic effects in humans will therefore have to take into account the timing and magnitude of the intervention.

13.6 Transgenerational Effects

Non-genomic transmission of induced phenotypes between generations may constitute an important mechanism in human disease.[49] Records from Överkalix in northern Sweden for individuals born in 1890, 1905, and 1920 have shown that diabetes mortality increased in men if the paternal grandfather was exposed to abundant nutrition during his pre-pubertal growth period.[75] This was later found to apply also to paternal grandmother/granddaughter pairs and to be transmitted in a gender-specific fashion.[108] Poor maternal nutrition has been associated with an increased risk of type 2 diabetes mellitus over several generations in North American Indians,[10] and individuals whose grandparents were in utero during the Dutch Hunger Winter had lower birth weight.[124] Exposure of pregnant women to diethylstilbestrol led to a marked increase in reproductive abnormalities and uterine fibroids,[4] an earlier menopause,[60] to an increase in breast cancer,[107] and to rare genital tract cancers in their children; there is also evidence of third-generational effects transmitted through the maternal line.[18]

Emerging evidence from small animal models suggests that induced phenotypes can be transmitted to one or more generations by a non-genomic mechanism.

In rats, feeding a PR diet to the F_0 generation during pregnancy leads to elevated blood pressure, endothelial dysfunction, and insulin resistance in the F_1 and F_2 generations[98,131,153] despite adequate nutrition during pregnancy in the F_1 generation. The feeding of a PR regimen during pregnancy in the F_0 generation has led to adverse effects on glucose homeostasis in the offspring up to the F_3 generation.[9] The administration of dexamethasone to dams in late pregnancy led to increased expression of the glucocorticoid receptor (GR) and its target gene phosphoenolpyruvate carboxykinase (PEPCK) in the livers of the F_1 and F_2 offspring, but not in the F_3 generation,[32] and it is now evident that assessment of true non-genomic transmission between generations requires studies that continue to at least the F_3 generation.[120] Transmission to the F_2 generation may be due either to changes induced in the F_1 generation that are transmitted to the F_2 generation or because F_1 germ line cells have been altered directly by environmental cues from the grandmother. Loss of transmission of the altered phenotype from the F_2 to the F_3 generation suggests that the phenotype present in the F_2 generation may have resulted from exposure of the F_1 germ line to dexamethasone.

There is substantial evidence for transgenerational epigenetic inheritance in non-mammalian species and its role in evolutionary biology,[49,71] however few studies have reported transmission of nutritionally induced epigenetic marks between generations. Compared to controls, GR and PPARα promoters were hypomethylated in 80-day-old male grand-offspring of rats exposed to maternal PR diet during gestation, even though F1 dams received adequate nutrition throughout pregnancy.[24] These findings imply that the female line can transmit such epigenetic information between generations. However, phenotypes induced in the offspring by maternal exposure to dexamethasone in pregnancy are transmitted to both males and females through the F_1 and F_2, but not the F_3 generation.[32] The tendency toward obesity in A^{vy} mice is exacerbated through successive generations,[142] but was prevented by supplementation of females with methyl donors and cofactors, although this was not associated with a change in the methylation status of the A^{vy} locus. The mechanism by which induced epigenetic marks are transmitted to subsequent generations is not known. Given transmission was only to the F_2 generation, a direct effect of the diet fed to the F_0 dams on germ cells that give rise to the F_2 offspring cannot be ruled out. Sequential transmission from F_1 to F_2, and possibly beyond, would involve induction in the germ line of altered epigenetic marks. Moreover, such changes in DNA methylation would have to be preserved during genome-wide demethylation that takes place in fertilization, possibly by a mechanism similar to that which preserves the methylation of imprinted genes.[84] It may also involve targeted preservation of nucleosome structure as occurs for specific developmental genes during spermatogenesis.[58] Alternatively, prenatal nutritional constraint may induce physical or physiological changes in the female which, in turn, restrict the intra-uterine environment in which her offspring develop. In this case, transmission of an altered phenotype between generations would involve induction of changes in gene methylation de novo in each generation. If that is so, the magnitude of the induced effect, epigenetic or phenotypic, may differ between generations. However, this possibility is not supported by the fact that reduction in birth size and blood glucose concentration in the F_1 and F_2 generations is similar in offspring of rat dams exposed to dexamethasone in late gestation.[32] Furthermore, the degree of hypomethylation of the hepatic GR and PPARα is similar in the F_1 and F_2 offspring of dams fed a PR diet in pregnancy.[24] Finally, de novo methylation in each generation would not pass phenotypic traits through the male line.[32]

13.7 Life Course and Evolutionary Implications of Induced Phenotypic and Epigenetic Change

13.7.1 Developmental Epigenetic Processes and the Adaptive Value of Non-genomic Inheritance

Developmental plasticity provides a mechanism by which epigenetic processes can lead to different phenotypes from the same genome. Many examples quoted from natural history, such as Royal Jelly feeding in honey bees, represent polyphenisms. Yet phenotypes induced by variations in maternal nutrition in mammals appear pleiotropic, such as the graded dose–response relationship between maternal protein intake and blood pressure in the offspring.[85] Experimental studies have shown that during early life, rodents are sensitive to a range of environmental cues and that variations in the duration and timing of nutritional changes can induce a range of phenotypes in the offspring. Many of the induced phenotypes are cryptic and become only manifest when an individual is exposed to a particular environmental challenge in later life.

The increasing evidence for non-genomic inheritance, and particularly epigenetic inheritance, raises the question of why the processes underpinning non-genomic inheritance have been preserved through evolution. Natural selection is a process by which a species and its environment become well matched. Developmental plasticity utilizes environmental cues to adjust individual phenotype to the current and predicted environment.[51,52] Developmental plasticity leading to non-genomic inheritance may have evolved to enhance fitness during shorter-term environmental shifts than Darwinian selection can necessarily cope with; it ensures a greater match to a variable environment than selection alone can generate. In addition, developmental plasticity leads to a wide range of phenotypes, permitting survival in a broader range of environments. Such strategies may have been important in the evolution of mammalian generalist species.[93] Theoretical models have been developed that show the circumstances under which fitness is enhanced if parents transmit information about the environment to their progeny. Factors to consider include the fidelity of transmission, the degree of predictability of environmental conditions, and the costs of incorrect prediction.[51,70,102,126]

After an initial period of plasticity, phenotype and epigenotype are likely to be resistant to further change in order to transmit the fitness advantage of adaptation to the predicted environment. This is not always the case, however, inasmuch as increased exposure to folic acid during the juvenile-pubertal period overcomes epigenetic and phenotypic changes induced before birth.[22] Neonatal administration of leptin also reverses developmental programming.[136] Such resetting of an induced phenotype and epigenotype by a modest nutritional intervention may seem to disprove the hypothesis that induced changes predict the future environment. However, plasticity before and during puberty allows further adaptation or could correct incorrect prediction before the onset of reproduction.

In addition to providing a means of increasing fitness, phenotypes altered in response to environmental cues and their transmission can also contribute to speciation and phylogeny.[144] The epigenetic changes which underlie induced phenotypes constitute one mechanism for increasing the susceptibility of the genome to mutation. This includes the instability of the genome due to either hyper- or hypo-DNA methylation or the deamination of methylcytosine to uracil. This reaction induces a single nucleotide polymorphism (SNP) by allowing insertion of thymidine in the place of guanine in the complementary strand during DNA replication. This is supported by the enrichment of SNPs in highly methylated sequences.[152] In mice, feeding a methyl donor-deficient diet increased the rate of DNA damage when they were subjected to irradiation.[6] Furthermore, in the white-cheeked gibbon (*Nomascus leucogenys*), hypomethylation of the genome is associated with a higher rate of chromosomal rearrangement and may have contributed to evolution of the species.[25] Thus, changes in DNA methylation in response to environmental cues contribute to genetic variation, thereby enhancing evolution. However, it is not yet known whether these genetic changes fix induced phenotypes or would tend to produce random changes in the genome.

13.8 Relevance of Epigenetic Processes to the Risk of Adult Disease

We now live much longer than our human ancestors. Mechanisms that enhanced fitness in early evolution may no longer have an advantage, or may have advantage for the young, but not for the health of the elderly. Fitness and health are not identical. There are limits to the environment that the fetus can sense and use to adjust its development.[43] Non-genomic transmission of environmental information to later generations may have evolved to assist humans in their migration and occupation of very different environments. Non-genomic epigenetic processes may also have buffered against short-term environmental changes occurring between generations, especially during the vulnerable period of weaning.[79] Such processes were not "designed," however, to deal with the mismatch between the generally constrained fetal environment and the modern postnatal environment of high-energy intake and low-energy expenditure,[46] as a result of which disease risk is amplified by the mismatch between the predicted and actual adult environments. This is particularly true for societies in rapid economic development.[14,44,111,112] Epigenetic and other non-genomic inheritance processes that conferred survival advantage on evolving humans now exacerbate disease risk for successive generations and play a major role in the current epidemics of metabolic and cardiovascular diseases.[42,44] In addition, exposure to xenobiotics, such as endocrine disruptors, may have multigenerational effects through similar epigenetic actions in both female and male lines.[2]

Finally, there is increasing evidence that the risks of metabolic and cardiovascular disease are increased in the offspring of obese or diabetic mothers,[37,38,119] as has also been shown in animal models.[1,39] These effects may amplify across generations, contributing to the rising incidence of metabolic and cardiovascular diseases in both developed and developing societies. Whether these risks are due to epigenetic processes is not known, but the offspring of high fat-fed dams show changes in the pattern of micro-RNA expression, particularly in those associated with IGF expression and methyl transferases.[154]

13.9 Summary and Conclusions

Variation in the expression of the genome leads to novel phenotypes, with implications for understanding evolutionary biology and the risk of disease. Epigenetic changes, DNA methylation in particular, provide a "memory" of plastic responses to the early environment during the course of development. These are central to the generation and lifetime stability of novel phenotypes. Effects may become manifest only later in life, for example, as altered responses to environmental challenges. Research is needed to determine whether epigenetic marks in early life constitute biomarkers to identify individuals who in development have experienced environmental perturbations, and thus are more likely to develop premature cardiovascular and metabolic diseases or other sequelae. Epigenetically induced change is central to phenotypic variation but may be secondary to the pathway by which signals from the future environment are transmitted to the embryo/fetus and subsequently give rise to epigenetic modifications. Understanding this process will provide a substantial step forward in biological research, including the understanding of the developmental origins of health and disease.

Acknowledgments Research by the authors is supported by the Biotechnology and Biological Sciences Research Council, the Medical Research Council, the British Heart Foundation, the National Institute for Health Research Southampton Nutrition, Diet and Lifestyle Biomedical Research Unit, Wessex Medical Research, and the Gerald Kerkut Charitable Trust. The authors are not aware of any conflicts of interest that may be perceived as affecting the objectivity of this review.

References

1. Aerts L, Van Assche FA. Animal evidence for the transgenerational development of diabetes mellitus. *Int J Biochem Cell Biol*. 2006;38:894-903.
2. Anway MD, Cupp AS, Uzumcu M, Skinner MK. Epigenetic transgenerational actions of endocrine disruptors and male fertility. *Science*. 2005;308:1466-1469.
3. Armitage JA, Khan IY, Taylor PD, Nathanielsz PW, Poston L. Developmental programming of the metabolic syndrome by maternal nutritional imbalance: how strong is the evidence from experimental models in mammals? *J Physiol*. 2004; 561:355-377.
4. Baird DD, Newbold R. Prenatal diethylstilbestrol (DES) exposure is associated with uterine leiomyoma development. *Reprod Toxicol*. 2005;20:81-84.
5. Bateson P, Barker D, Clutton-Brock T, et al. Developmental plasticity and human health. *Nature*. 2004;430:419-421.
6. Batra V, Mishra KP. Modulation of DNA methyltransferase profile by methyl donor starvation followed by gamma irradiation. *Mol Cell Biochem*. 2007;294:181-187.
7. Bellinger L, Sculley DV, Langley-Evans SC. Exposure to undernutrition in fetal life determines fat distribution, locomotor activity and food intake in ageing rats. *Int J Obes Lond*. 2006;30:729-738.
8. Benvenisty N, Mencher D, Meyuhas O, Razin A, Reshef L. Sequential changes in DNA methylation patterns of the rat phosphoenolpyruvate carboxykinase gene during development. *Proc Natl Acad Sci USA*. 1985;82:267-271.
9. Benyshek DC, Johnston CS, Martin JF. Glucose metabolism is altered in the adequately-nourished grand-offspring (F-3 generation) of rats malnourished during gestation and perinatal life. *Diabetologia*. 2006;49:1117-1119.
10. Benyshek DC, Martin JF, Johnston CS. A reconsideration of the origins of the type 2 diabetes epidemic among Native Americans and the implications for intervention policy. *Med Anthropol*. 2001;20:25-64.
11. Bertram CE, Hanson MA. Animal models and programming of the metabolic syndrome. *Br Med Bull*. 2001;60:103-121.
12. Bertram C, Khan O, Ohri S, Phillips DI, Matthews SG, Hanson MA. Transgenerational effects of prenatal nutrient restriction on cardiovascular and hypothalamic-pituitary-adrenal function. *J Physiol*. 2008;586:2217-2229.
13. Bertram C, Trowern AR, Copin N, Jackson AA, Whorwood CB. The maternal diet during pregnancy programs altered expression of the glucocorticoid receptor and type 2 11beta-hydroxysteroid dehydrogenase: potential molecular mechanisms underlying the programming of hypertension in utero. *Endocrinology*. 2001;142:2841-2853.

14. Bhargava SK, Sachdev HS, Fall CH, et al. Relation of serial changes in childhood body-mass index to impaired glucose tolerance in young adulthood. *N Engl J Med*. 2004;350: 865-875.
15. Bogdarina I, Murphy HC, Burns SP, Clark AJ. Investigation of the role of epigenetic modification of the rat glucokinase gene in fetal programming. *Life Sci*. 2004;74:1407-1415.
16. Brandeis M, Kafri T, Ariel M, et al. The ontogeny of allele-specific methylation associated with imprinted genes in the mouse. *EMBO J*. 1993;12:3669-3677.
17. Brennan KA, Gopalakrishnan GS, Kurlak L, Rhind SM, Kyle CE, et al. Impact of maternal undernutrition and fetal number on glucocorticoid, growth hormone and insulin-like growth factor receptor mRNA abundance in the ovine fetal kidney. *Reproduction*. 2005;129:151-159.
18. Brouwers MM, Feitz WF, Roelofs LA, Kiemeney LA, de Gier RP, Roeleveld N. Hypospadias: a transgenerational effect of diethylstilbestrol? *Hum Reprod*. 2006;21:666-669.
19. Burdge GC, Hanson MA, Slater-Jefferies JL, Lillycrop KA. Epigenetic regulation of transcription: a mechanism for inducing variations in phenotype (fetal programming) by differences in nutrition during early life? *Br J Nutr*. 2007;97:1036-1046.
20. Burdge GC, Lillycrop KA, Jackson AA. Nutrition in early life, and risk of cancer and metabolic disease: alternative endings in an epigenetic tale? *Br J Nutr*. 2009;101:619-630.
21. Burdge GC, Lillycrop KA, Jackson AA, Gluckman PD, Hanson MA. The nature of the growth pattern and of the metabolic response to fasting in the rat are dependent upon the dietary protein and folic acid intakes of their pregnant dams and post-weaning fat consumption. *Br J Nutr*. 2008;99:540-549.
22. Burdge GC, Lillycrop KA, Phillips ES, Slater-Jefferies JL, Jackson AA, Hanson MA. Folic acid supplementation during the juvenile-pubertal period in rats modifies the phenotype and epigenotype induced by prenatal nutrition. *J Nutr*. 2009;139:1054-1060.
23. Burdge GC, Phillips ES, Dunn RL, Jackson AA, Lillycrop KA. Effect of reduced maternal protein consumption during pregnancy in the rat on plasma lipid concentrations and expression of peroxisomal proliferator-activated receptors in the liver and adipose tissue of the offspring. *Nutr Res*. 2004;24:639-646.
24. Burdge GC, Slater-Jefferies J, Torrens C, Phillips ES, Hanson MA, Lillycrop KA. Dietary protein restriction of pregnant rats in the F0 generation induces altered methylation of hepatic gene promoters in the adult male offspring in the F1 and F2 generations. *Br J Nutr*. 2007;97:435-439.
25. Carbone L, Harris RA, Vessere GM, et al. Evolutionary breakpoints in the gibbon suggest association between cytosine methylation and karyotype evolution. *PLoS Genet*. 2009;5:e1000538.
26. Catalano PM, Kirwan JP. Maternal factors that determine neonatal size and body fat. *Curr Diab Rep*. 2001;1:71-77.
27. Cleal JK, Poore KR, Boullin JP, et al. Mismatched pre- and postnatal nutrition leads to cardiovascular dysfunction and altered renal function in adulthood. *Proc Natl Acad Sci USA*. 2007;104:9529-9533.
28. Davies MJ. Evidence for effects of weight on reproduction in women. *Reprod Biomed Online*. 2006;12:552-561.
29. Delaval K, Feil R. Epigenetic regulation of mammalian genomic imprinting. *Curr Opin Genet Dev*. 2004;14:188-195.
30. Dolinoy DC, Das R, Weidman JR, Jirtle RL. Metastable epialleles, imprinting, and the fetal origins of adult diseases. *Pediatr Res*. 2007;61:30R-37R.
31. Dolinoy DC, Huang D, Jirtle RL. Maternal nutrient supplementation counteracts bisphenol A-induced DNA hypomethylation in early development. *Proc Natl Acad Sci USA*. 2007;104:13056-13061.
32. Drake AJ, Walker BR, Seckl JR. Intergenerational consequences of fetal programming by in utero exposure to glucocorticoids in rats. *Am J Physiol Regul Integr Comp Physiol*. 2005;288:R34-R38.
33. Druker R, Bruxner TJ, Lehrbach NJ, Whitelaw E. Complex patterns of transcription at the insertion site of a retrotransposon in the mouse. *Nucleic Acids Res*. 2004;32:5800-5808.
34. Duhl DMJ, Vrieling H, Miller KA, Wolff GL, Barsh GS. Neomorphic agouti mutations in obese yellow mice. *Nat Genet*. 1994;8:59-65.
35. Ehrich M, Nelson MR, Stanssens P, et al. Quantitative high-throughput analysis of DNA methylation patterns by base-specific cleavage and mass spectrometry. *Proc Natl Acad Sci USA*. 2005;102:15785-15790.
36. Fernandez-Twinn DS, Wayman A, Ekizoglou S, Martin MS, Hales CN, Ozanne SE. Maternal protein restriction leads to hyperinsulinemia and reduced insulin-signaling protein expression in 21-mo-old female rat offspring. *Am J Physiol Regul Integr Comp Physiol*. 2005;288:R368-R373.
37. Forsen T, Eriksson JG, Tuomilehto J, Teramo K, Osmond C, Barker DJP. Mother's weight in pregnancy and coronary heart disease in a cohort of Finnish men: follow up study. *BMJ*. 1997;315:837-840.
38. Gale CR, Javaid MK, Robinson SM, Law CM, Godfrey KM, Cooper C. Maternal size in pregnancy and body composition in children. *J Clin Endocrinol Metab*. 2007;92:3904-3911.
39. Galili O, Versari D, Sattler KJ, et al. Early experimental obesity is associated with coronary endothelial dysfunction and oxidative stress. *Am J Physiol Heart Circ Physiol*. 2007;292:H904-H911.
40. Gheorghe CP, Goyal R, Holweger JD, Longo LD. Placental gene expression responses to maternal protein restriction in the mouse. *Placenta*. 2009;30:411-417.
41. Gidekel S, Bergman Y. A unique developmental pattern of Oct-3/4 DNA methylation is controlled by a cis-demodification element. *J Biol Chem*. 2002;277:34521-34530.
42. Gluckman PD, Hanson MA. Living with the past: evolution, development, and patterns of disease. *Science*. 2004;305: 1733-1736.
43. Gluckman PD, Hanson MA. Maternal constraint of fetal growth and its consequences. *Semin Fetal Neonatal Med*. 2004;9:419-425.
44. Gluckman PD, Hanson MA. The developmental origins of the metabolic syndrome. *Trends Endocrinol Metab*. 2004; 15:183-187.
45. Gluckman PD, Hanson MA. *The fetal matrix: evolution, development, and disease*. Cambridge, UK: Cambridge University Press; 2005.
46. Gluckman PD, Hanson MA. *Mismatch; how our world no longer fits our bodies*. Oxford: Oxford University Press; 2006.

47. Gluckman PD, Hanson MA. Evolution, development and timing of puberty. *Trends Endocrinol Metab*. 2006;17: 7-12.
48. Gluckman PD, Hanson MA, Beedle AS. Early life events and their consequences for later disease; a life history and evolutionary perspective. *Am J Hum Biol*. 2007;19:1-19.
49. Gluckman PD, Hanson MA, Beedle AS. Non-genomic transgenerational inheritance of disease risk. *BioEssays*. 2007;29:145-154.
50. Gluckman PD, Hanson MA, Cooper C, Thornburg KL. Effect of in utero and early-life conditions on adult health and disease. *N Engl J Med*. 2008;359:61-73.
51. Gluckman PD, Hanson MA, Spencer HG. Predictive adaptive responses and human evolution. *Trends Ecol Evol*. 2005;20:527-533.
52. Gluckman PD, Hanson MA, Spencer HG, Bateson P. Environmental influences during development and their later consequences for health and disease: implications for the interpretation of empirical studies. *Proc R Soc Lond B*. 2005;272:671-677. Containing papers of a Biological character. Royal Society (Great Britain).
53. Gluckman PD, Lillycrop KA, Vickers MH, et al. Metabolic plasticity during mammalian development is directionally dependent on early nutritional status. *Proc Natl Acad Sci USA*. 2007;104:12796-12800.
54. Gnanalingham MG, Mostyn A, Dandrea J, Yakubu DP, Symonds ME, Stephenson T. Ontogeny and nutritional programming of uncoupling protein-2 and glucocorticoid receptor mRNA in the ovine lung. *J Physiol*. 2005;565:159-169.
55. Godfrey KM. The "Developmental Origins" hypothesis: epidemiology. In: Hanson MA, Gluckman PD, eds. *Developmental Origins of Health and Disease – A Biomedical Perspective*. Cambridge, UK: Cambridge University Press; 2006:6-32.
56. Godfrey KM, Gluckman PD, Hanson MA. Developmental origins of metabolic disease: life course and intergenerational perspectives. *Trends Endocrinol Metab*. 2010;21(4): 199-205.
57. Grainger RM, Hazard-Leonards RM, Samaha F, Hougan LM, Lesk MR, Thomsen GH. Is hypomethylation linked to activation of delta-crystallin genes during lens development? *Nature*. 1983;306:88-91.
58. Hammoud SS, Nix DA, Zhang H, Purwar J, Carrell DT, Cairns BR. Distinctive chromatin in human sperm packages genes for embryo development. *Nature*. 2009;460:473-478.
59. Hanson MA, Gluckman PD. Developmental processes and the induction of cardiovascular function: conceptual aspects. *J Physiol*. 2005;565:27-34.
60. Hatch EE, Troisi R, Wise LA, et al. Age at natural menopause in women exposed to diethylstilbestrol in utero. *Am J Epidemiol*. 2006;164:682-688.
61. Hawkins P, Steyn C, McGarrigle HH, et al. Effect of maternal nutrient restriction in early gestation on responses of the hypothalamic-pituitary-adrenal axis to acute isocapnic hypoxaemia in late gestation fetal sheep. *Exp Physiol*. 2000;85:85-96.
62. Heijmans BT, Tobi EW, Stein AD, et al. Persistent epigenetic differences associated with prenatal exposure to famine in humans. *Proc Natl Acad Sci USA*.2008;105: 17046-17049.
63. Herrick K, Phillips DI, Haselden S, Shiell AW, Campbell-Brown M, Godfrey KM. Maternal consumption of a high-meat, low-carbohydrate diet in late pregnancy: relation to adult cortisol concentrations in the offspring. *J Clin Endocrinol Metab*. 2003;88:3554-3560.
64. Hershko AY, Kafri T, Fainsod A, Razin A. Methylation of HoxA5 and HoxB5 and its relevance to expression during mouse development. *Gene*. 2003;302:65-72.
65. Hofman PL, Regan F, Jackson WE, et al. Premature birth and later insulin resistance. *N Engl J Med*. 2004;351: 2179-2186.
66. Hollingsworth JW, Maruoka S, Boon K, et al. In utero supplementation with methyl donors enhances allergic airway disease in mice. *J Clin Invest*. 2008;118:3462-3469.
67. Hoppe CC, Evans RG, Moritz KM, et al. Combined prenatal and postnatal protein restriction influences adult kidney structure, function and arterial pressure. *Am J Physiol*. 2007;292:R462-R469.
68. Hovi P, Andersson S, Eriksson JG, et al. Glucose regulation in young adults with very low birth weight. *N Engl J Med*. 2007;356:2053-2063.
69. Huxley R, Owen C, Whincup P, Cook D, Rich-Edwards J, Smith G. Is birthweight a risk factor for coronary heart disease in later life? *J Epidemiol*. 2007;85:1244-1250.
70. Jablonka E, Oborny B, Molnar I, Kisdi E, Hofbauer J, Czaran T. The adaptive advantage of phenotypic memory in changing environments. *Philos Trans R Soc Lond B Biol Sci*. 1995; 350:133-141.
71. Jablonka E, Raz G. Transgenerational epigenetic inheritance: prevalence, mechanisms, and implications for the study of heredity and evolution. *Q Rev Biol*. 2009;84: 131-176.
72. Jackson AA, Dunn RL, Marchand MC, Langley-Evans SC. Increased systolic blood pressure in rats induced by a maternal low-protein diet is reversed by dietary supplementation with glycine. *Clin Sci (Lond)*. 2002;103:633-639.
73. Jackson-Grusby L, Beard C, Possemato R, et al. Loss of genomic methylation causes p53-dependent apoptosis and epigenetic deregulation. *Nat Genet*. 2001;27:31-39.
74. Jaenisch R, Bird A. Epigenetic regulation of gene expression: how the genome integrates intrinsic and environmental signals. *Nat Genet*. 2003;33(Suppl):245-254.
75. Kaati G, Bygren LO, Edvinsson S. Cardiovascular and diabetes mortality determined by nutrition during parents' and grandparents' slow growth period. *Eur J Hum Genet*. 2002;10:682-688.
76. Khan I, Dekou V, Hanson M, Poston L, Taylor P. Predictive adaptive responses to maternal high-fat diet prevent endothelial dysfunction but not hypertension in adult rat offspring. *Circulation*. 2004;110:1097-1102.
77. Kind KL, Clifton PM, Katsman AI, Tsiounis M, Robinson JS, Owens JA. Restricted fetal growth and the response to dietary cholesterol in the guinea pig. *Am J Physiol*. 1999;277:R1675-R1682.
78. Kucharski R, Maleszka J, Foret S, Maleszka R. Nutritional control of reproductive status in honeybees via DNA methylation. *Science*. 2008;319:1827-1830.
79. Kuzawa CW. Adipose tissue in human infancy and childhood: an evolutionary perspective. *Yearb Phys Anthropol*. 1998;41:177-209.

80. Kuzawa CW. Developmental perspectives on the origin of obesity. In: Fantuzzi G, Mazzone T, eds. *Adipose Tissue and Adipokines in Health and Disease*. Totowa, NJ: Humana; 2007:207-219.
81. Kwong WY, Wild AE, Roberts P, Willis AC, Fleming TP. Maternal undernutrition during the preimplantation period of rat development causes blastocyst abnormalities and programming of postnatal hypertension. *Development*. 2000;127:4195-4202.
82. Laforsch C, Tollrian R. Embryological aspects of inducible morphological defenses in Daphnia. *J Morphol*. 2004;262: 701-707.
83. Laird PW. Cancer epigenetics. *Hum Mol Genet*. 2005;14: R65-R76.
84. Lane N, Dean W, Erhardt S, et al. Resistance of IAPs to methylation reprogramming may provide a mechanism for epigenetic inheritance in the mouse. *Genesis*. 2003;35: 88-93.
85. Langley SC, Jackson AA. Increased systolic blood pressure in adult rats induced by fetal exposure to maternal low protein diets. *Clin Sci (Lond)*. 1994;86:217-222.
86. Lee TM, Zucker I. Vole infant development is influenced perinatally by maternal photoperiodic history. *Am J Physiol*. 1988;255:R831-R838.
87. Li E. Chromatin modification and epigenetic reprogramming in mammalian development. *Nat Rev Genet*. 2002;3:662-673.
88. Lillycrop KA, Phillips ES, Jackson AA, Hanson MA, Burdge GC. Dietary protein restriction of pregnant rats induces and folic acid supplementation prevents epigenetic modification of hepatic gene expression in the offspring. *J Nutr*. 2005;135:1382-1386.
89. Lillycrop KA, Phillips ES, Jackson AA, Hanson MA, Burdge GC. Dietary protein restriction in the pregnant rat induces altered epigenetic regulation of the glucocorticoid receptor and peroxisomal proliferator-activated receptor alpha in the heart of the offspring which is prevented by folic acid. *Proc Nutr Soc*. 2006;65:65A.
90. Lillycrop KA, Phillips ES, Torrens C, Hanson MA, Jackson AA, Burdge GC. Feeding pregnant rats a protein-restricted diet persistently alters the methylation of specific cytosines in the hepatic PPARalpha promoter of the offspring. *Br J Nutr*. 2008;100:278-282.
91. Lillycrop KA, Rodford J, Garratt ES, et al. Maternal protein restriction with or without folic acid supplementation during pregnancy alters the hepatic transcriptome in adult male rats. *Br J Nutr*. 2010;103(12):1711-1719.
92. Lillycrop KA, Slater-Jefferies JL, Hanson MA, Godfrey KM, Jackson AA, Burdge GC. Induction of altered epigenetic regulation of the hepatic glucocorticoid receptor in the offspring of rats fed a protein-restricted diet during pregnancy suggests that reduced DNA methyltransferase-1 expression is involved in impaired DNA methylation and changes in histone modifications. *Br J Nutr*. 2007;97: 1064-1073.
93. Lister AM. The impact of quaternary ice ages on mammalian evolution. *Philos Trans R Soc Lond B Biol Sci*. 2004; 359:221-241.
94. Lopatina N, Haskell JF, Andrews LG, Poole JC, Saldanha S, Tollefsbol T. Differential maintenance and de novo methylating activity by three DNA methyltransferases in aging and immortalized fibroblasts. *J Cell Biochem*. 2002;84:324-334.
95. Maleszka R. Epigenetic integration of environmental and genomic signals in honey bees. *Epigenetics*. 2008;3:188-192.
96. Maloney CA, Gosby AK, Phuyal JL, Denyer GS, Bryson JM, Caterson ID. Site-specific changes in the expression of fat-partitioning genes in weanling rats exposed to a low-protein diet in utero. *Obes Res*. 2003;11:461-468.
97. Manolio TA, Collins FS, Cox NJ, et al. Finding the missing heritability of complex diseases. *Nature*. 2009;461:747-753.
98. Martin JF, Johnston CS, Han CT, Benyshek DC. Nutritional origins of insulin resistance: a rat model for diabetes-prone human populations. *J Nutr*. 2000;130:741-744.
99. Mayer W, Niveleau A, Walter J, Fundele R, Haaf T. Demethylation of the zygotic paternal genome. *Nature*. 2000;403:501-502.
100. McGowan PO, Sasaki A, Huang TC, et al. Promoter-wide hypermethylation of the ribosomal RNA gene promoter in the suicide brain. *PLoS ONE*. 2008;3:e2085.
101. Milutinovic S, Zhuang Q, Niveleau A, Szyf M. Epigenomic stress response. Knockdown of DNA methyltransferase 1 triggers an intra-S-phase arrest of DNA replication and induction of stress response genes. *J Biol Chem*. 2003; 278:14985-14995.
102. Moran NA. The evolutionary maintenance of alternative phenotypes. *Am Nat*. 1992;139:971-989.
103. Morison IM, Paton CJ, Cleverley SD. The imprinted gene and parent-of-origin effect database. *Nucleic Acids Res*. 2001;29:275-276.
104. Okada Y, Yamagata K, Hong K, Wakayama T, Zhang Y. A role for the elongator complex in zygotic paternal genome demethylation. *Nature*. 2010;463(7280):554-558. doi:10.1038/nature08732.
105. Okano M, Bell DW, Haber DA, Li E. DNA methyltransferases Dnmt3a and Dnmt3b are essential for de novo methylation and mammalian development. *Cell*. 1999;99: 247-257.
106. Oswald J, Engemann S, Lane N, et al. Active demethylation of the paternal genome in the mouse zygote. *Curr Biol*. 2000;10:475-478.
107. Palmer JR, Wise LA, Hatch EE, et al. Prenatal diethylstilbestrol exposure and risk of breast cancer. *Cancer Epidemiol Biomarkers Prev*. 2006;15:1509-1514.
108. Pembrey ME, Bygren LO, Kaati G, et al. Sex-specific, male-line transgenerational responses in humans. *Eur J Hum Genet*. 2006;14:159-166.
109. Pener MP, Yerushalmi Y. The physiology of locust phase polymorphism: an update. *J Insect Physiol*. 1998;44:365-377.
110. Pham TD, MacLennan NK, Chiu CT, Laksana GS, Hsu JL, Lane RH. Uteroplacental insufficiency increases apoptosis and alters p53 gene methylation in the full-term IUGR rat kidney. *Am J Physiol*. 2003;285:R962-R970.
111. Popkin BM. Nutrition in transition: the changing global nutrition challenge. *Asia Pac J Clin Nutr*. 2001;10: S13-S18.
112. Prentice AM, Moore SE. Early programming of adult diseases in resource poor countries. *Arch Dis Child*. 2005;90:429-432.
113. Reik W, Walter J. Genomic imprinting: parental influence on the genome. *Nat Rev Genet*. 2001;2:21-32.
114. Rhee I, Jair KW, Yen RW, et al. CpG methylation is maintained in human cancer cells lacking DNMT1. *Nature*. 2000;404:1003-1007.

115. Richardson B. Impact of aging on DNA methylation. *Ageing Res Rev*. 2003;2:245-261.
116. Roseboom T, de Rooji S, Painter R. The Dutch famine and its long-term consequences for adult health. *Early Hum Dev*. 2006;82:485-491.
117. Sellayah D, Sek K, Anthony FW, et al. Appetite regulatory mechanisms and food intake in mice are sensitive to mismatch in diets between pregnancy and postnatal periods. *Brain Res*. 2008;1237:146-152.
118. Shiell AW, Campbell-Brown M, Haselden S, Robinson S, Godfrey KM, Barker DJ. High-meat, low-carbohydrate diet in pregnancy: relation to adult blood pressure in the offspring. *Hypertension*. 2001;38:1282-1288.
119. Silverman BL, Purdy LP, Metzger BE. The intrauterine environment: implications for the offspring of diabetic mothers. *Diab Rev*. 1996;4:21-35.
120. Skinner MK. What is an epigenetic transgenerational phenotype? F3 or F2. *Reprod Toxicol*. 2008;25:2-6.
121. Sloboda DM, Hart R, Doherty DA, Pennell CE, Hickey M. Age at menarche: influences of prenatal and postnatal growth. *J Clin Endocrinol Metab*. 2007;92:46-50.
122. Sloboda DM, Howie GJ, Pleasants A, Gluckman PD, Vickers MH. Pre- and postnatal nutritional histories influence reproductive maturation and ovarian function in the rat. *PLoS ONE*. 2009;4:e6744.
123. Steegers-Theunissen RP, Obermann-Borst SA, Kremer D, et al. Periconceptional maternal folic acid use of 400 microg per day is related to increased methylation of the IGF2 gene in the very young child. *PLoS ONE*. 2009;4:e7845.
124. Stein AD, Lumey LH. The relationship between maternal and offspring birth weights after maternal prenatal famine exposure: the Dutch Famine Birth Cohort Study. *Hum Biol*. 2000;72:641-654.
125. Suetake I, Shi LH, Watanabe D, Nakamura M, Tajima S. Proliferation stage-dependent expression of DNA methyltransferase (Dnmt1) in mouse small intestine. *Cell Struct Funct*. 2001;26:79-86.
126. Sultan SE, Spencer HG. Metapopulation structure favors plasticity over local adaptation. *Am Nat*. 2002;160: 271-283.
127. Temple IK. Imprinting in human disease with special reference to transient neonatal diabetes and Beckwith–Wiedemann syndrome. *Endocr Dev*. 2007;12:113-123.
128. Tobi EW, Lumey LH, Talens RP, et al. DNA methylation differences after exposure to prenatal famine are common and timing- and sex-specific. *Hum Mol Genet*. 2009; 18(21):4046-4053.
129. Tollefsbol TO, Andrews LG. Mechanisms for telomerase gene control in aging cells and tumorigenesis. *Med Hypotheses*. 2001;56:630-637.
130. Torrens C, Brawley L, Anthony FW, et al. Folate supplementation during pregnancy improves offspring cardiovascular dysfunction induced by protein restriction. *Hypertension*. 2006;47:982-987.
131. Torrens C, Poston L, Hanson MA. Transmission of raised blood pressure and endothelial dysfunction to the F2 generation induced by maternal protein restriction in the F0, in the absence of dietary challenge in the F1 generation. *Br J Nutr*. 2008;100:760-766.
132. Tycko B, Morison IM. Physiological functions of imprinted genes. *J Cell Physiol*. 2002;192:245-258.
133. Uller T. Developmental plasticity and the evolution of parental effects. *Trends Ecol Evol*. 2008;23:432-438.
134. Vasicek TJ, Zeng L, Guan XJ, Zhang T, Costantini F, Tilghman SM. Two dominant mutations in the mouse fused gene are the result of transposon insertions. *Genetics*. 1997;147:777-786.
135. Vickers MH, Breier BH, Cutfield WS, Hofman PL, Gluckman PD. Fetal origins of hyperphagia, obesity, and hypertension and postnatal amplification by hypercaloric nutrition. *Am J Physiol Endocrinol Metab*. 2000;279: E83-E87.
136. Vickers MH, Gluckman PD, Coveny AH, et al. Neonatal leptin treatment reverses developmental programming. *Endocrinology*. 2005;146:4211-4216.
137. Waddington CH. Canalization of development and the inheritance of acquired characters. *Nature*. 1942;150: 563-565.
138. Waddington CH. Canalization of development and genetic assimilation of acquired characters. *Nature*. 1959;183: 1654-1655.
139. Waterland RA, Dolinoy DC, Lin JR, Smith CA, Shi X, Tahiliani KG. Maternal methyl supplements increase offspring DNA methylation at axin fused. *Genesis*. 2006; 44:401-406.
140. Waterland RA, Jirtle RL. Transposable elements: targets for early nutritional effects on epigenetic gene regulation. *Mol Cell Biol*. 2003;23:5293-5300.
141. Waterland RA, Jirtle RL. Early nutrition, epigenetic changes at transposons and imprinted genes, and enhanced susceptibility to adult chronic diseases. *Nutrition*. 2004; 20:63-68.
142. Waterland RA, Travisano M, Tahiliani KG, Rached MT, Mirza S. Methyl donor supplementation prevents transgenerational amplification of obesity. *Int J Obes (Lond)*. 2008;32:1373-1379.
143. Weaver IC, Cervoni N, Champagne FA, et al. Epigenetic programming by maternal behavior. *Nat Neurosci*. 2004;7:847-854.
144. West-Eberhard MJ. Alternative adaptations, speciation, and phylogeny (A Review). *Proc Natl Acad Sci USA*. 1986;83:1388-1392.
145. West-Eberhard MJ. *Developmental plasticity and evolution*. New York: Oxford University Press; 2003.
146. Whorwood CB, Firth KM, Budge H, Symonds ME. Maternal undernutrition during early to midgestation programs tissue-specific alterations in the expression of the glucocorticoid receptor, 11beta-hydroxysteroid dehydrogenase isoforms, and type 1 angiotensin II receptor in neonatal sheep. *Endocrinology*. 2001;142:2854-2864.
147. Winick M, Noble A. Cellular response in rats during malnutrition at various ages. *J Nutr*. 1966;89:300-306.
148. Wolff GL, Kodell RL, Moore SR, Cooney CA. Maternal epigenetics and methyl supplements affect agouti gene expression in Avy/a mice. *FASEB J*. 1998;12:949-957.
149. Woodall SM, Johnston BM, Breier BH, Gluckman PD. Chronic maternal undernutrition in the rat leads to delayed postnatal growth and elevated blood pressure of offspring. *Pediatr Res*. 1996;40:438-443.
150. World Health Organization. *Promoting optimal fetal development*. Geneva: World Health Organization; 2006. http://www.who.int/nutrition/topics/fetal_dev_report_EN.pdf.

151. World Health Organization. *2008–2013 Action plan for the global strategy for the prevention and control of noncommunicable diseases*. Geneva: World Health Organization; 2008.
152. Xie H, Wang M, Bonaldo Mde F, et al. High-throughput sequence-based epigenomic analysis of Alu repeats in human cerebellum. *Nucleic Acids Res*. 2009;37:4331-4340.
153. Zambrano E, Martínez-Samayoa PM, Bautista CJ, et al. Sex differences in transgenerational alterations of growth and metabolism in progeny (F-2) of female offspring (F-1) of rats fed a low protein diet during pregnancy and lactation. *J Physiol (Lond)*. 2005;566:225-236.
154. Zhang J, Zhang F, Didelot X, et al. Maternal high fat diet during pregnancy and lactation alters hepatic expression of insulin like growth factor-2 and key microRNAs in the adult offspring. *BMC Genomics*. 2009;10:478.

Targeting Histone Deacetylases in Chronic Obstructive Pulmonary Disease

14

Peter J. Barnes

14.1 Introduction

Histone deacetylases (HDAC) play a critical role in gene suppression through the recruitment of corepressor proteins to switch off gene transcription, which is activated by histone acetylation. Activated inflammatory genes are suppressed by HDAC2. Corticosteroids recruit HDAC2 to switch off inflammatory genes. HDAC2 may therefore be a target for the development of anti-inflammatory treatments, particularly in diseases where there is active corticosteroid resistance, as in chronic obstructive pulmonary disease (COPD).

14.2 Chronic Obstructive Pulmonary Disease

COPD is a common disease of progressive airway obstruction that is the result of emphysema and small airway disease (chronic obstructive bronchiolitis). It gradually leads to increasing shortness of breath and limitation of exercise.[4,79] COPD is a global health problem that is the fourth commonest cause of death in developed countries and an increasing cause of death in developing countries[11] COPD is also a common cause of morbidity and exacerbation of COPD is a common reason for hospital admission, imposing a large burden on health resources. COPD has a worldwide prevalence of over 10% in men over the age of 40 years and is moving toward this percentage in women.[65] Because of the high incidence and escalating health care costs, there is renewed interest in the underlying cellular and molecular mechanisms of COPD[5] and in the search for new therapies.[13,22,47] The definition of COPD adopted by the Global initiative on Obstructive Lung Disease (GOLD) is that COPD is a chronic inflammatory disease. Much of recent research on disease mechanisms has focused on the nature of this inflammatory response.[79] COPD progresses slowly over many decades and leads to death from respiratory failure. Most patients however die of co-morbidities such as heart failure and myocardial infarction, or lung cancer. The commonest cause of COPD is chronic cigarette smoking. Some patients, however, particularly in developing countries, develop the disease from inhalation of wood smoke or other inhaled irritants.[65,83] Given only about 25% of smokers develop COPD,[62] genetic or host factors, not yet identified, may predispose to its development. Airway obstruction is relentlessly progressive and only smoking cessation reduces the rate of decline in lung function. As the disease becomes more severe, the smoking cessation has less benefit and lung inflammation persists in ex-smokers.[46,95]

As COPD, in sharp contrast to asthma, responds poorly to currently available therapies there is pressing need for the development of effective anti-inflammation treatments.[25] A major barrier to therapy is resistance of the disease to the anti-inflammatory effects of corticosteroids. With the molecular basis for the anti-inflammatory effects of corticosteroids better understood,[8] it has become clear that an important mechanism of corticosteroid resistance in COPD is the reduction in the critical nuclear enzyme histone deacetylase-2 (HDAC2).[9] Resistance to corticosteroids is also linked to amplification of the inflammatory process.

P.J. Barnes
National Heart and Lung Institute,
Imperial College London, London, UK
e-mail: p.j.barnes@imperial.ac.uk

H.I. Roach et al. (eds.), *Epigenetic Aspects of Chronic Diseases*,
DOI: 10.1007/978-1-84882-644-1_14, © Springer-Verlag London Limited 2011

14.3 Inflammation in COPD

The progressive airflow limitation in COPD is due to two major pathological processes: remodeling and narrowing of small airways and destruction of the lung parenchyma with consequent destruction of the alveolar attachments of these airways. This results in diminished lung recoil, higher resistance to flow, and closure of small airways at higher lung volumes during expiration, which results in air being trapped in the lung. This in turn leads to the characteristic hyperinflation of the lungs, which gives rise to the sensation of dyspnea and limits exercise capacity.[73] Remodeling and narrowing of the small airway and emphysema are the likely result of chronic inflammation in the lung periphery.[50] Quantitative studies have shown that the inflammatory response in small airways increases as the disease progresses, leading to peribronchiolar fibrosis [51] Inflammation in COPD airways and of lung parenchyma follows a specific pattern, with macrophages playing an important inflammatory role through the release of proinflammatory mediators [5] These include T-lymphocytes, with $CD8^+$ (cytotoxic) T-cells predominating, plus B-lymphocytes in more severe disease, and many neutrophils in the lumen [24].

The inflammatory response in COPD involves both innate and adaptive immune responses.[14,40] The inflammatory cells and structural cells of the airways and lungs give rise to inflammatory mediators,[6] with many cytokines and chemokines orchestrating the chronic inflammatory process.[12,18] The pattern of inflammation in smokers without airflow limitation is similar, but in COPD the number of inflammatory cells and the concentration of mediators is higher. During acute exacerbations of the disease, usually precipitated by bacterial and viral infections, inflammation is further increased.[31,94,43] The molecular basis of this increase is not understood, but reduced HDAC activity in COPD may be a potential mechanism. The amplification of inflammation and susceptibility to develop airflow limitation may have a genetic basis, but the specific genes have not been identified.[86]

Cigarette smoke and other irritants in the respiratory tract may activate surface macrophages and airway epithelial cells to release chemokines that then attract circulating leukocytes into the lungs causing the initial inflammatory events that occur in all smokers.[23,24,28] However, in smokers who develop COPD this inflammation progresses into a more complicated inflammatory pattern of adaptive immunity and involves T- and B-lymphocyte infiltration and possibly dendritic cells, along with a complicated interacting array of cytokines and other mediators.[12,14]

Increased expression of inflammatory proteins is regulated by proinflammatory transcription factors, such as nuclear factor-κB (NF-κB)[10] which is activated in COPD lungs, in alveolar macrophages, and airway epithelial cells.[30,42]

14.4 Histone Acetylation and Deacetylation

14.4.1 *Histone Acetylation*

Gene expression is regulated by acetylation of core histones which open up the chromatin structure (chromatin remodeling) to allow transcription factors and RNA polymerase to bind to DNA, thus initiating gene transcription.[61,84,91] Gene expression is regulated by various coactivator molecules, such as the CREB-binding protein and p300, which have intrinsic histone acetyltransferase (HAT) activity. Expression of inflammatory genes is regulated by increased acetylation of histone-4.[21,52] Thus epigenetic factors play a critical role in chronic inflammation.[2]

In COPD peripheral lung, airway biopsies, and alveolar macrophages the increase in the acetylation of histones is associated with the promoter region of inflammatory genes, such as CXCL8 (interleukin-8) that are regulated by NF-κB; the degree of acetylation increases with disease severity.[54] The increase in acetylation of histones is not due to an increase in HAT activity in lungs or in macrophages as in asthma. Rather it is due to a decrease in HDAC activity. Many of the inflammatory genes are subject to regulation by NF-κB, which in turn may also be acetylated.

Asthma patients, on the other hand, have an increase in acetylation of histone-4. This is consistent with an increase in expression of inflammatory genes, but with little reduction in HDAC activity.[53]

14.4.2 *Histone Deactylases*

Histone acetylation is reversed by HDACs and there are 11 HDAC isoenzymes that deacetylate histones and other proteins within the nucleus.[92] Specific

HDACs appear to be differentially regulated and to regulate different groups of genes. HDACs play a critical role in the suppression of gene expression by reversing the hyperacetylation of core histones. For the regulation of inflammatory genes HDAC2 appears to be of critical importance.[52,58] The role of other HDACs in inflammatory diseases is less certain. HDAC1 is often associated with HDAC2 in the nucleus, but may not be directly involved with regulation of inflammatory genes or the anti-inflammatory actions of corticosteroids. Inflammatory gene expression is determined by the balance between histone acetylation, which activates transcription, and deacetylation, which switches off transcription.

Many regulatory proteins, particularly transcription factors and nuclear receptors, are regulated by acetylation which in turn is controlled by HATs and HDACs.[78] Acetylation plays a key role in the regulation of androgen and estrogen receptors, and of glucocorticoid receptors (GR).[58] GR is acetylated in the nucleus at specific lysine residues close to the hinge region of the receptor and only binds to its DNA binding site in its acetylated form. However, in order to inhibit NF-κB-activated genes, it is necessary to deacetylate the receptor and this is achieved by HDAC2 (Fig. 14.1). Binding of GR to DNA requires acetylation of the receptor, in order for GR acetylation to activate genes, including those that mediate the metabolic and endocrine side effects of corticosteroids.

14.5 HDAC2 in COPD

The reduced HDAC activity in alveolar macrophages of cigarette smokers, compared to that of nonsmokers, is correlated with increased expression of inflammatory genes.[16,55] A comparable reduction in total HDAC activity in peripheral lung, bronchial biopsies, and alveolar macrophages from COPD patients is correlated with disease severity, with increased gene expression of CXCL8, and increased acetylation of histone, associated with the NF-κB binding site on the CXCL8 promoter[54,90] (Fig. 14.2). The expression of HDAC2 is also selectively reduced with lesser reductions in HDAC3 and HDAC5. In patients with very severe COPD (GOLD stage 4), the expression of HDAC2 was less than 5% of that seen in normal lung. Because both protein and messenger RNA expressions of HDAC2 are reduced, HDAC2 gene transcription is likely reduced as well. Alternatively the stability of its mRNA may be reduced. However, almost nothing is known about the transcriptional regulation of HDAC genes.

Restoration in alveolar macrophages from COPD patients of HDAC2 expression to normal by transfection with a plasmid vector of HDAC2 reverses corticosteroid resistance in these cells, whereas transfection with an HDAC1 vector is without effect.[55] This provides compelling evidence that the reduction in HDAC2 seen in COPD is linked to reduced corticosteroid responsiveness.

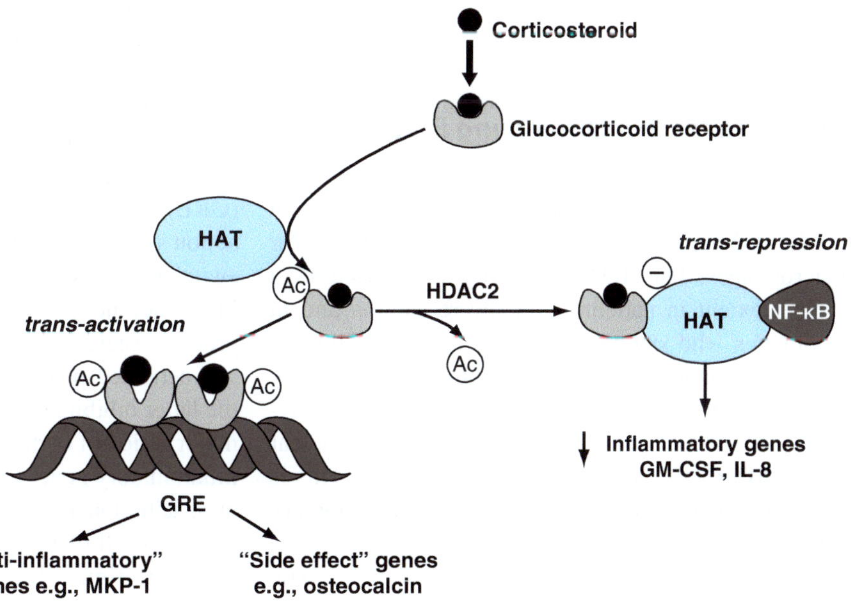

Fig. 14.1 Acetylation of the glucocorticoid receptor (*GR*). After corticosteroid binds to the GR, the receptor translocates to the nucleus, where it is acetylated by a histone acetyltransferase (*HAT*). This is necessary for GR to bind to its glucocorticoid receptor recognition element (*GRE*) in the promoter region of steroid-sensitive genes. These include genes that mediate the side effects of corticosteroids, such as osteocalcin. It is necessary for the acetylated GR to be deacetylated by histone deacetylase 2 (*HDAC2*) in order to inhibit the activated nuclear factor-κB (*NF-κB*) to suppress activated inflammatory genes

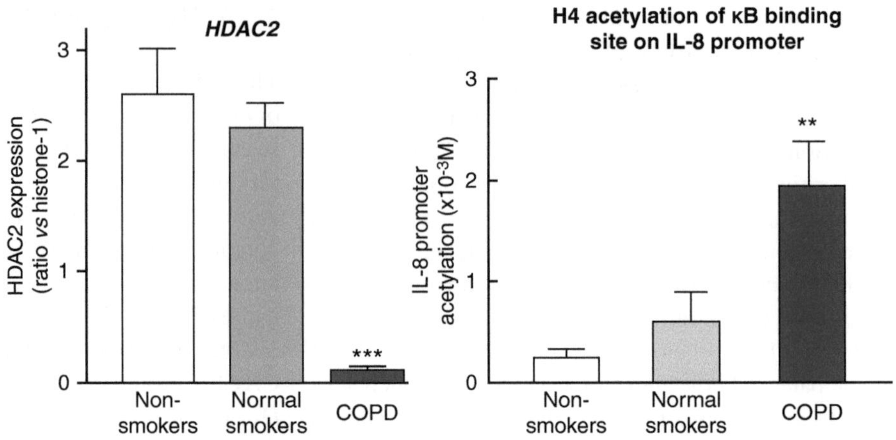

Fig. 14.2 Reduced histone deacetylase-2 (*HDAC2*) in peripheral lung of patients with severe COPD compared to smokers with normal lung function and age-matched nonsmokers (*left panel*). Right panel shows the increased acetylation of histones at the nuclear factor-κB (*NF-κB*) binding site on the promoter region of the CXCL8 (IL-8) gene) (Adapted from Culpitt[38])

As discussed above, HDAC2 is required for the deacetylation-activated nuclear GR in order for GR to inhibit NF-κB activity, and inflammatory gene expression. Reduced activity of HDAC2 in COPD patients which is associated with the increased acetylation of GR may be a major mechanism that accounts for corticosteroid resistance in COPD.[41] In addition, the increase in acetylated GR may promote gene activation and gene suppression by binding to GR recognition sequences (GRE) in steroid-sensitive genes. There may be genes involved in side effects of corticosteroids; examples are osteocalcin in osteoporosis and suppression of the hypothalamo-pituitary–adrenal axis by pro-opiomelanocortin. The reduction in HDAC2 in COPD may therefore lead not only to greater inflammation and corticosteroid resistance, but may also raise the risk of corticosteroid-induced side effects.

14.5.1 HDAC2 in Other Lung Diseases

In patients with mild asthma there is a small reduction in HDAC2 activity in bronchial biopsies and alveolar macrophages, but HDAC2 activity is reduced to a greater extent in patients with severe asthma and in asthmatics who smoke.[26,53] The reduction in HDAC2 activity in patients with mild asthma is restored by treatment with inhaled corticosteroids[53]. HDAC activity is also reduced in peripheral blood mononuclear cells (PBMC) of patients with severe asthma. This is correlated with resistance to the anti-inflammatory effects of corticosteroids in vitro.[49]

Peripheral lung tissue from patients with cystic fibrosis and interstitial pulmonary fibrosis has normal HDAC2 expression and activity.[54] However, the role of HDAC2 in these diseases has not been carefully investigated and preliminary findings from our laboratory suggest that HDAC2 is reduced in sputum macrophages and in PBMC of patients with cystic fibrosis.

14.6 Effect of Corticosteroids on HDAC2

Advances in understanding the molecular mechanisms of how glucocorticoids suppress inflammation also provide insights how corticosteroid-resistance arises.[8,20]

Corticosteroids diffuse across the cell membrane and bind to glucocorticoid receptors (GR) in the cytoplasm. Binding activates GR and release them from chaperone proteins (heat shock protein 90 and others). This allows a rapid translocation to the nucleus where GRs activate and suppress the genes involved in the inflammatory process. GRs do so by homodimerizing and binding to glucocorticoid response elements (GRE) in the promoter region of the genes. This interaction switches on gene transcription and, on occasion, switches it off. Glucocorticoids switch on genes encoding β_2-adrenergic receptors and the anti-inflammatory proteins secretory leukoprotease inhibitor and mitogen-activated protein kinase phosphatase-1 (MKP-1); the latter inhibits MAP kinase pathways. However, the major action of corticosteroids is to switch off activated inflammatory genes that encode for cytokines, chemokines, adhesion molecules, inflammatory enzymes, and receptors, all of which are regulated by proinflammatory transcription factors, such as NF-κB and activator protein-1 (AP-1). These transcription factors activate inflammatory genes through histone acetylation, whereas activated GR reverse this process by interacting with corepressor molecules to attenuate NF-κB-associated coactivator activity; this reduces histone acetylation.[8,52] Reduction of histone acetylation occurs mainly through specific recruitment of HDAC2 to the activated inflammatory gene complex by activated GR. This then effectively suppresses activated inflammatory genes within the nucleus. GRs become acetylated and then bind to GREs. HDAC2 targets the acetylated GRs and then can associate with the NF-κB complex[58] (Fig. 14.1).

14.7 Corticosteroid Resistance in COPD

Corticosteroid resistance is also important in severe asthma and in asthma of smokers.[1,20] Recruitment by ligand-bound GR of HDAC2 to activated inflammatory genes reverses the acetylation of the inflammatory genes and silences their transcription.[8,19,52] The reduction in HDAC2 expression in patients with COPD may therefore account for their corticosteroid insensitivity.[23] The fact that even high systemic doses of corticosteroids do not suppress lung inflammation is in marked contrast to their high level of efficacy in asthma.[38,59,63] In COPD, corticosteroids fail to suppress secretion from macrophages of inflammatory proteins, such as CXCL8 and matrix metalloproteinase-9.[39,82] The reduction in HDAC2 expression in COPD cells may therefore not only account for the amplification of inflammation, but also for the insensitivity to the anti-inflammatory effects of corticosteroids[9,16] and the lack of their clinical benefits[29,88,96] (Fig. 14.3).

14.7.1 Role in Asthma

Patients with severe asthma also have reduced responsiveness to corticosteroids; this may be manifest in circulating PBMCs and in alveolar macrophages.[29,49,69] Reduced HDAC2 activity accounts for at least part of corticosteroid resistance, although other mechanisms may also be involved.[1] Smoking asthmatics are also relatively resistant to the anti-inflammatory effects of

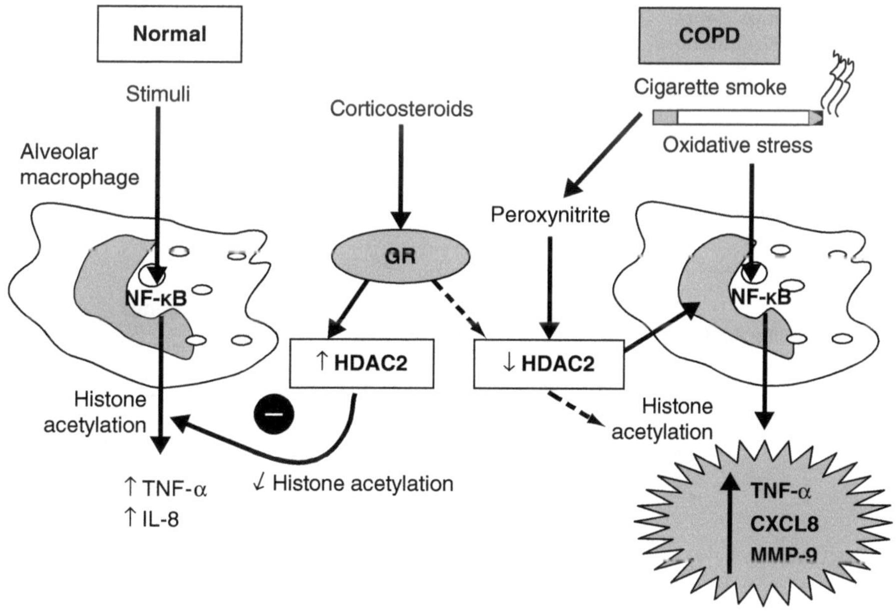

Fig. 14.3 Proposed mechanism of corticosteroid resistance in COPD patients. Stimulation of normal alveolar macrophages activates nuclear factor-κB (*NF-κB*) and other transcription factors to switch on histone acetyltransferase leading to histone acetylation and subsequently to transcription of genes encoding inflammatory proteins, such as tumor necrosis factor-α (*TNF-α*) and CXCL8 (IL-8). Corticosteroids reverse this by binding to glucocorticoid receptors (*GR*) and recruiting histone deacetylase-2 (*HDAC2*). This reverses the histone acetylation induced by NF-κB and switches off the activated inflammatory genes. In COPD patients, cigarette smoke activates macrophages as in normal subjects but oxidative stress (acting through the formation of peroxynitrite) impairs the activity of HDAC2. This amplifies the inflammatory response to NF-κB activation, but also reduces the anti-inflammatory effect of corticosteroids as HDAC2 is now unable to reverse histone acetylation

corticosteroids. This makes them more difficult to control than nonsmoking asthmatics.[34] Because cigarette smoking is an oxidative stress, the difficulty in controlling these patients is quite possibly due to a reduction in HDAC2 activity.[3]

14.8 Mechanisms of HDAC Reduction

The reasons why HDAC, particularly HDAC2, are reduced in COPD and in severe/smoking asthma are not fully understood. It may be due to inactivation of the enzyme by oxidative and nitrative stress[23] (Fig. 14.4). Oxidative stress is increased in COPD and increases with disease severity.[27,72,77] Nitrative stress is also increased in the peripheral lung of COPD patients[66] and there is increased expression of inducible nitric oxide synthase in small airways and the lung parenchyma.[80] Oxidative stress is also increased in severe and smoking asthma.[71,76] Oxidative and nitrative stress lead to the rapid formation of peroxynitrite, which is increased in exhaled breath condensate of patients with COPD.[75] Peroxynitrite nitrates bind to tyrosine residues on certain proteins. Tyrosine nitration is increased in HDAC2, but not other isoforms of HDAC, and has increased in macrophages and peripheral lung of COPD patients. The increase in nitration is correlated with an increase in expression of CXCL8.[71,57] Oxidative and nitrative stress induce corticosteroid resistance in macrophage-like cells (U937 cells) in vitro. This is mimicked by cigarette smoke extract and reversed by the antioxidant *N*-acetylcysteine. In vivo exposure of mice to cigarette smoke reduces HDAC activity in the lungs and induces neutrophilic inflammation that is steroid-resistant.[72,73,45,67] Nitration of HDAC2 not only inactivates the catalytic activity of this enzyme, but also leads to its ubiquitination, which marks it for degradation by a proteasome. As a result, HDAC2 protein concentration in the lungs of patients with severe

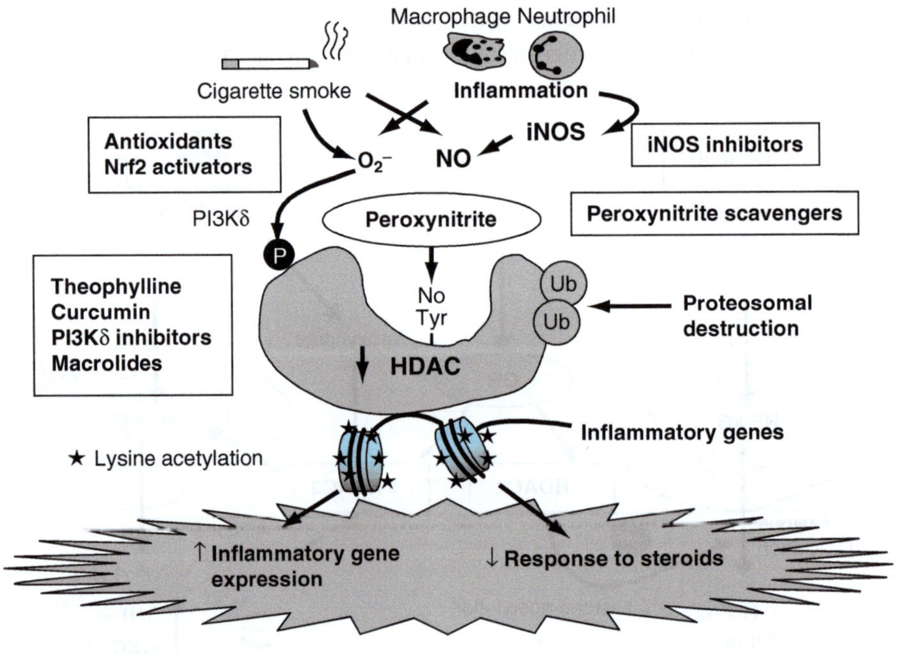

Fig. 14.4 Possible mechanisms for decreased histone deacetylase (*HDAC*) in chronic obstructive pulmonary disease (*COPD*). Superoxide anions (O_{2-}) and nitric oxide (*NO*) generated by cigarette smoke and inflammatory cells combine to form peroxynitrite. NO production from inflammatory cells is derived from inducible NO synthase (*iNOS*) in response to inflammatory stimuli. Peroxynitrite nitrates histone deacetylase-2 (*HDAC2*) at a tyrosine (*Tyr*) residue within the catalytic site. This inactivates HDAC2 and leads to ubiquitination (*Ub*) of the enzyme which labels HDAC2 for degradation by the proteasome, resulting in reduced expression. Oxidative stress also activates a phosphoinositide-3-kinase (*PI3K*) pathway that phosphorylates (*P*) and inactivates HDAC2. Loss of HDAC function then results in enhanced inflammatory gene expression and blocks the anti-inflammatory action of corticosteroids. HDAC function may be restored by antioxidants, iNOS inhibitors, or peroxynitrite scavengers which reduce tyrosine nitration or by theophylline, curcumin, or PI3K inhibitors which restore HDAC function to normal

COPD is very low.[28,74] HDAC2 mRNA is also reduced in COPD peripheral lungs due apparently to a reduction in gene transcription. Little is known about the regulation of HDAC2 gene expression. Charon et al.[33] have shown that hypoxia reduces HDAC2 expression, because the transcription factor, hypoxia-inducible factor-1α (HIF-1α), binds to a recognition sequence in the promoter region, thereby causing transcription to be reduced.

Oxidative stress also activates the phosphoinositide-3-kinase (PI3K) pathway; as a result serine residues are phosphorylated and HDAC2 is inactivated.[44] It thus seems that a high level of oxidative/nitrative stress in COPD lungs, particularly as the disease progresses, leads to an increase in tyrosine nitration, serine phosphorylation, and ubiquitination. HDAC2 catalytic activity is therefore impaired and HDAC2 levels are reduced. As a result, inflammatory gene expression is increased and the response to corticosteroids is impaired.

14.9 Theophylline as an HDAC Activator

An attractive therapeutic option is to reverse corticosteroid resistance by interfering with specific signal pathways. Selective activation of HDAC2 can be achieved with low concentrations of theophylline ($\sim 10^{-6}$ M) that restore HDAC2 activity in COPD macrophages to normal and reverse corticosteroid resistance.[37,56] In cigarette smoke exposed mice, which develop corticosteroid-resistant inflammation, oral theophylline reverses resistance. Clinical trials to test this in COPD patients are under way. In smoking asthmatics who are corticosteroid resistant, a low dose of theophylline reverses resistance.[87] Theophylline accelerates recovery from acute COPD exacerbation. Recovery is accompanied by increased HDAC activity in sputum macrophages and reduced inflammatory mediators.[36]

14.9.1 Molecular Mechanisms of Action

Therapeutic concentrations of theophylline inhibit phosphodiesterase and overcome receptor antagonism to adenosine. However, theophylline restores HDAC2 via selective inhibition of PI3Kδ, which in turn is activated by oxidative stress.[66,68] Selective PI3Kδ inhibitors may therefore also have therapeutic value. Drugs to inhibit PI3Kδ are currently in clinical development for other diseases. PI3K-δ also modulates expression of inflammatory genes. Several PI3K-δ or mixed PI3K-γ/δ inhibitors are now in development.[93] As oxidative stress appears to be important in reducing HDAC2 and leads to corticosteroid resistance, antioxidants should also be effective. Once molecular signaling pathways that regulate HDAC2 are better understood, it may be possible to develop drugs that increase HDAC2 expression.

14.10 Future Directions

Inasmuch as histone acetylation activates inflammatory genes and can be modified by anti-inflammatory drugs, such as corticosteroids, it may be possible to identify new targets that in turn may lead to novel therapies.[15,17] Understanding these mechanisms may also lead to the development of treatments that reverse corticosteroid resistance in airway disease.

14.10.1 Antioxidants

Given oxidative/nitrative stress appears to be a mechanism that leads to corticosteroid resistance, antioxidants and inhibitors of inducible nitric oxide synthase (iNOS) may inhibit the generation of peroxynitrite. Currently available antioxidants, such as vitamins C and E and *N*-acetylcysteine, are not very potent and do not sufficiently reduce oxidative stress in the lung. Oral *N*-acetyl cysteine has failed to reduce disease exacerbations or the decline in lung function in COPD patients who have been treated with or without inhaled corticosteroids.[41] Selective iNOS inhibitors are now in clinical development and may be effective in reducing peroxynitrite formation,[48] but may not prevent the generation of peroxynitrite from NO in cigarette smoke. More potent antioxidants and peroxynitrite scavenger drugs are now in development.[60] The transcription factor Nrf2 (nuclear factor erythroid-derived 2-related factor-2) plays a key role in the regulation of endogenous antioxidant genes and is defective in COPD patients.[64] Several Nrf2 activators, such as sulforaphane (which occurs naturally in broccoli) and the synthetic triterpenoid, 1-[2-cyano-3-,12-dioxooleana-1,9-dien-28-oyl]imidazole-methyl] ester, have now been identified as potential therapies.[89]

14.10.2 New Theophylline Derivatives

Theophylline increases HDAC activity and expression through a mechanism that is independent of phosphodiesterase (PDE) inhibition or adenosine receptor antagonism, which together account for all known side effects of theophylline.[56] It may therefore be possible to design molecules that preserve the HDAC activating effect without effects on PDE or adenosine receptors. Given, as discussed above, low concentrations of theophylline increase HDAC2 expression via inhibition of oxidant-activated PI3Kδ, and because its effects are mimicked by PI3Kδ inhibitors that have been developed for other indications, drugs that act further down the PI3Kδ pathway may be similarly effective.

14.10.3 Curcumin

Curcumin, a polyphenol found in curry powder, also reverses corticosteroid sensitivity by restoring HDAC2 expression to normal.[70] However, curcumin is not very specific and inhibits HAT activity and NF-κB. Its effects are therefore difficult to interpret.[35]

14.10.4 Macrolides

Macrolide antibiotics have anti-inflammatory effects, although the molecular mechanisms are poorly understood.[81] Several non-antibiotic macrolides have been developed as possible anti-inflammatory treatments. Macrolides also reverse corticosteroid resistance, but a different molecular mechanism appears involved.[32]

14.10.5 HDAC2 Activators

With high-throughput screening it may be possible to identify drugs that directly activate HDAC2.[7] Several non-selective HDAC inhibitors, such as trichostatin A and valproate, have been developed to treat malignancies,[85] but activators have so far proved to be elusive.

14.11 Conclusion

The elucidation of HDAC2 as a critical mechanism in the regulation of inflammatory genes has identified several novel molecular targets, particularly in inflammatory diseases where corticosteroids are relatively ineffective due to reduced HDAC2 activity and expression, as occurs in COPD. Understanding the role of HDAC2 may lead to the development of novel anti-inflammatory therapies, including drugs that reverse corticosteroid resistance in COPD, severe asthma, and cystic fibrosis, diseases currently difficult to manage.

References

1. Adcock IM, Barnes PJ. Molecular mechanisms of corticosteroid resistance. *Chest*. 2008;134:394-401.
2. Adcock IM, Ford P, Barnes PJ, Ito K. Epigenetics and airways disease. *Respir Res*. 2006;7:21.
3. Ahmad T, Barnes PJ, Adcock IM. Overcoming steroid insensitivity in smoking asthmatics. *Curr Opin Investig Drugs*. 2008;9:470-477.
4. Barnes PJ. Chronic obstructive pulmonary disease. *New Engl J Med*. 2000;343:269-280.
5. Barnes PJ. Macrophages as orchestrators of COPD. *J COPD*. 2004;1:59-70.
6. Barnes PJ. Mediators of chronic obstructive pulmonary disease. *Pharm Rev*. 2004;56:515-548.
7. Barnes PJ. Targeting histone deacetylase 2 in chronic obstructive pulmonary disease treatment. *Expert Opin Ther Targets*. 2005;9:1111-1121.
8. Barnes PJ. How corticosteroids control inflammation. *Br J Pharmacol*. 2006;148:245-254.
9. Barnes PJ. Reduced histone deacetylase in COPD: clinical implications. *Chest*. 2006;129:151-155.
10. Barnes PJ. Transcription factors in airway diseases. *Lab Invest*. 2006;86:867-872.
11. Barnes PJ. Chronic obstructive pulmonary disease: a growing but neglected epidemic. *PLoS Med*. 2007;4:e112.
12. Barnes PJ. Cytokine networks in asthma and chronic obstructive pulmonary disease. *J Clin Invest*. 2008;118:3546-3556.
13. Barnes PJ. Frontrunners in novel pharmacotherapy of COPD. *Curr Opin Pharmacol*. 2008;8:300-307.
14. Barnes PJ. Immunology of asthma and chronic obstructive pulmonary disease. *Nat Immunol Rev*. 2008;8:183-192.
15. Barnes PJ. Histone deacetylase-2 and airway disease. *Ther Adv Respir Dis*. 2009;3:235-243.
16. Barnes PJ. Role of HDAC2 in the pathophysiology of COPD. *Ann Rev Physiol*. 2009;71:451-464.

17. Barnes PJ. Targeting the epigenome in the treatment of asthma and COPD. *Proc Am Thorac Soc*. 2009;6(8):693-696.
18. Barnes PJ. The cytokine network in COPD. *Am J Respir Cell Mol Biol*. 2009;41:631-638.
19. Barnes PJ, Adcock IM. How do corticosteroids work in asthma? *Ann Intern Med*. 2003;139:359-370.
20. Barnes PJ, Adcock IM. Glucocorticoid resistance in inflammatory diseases. *Lancet*. 2009;342:1905-1917.
21. Barnes PJ, Adcock IM, Ito K. Histone acetylation and deacetylation: importance in inflammatory lung diseases. *Eur Respir J*. 2005;25:552-563.
22. Barnes PJ, Hansel TT. Prospects for new drugs for chronic obstructive pulmonary disease. *Lancet*. 2004;364:985-996.
23. Barnes PJ, Ito K, Adcock IM. A mechanism of corticosteroid resistance in COPD: inactivation of histone deacetylase. *Lancet*. 2004;363:731-733.
24. Barnes PJ, Shapiro SD, Pauwels RA. Chronic obstructive pulmonary disease: molecular and cellular mechanisms. *Eur Respir J*. 2003;22:672-688.
25. Barnes PJ, Stockley RA. COPD: current therapeutic interventions and future approaches. *Eur Respir J*. 2005;25: 1084-1106.
26. Bhavsar P, Hew M, Khorasani N, et al. Relative corticosteroid insensitivity of alveolar macrophages in severe asthma compared to non-severe asthma. *Thorax*. 2008;63:784-790.
27. Bowler RP, Barnes PJ, Crapo JD. The role of oxidative stress in chronic obstructive pulmonary disease. *J COPD*. 2004;2: 255-277.
28. Brindicci C, Ito K, Resta O, Pride NB, Barnes PJ, Kharitonov SA. Exhaled nitric oxide from lung periphery is increased in COPD. *Eur Respir J*. 2005;26:52-59.
29. Calverley PM, Anderson JA, Celli B, et al. Salmeterol and fluticasone propionate and survival in chronic obstructive pulmonary disease. *N Engl J Med*. 2007;356:775-789.
30. Caramori G, Romagnoli M, Casolari P, et al. Nuclear localisation of p65 in sputum macrophages but not in sputum neutrophils during COPD exacerbations. *Thorax*. 2003;58: 348-351.
31. Celli BR, Barnes PJ. Exacerbations of chronic obstructive pulmonary disease. *Eur Respir J*. 2007;29:1224-1238.
32. Charron C, Sumakuza T, Oomura S, Ito K. EM-703, a non-antibacterial erythromycin derivative, restores HDAC2 activation diminished by hypoxia and oxidative stress. *Proc Am Thorac Soc*. 2007;175:A640.
33. Charron CE, Chou PC, Coutts DJ, et al. Hypoxia inducible factor 1 alpha (HIF-1a) induces corticosteroid-insensitive inflammation via reduction of histone deacetylase-2 (HDAC2) transcription. *J Biol Chem*. 2009;284(52): 36047-36054.
34. Chaudhuri R, Livingston E, McMahon AD, et al. Effects of smoking cessation on lung function and airway inflammation in smokers with asthma. *Am J Respir Crit Care Med*. 2006;174:127-133.
35. Chen Y, Shu W, Chen W, Wu Q, Liu H, Cui G. Curcumin, both histone deacetylase and p300/CBP-specific inhibitor, represses the activity of nuclear factor kappa B and Notch 1 in Raji cells. *Basic Clin Pharmacol Toxicol*. 2007;101: 427-433.
36. Cosio BG, Iglesias A, Rios A, et al. Low-dose theophylline enhances the anti-inflammatory effects of steroids during exacerbations of chronic obstructive pulmonary disease. *Thorax*. 2009;64:424-429.
37. Cosio BG, Tsaprouni L, Ito K, Jazrawi E, Adcock IM, Barnes PJ. Theophylline restores histone deacetylase activity and steroid responses in COPD macrophages. *J Exp Med*. 2004;200:689-695.
38. Culpitt SV, Nightingale JA, Barnes PJ. Effect of high dose inhaled steroid on cells, cytokines and proteases in induced sputum in chronic obstructive pulmonary disease. *Am J Respir Crit Care Med*. 1999;160:1635-1639.
39. Culpitt SV, Rogers DF, Shah P, et al. Impaired inhibition by dexamethasone of cytokine release by alveolar macrophages from patients with chronic obstructive pulmonary disease. *Am J Respir Crit Care Med*. 2003;167:24-31.
40. Curtis JL, Freeman CM, Hogg JC. The immunopathogenesis of chronic obstructive pulmonary disease: insights from recent research. *Proc Am Thorac Soc*. 2007;4:512-521.
41. Decramer M, Rutten-van MM, Dekhuijzen PN, et al. Effects of *N*-acetylcysteine on outcomes in chronic obstructive pulmonary disease (Bronchitis Randomized on NAC Cost-Utility Study, BRONCUS): a randomised placebo-controlled trial. *Lancet*. 2005;365:1552-1560.
42. Di Stefano A, Caramori G, Capelli A, et al. Increased expression of NF-kB in bronchial biopsies from smokers and patients with COPD. *Eur Resp J*. 2002;20:556-563.
43. Donnelly LE, Barnes PJ. Chemokine receptors as therapeutic targets in chronic obstructive pulmonary disease. *Trends Pharmacol Sci*. 2006;27:546-553.
44. Failla M, To Y, Ito M, Adcock IM, Barnes PJ, Ito K. Oxidative stress-induced PI3-kinase activation reduces HDAC activity and is inhibited by theophylline. *Proc Am Thorac Soc*. 2007;2:A45.
45. Fox JC, Spicer D, Ito K, Barnes PJ, Fitzgerald MF. Oral or inhaled corticosteroid combination therapy with low dose theophylline reverses corticosteroid insensitivity in a smoking mouse model. *Proc Am Thorac Soc*. 2007;2:A637.
46. Gamble E, Grootendorst DC, Hattotuwa K, et al. Airway mucosal inflammation in COPD is similar in smokers and ex-smokers: a pooled analysis. *Eur Respir J*. 2007;30: 467-471.
47. Hansel TT, Barnes PJ. New drugs for exacerbations of chronic obstructive pulmonary disease. *Lancet*. 2009;374: 744-755.
48. Hansel TT, Kharitonov SA, Donnelly LE, et al. A selective inhibitor of inducible nitric oxide synthase inhibits exhaled breath nitric oxide in healthy volunteers and asthmatics. *FASEB J*. 2003;17:1298-1300.
49. Hew M, Bhavsar P, Torrego A, et al. Relative corticosteroid insensitivity of peripheral blood mononuclear cells in severe asthma. *Am J Respir Crit Care Med*. 2006;174:134-141.
50. Hogg JC. Pathophysiology of airflow limitation in chronic obstructive pulmonary disease. *Lancet*. 2004;364:709-721.
51. Hogg JC, Chu F, Utokaparch S, et al. The nature of small-airway obstruction in chronic obstructive pulmonary disease. *New Engl J Med*. 2004;350:2645-2653.
52. Ito K, Barnes PJ, Adcock IM. Glucocorticoid receptor recruitment of histone deacetylase 2 inhibits IL-1b-induced histone H4 acetylation on lysines 8 and 12. *Mol Cell Biol*. 2000;20:6891-6903.
53. Ito K, Caramori G, Lim S, et al. Expression and activity of histone deacetylases (HDACs) in human asthmatic airways. *Am J Respir Crit Care Med*. 2002;166:392-396.
54. Ito K, Ito M, Elliott WM, et al. Decreased histone deacetylase activity in chronic obstructive pulmonary disease. *New Engl J Med*. 2005;352:1967-1976.
55. Ito K, Lim S, Caramori G, Chung KF, Barnes PJ, Adcock IM. Cigarette smoking reduces histone deacetylase 2 expression, enhances cytokine expression and inhibits glucocorticoid

actions in alveolar macrophages. *FASEB J*. 2001;15: 1100-1102.
56. Ito K, Lim S, Caramori G, et al. A molecular mechanism of action of theophylline: Induction of histone deacetylase activity to decrease inflammatory gene expression. *Proc Natl Acad Sci USA*. 2002;99:8921-8926.
57. Ito K, Tomita T, Barnes PJ, Adcock IM. Oxidative stress reduces histone deacetylase (HDAC)2 activity and enhances IL-8 gene expression: role of tyrosine nitration. *Biochem Biophys Res Commun*. 2004;315:240-245.
58. Ito K, Yamamura S, Essilfie-Quaye S, et al. Histone deacetylase 2-mediated deacetylation of the glucocorticoid receptor enables NF-kB suppression. *J Exp Med*. 2006;203:7-13.
59. Keatings VM, Jatakanon A, Worsdell YM, Barnes PJ. Effects of inhaled and oral glucocorticoids on inflammatory indices in asthma and COPD. *Am J Respir Crit Care Med*. 1997;155:542-548.
60. Kirkham P, Rahman I. Oxidative stress in asthma and COPD: antioxidants as a therapeutic strategy. *Pharmacol Ther*. 2006;111:476-494.
61. Kouzarides T. Chromatin modifications and their function. *Cell*. 2007;128:693-705.
62. Lokke A, Lange P, Scharling H, Fabricius P, Vestbo J. Developing COPD: a 25 year follow up study of the general population. *Thorax*. 2006;61:935-939.
63. Loppow D, Schleiss MB, Kanniess F, Taube C, Jorres RA, Magnussen H. In patients with chronic bronchitis a four week trial with inhaled steroids does not attenuate airway inflammation. *Respir Med*. 2001;95:115-121.
64. Malhotra D, Thimmulappa R, Navas-Acien A, et al. Decline in NRF2 regulated antioxidants in COPD lungs due to loss of its positive regulator DJ-1. *Am J Respir Crit Care Med*. 2008;178:592-604.
65. Mannino DM, Buist AS. Global burden of COPD: risk factors, prevalence, and future trends. *Lancet*. 2007;370: 765-773.
66. Marwick JA, Caramori G, Stevenson CC, et al. Inhibition of PI3Kd restores glucocorticoid function in smoking-induced airway inflammation in mice. *Am J Respir Crit Care Med*. 2009;179:542-548.
67. Marwick JA, Kirkham PA, Stevenson CS, et al. Cigarette smoke alters chromatin remodelling and induces pro-inflammatory genes in rat lungs. *Am J Respir Cell Mol Biol*. 2004;31:633-642.
68. Marwick JA, Wallis G, Meja K, et al. Oxidative stress modulates theophylline effects on steroid responsiveness. *Biochem Biophys Res Commun*. 2008;377:797-802.
69. Matthews JG, Ito K, Barnes PJ, Adcock IM. Defective glucocorticoid receptor nuclear translocation and altered histone acetylation patterns in glucocorticoid-resistant patients. *J Allergy Clin Immunol*. 2004;113:1100-1108.
70. Meja KK, Rajendrasozhan S, Adenuga D, et al. Curcumin restores corticosteroid function in monocytes exposed to oxidants by maintaining HDAC2. *Am J Respir Cell Mol Biol*. 2008;39:312-323.
71. Montuschi P, Ciabattoni G, Corradi M, et al. Increased 8-Isoprostane, a marker of oxidative stress, in exhaled condensates of asthmatic patients. *Am J Respir Crit Care Med*. 1999;160:216-220.
72. Montuschi P, Collins JV, Ciabattoni G, et al. Exhaled 8-isoprostane as an *in vivo* biomarker of lung oxidative stress in patients with COPD and healthy smokers. *Am J Respir Crit Care Med*. 2000;162:1175-1177.
73. O'Donnell DE, Laveneziana P. Dyspnea and activity limitation in COPD: mechanical factors. *COPD*. 2007;4:225-236.
74. Osoata G, Yamamura S, Ito M, et al. Nitration of distinct tyrosine residues causes inactivation of histone deacetylase 2. *Biochem Biophy Res Commun*. 2009;384:366-371.
75. Osoata GO, Hanazawa T, Brindicci C, et al. Peroxynitrite elevation in exhaled breath condensate of COPD and its inhibition by fudosteine. *Chest*. 2009;135:1513-1520.
76. Paredi P, Kharitonov SA, Barnes PJ. Faster rise of exhaled breath temperature in asthma. A novel marker of airway inflammation? *Am J Respir Crit Care Med*. 2002;165:181-184.
77. Paredi P, Kharitonov SA, Leak D, Ward S, Cramer D, Barnes PJ. Exhaled ethane, a marker of lipid peroxidation, is elevated in chronic obstructive pulmonary disease. *Am J Respir Crit Care Med*. 2000;162:369-373.
78. Popov VM, Wang C, Shirley LA, et al. The functional significance of nuclear receptor acetylation. *Steroids*. 2007;72: 221-230.
79. Rabe KF, Hurd S, Anzueto A, et al. Global strategy for the diagnosis, management, and prevention of COPD – 2006 Update. *Am J Respir Crit Care Med*. 2007;176:532-555.
80. Ricciardolo FL, Caramori G, Ito K, et al. Nitrosative stress in the bronchial mucosa of severe chronic obstructive pulmonary disease. *J Allergy Clin Immunol*. 2005;116: 1028-1035.
81. Rubin BK, Henke MO. Immunomodulatory activity and effectiveness of macrolides in chronic airway disease. *Chest*. 2004;125:70S-78S.
82. Russell RE, Culpitt SV, DeMatos C, et al. Release and activity of matrix metalloproteinase-9 and tissue inhibitor of metalloproteinase-1 by alveolar macrophages from patients with chronic obstructive pulmonary disease. *Am J Respir Cell Mol Biol*. 2002;26:602-609.
83. Salvi SS, Barnes PJ. Chronic obstructive pulmonary disease in non-smokers. *Lancet*. 2009;374:733-743.
84. Shahbazian MD, Grunstein M. Functions of site-specific histone acetylation and deacetylation. *Annu Rev Biochem*. 2007;76:75-100.
85. Shankar S, Srivastava RK. Histone deacetylase inhibitors: mechanisms and clinical significance in cancer: HDAC inhibitor-induced apoptosis. *Adv Exp Med Biol*. 2008;615: 261-298.
86. Silverman EK. Progress in chronic obstructive pulmonary disease genetics. *Proc Am Thorac Soc*. 2006;3:405-408.
87. Spears M, Donnelly I, Jolly L, et al. Effect of theophylline plus beclometasone on lung function in smokers with asthma-a pilot study. *Eur Respir J*. 2009;33:1010-1017.
88. Suissa S, Barnes PJ. Inhaled corticosteroids in COPD: the case against. *Eur Respir J*. 2009;34:13-16.
89. Sussan TE, Rangasamy T, Blake DJ, et al. Targeting Nrf2 with the triterpenoid CDDO-imidazolide attenuates cigarette smoke-induced emphysema and cardiac dysfunction in mice. *Proc Natl Acad Sci USA*. 2009;106:250-255.
90. Szulakowski P, Crowther AJ, Jimenez LA, et al. The effect of smoking on the transcriptional regulation of lung inflammation in patients with chronic obstructive pulmonary disease. *Am J Respir Crit Care Med*. 2006;174:41-50.
91. Szyf M. Epigenetics, DNA methylation, and chromatin modifying drugs. *Annu Rev Pharmacol Toxicol*. 2009;49:243-263.

92. Thiagalingam S, Cheng KH, Lee HJ, Mineva N, Thiagalingam A, Ponte JF. Histone deacetylases: unique players in shaping the epigenetic histone code. *Ann NY Acad Sci*. 2003;983: 84-100.
93. Ward S, Sotsios Y, Dowden J, Bruce I, Finan P. Therapeutic potential of phosphoinositide 3-kinase inhibitors. *Chem Biol*. 2003;10:207-213.
94. Wilkinson TM, Hurst JR, Perera WR, Wilks M, Donaldson GC, Wedzicha JA. Effect of interactions between lower airway bacterial and rhinoviral infection in exacerbations of COPD. *Chest*. 2006;129:317-324.
95. Willemse BW, ten Hacken NH, Rutgers B, Lesman-Leegte IG, Postma DS, Timens W. Effect of 1-year smoking cessation on airway inflammation in COPD and asymptomatic smokers. *Eur Respir J*. 2005;26:835-845.
96. Yang IA, Fong KM, Sim EH, Black PN, Lasserson TJ. Inhaled corticosteroids for stable chronic obstructive pulmonary disease. *Cochrane Database Syst Rev*. 2007;CD002991.

15 Clinical Trials of Epigenetic Modifiers in the Treatment of Myelodysplastic Syndrome

Lauren C. Suarez and Steven D. Gore

15.1 Introduction

Myeloid malignancies, including acute myeloid leukemia (AML) and its precursor lesion myelodysplastic syndrome (MDS), are the result of bone marrow failure due to aberrant or arrested terminal differentiation. The prominent differentiation abnormalities in myeloid malignancies have led to a quest for pharmacologic agents that affect normal hematopoietic differentiation, leading to improved hematopoiesis with or without clonal extinction. The development of all-trans-retinoic acid for the treatment of acute promyelocytic leukemia has been the outstanding success of this effort.[17]

Piskala, Pliml, and Sorm introduced the cytidine analogs 5-azacytidine and 2′-deoxy-5-azacytidine as cytotoxic cancer agents in 1964, agents that clearly demonstrated early effectiveness in acute myeloid leukemia (AML).[67,107] In 1977, Jones and colleagues showed that non-myoblast mouse embryo cells differentiated into functional muscle cells following treatment with 5-azacytidine. However, differentiation was not immediate and did not occur until subsequent cellular divisions. Treatment was recorded to be most effective when cells were in early S phase.[13] The ability of 5-azacytidine to act as a differentiating agent in vitro[51,100–102] was subsequently confirmed and led to studies of 5-azacytidine in the treatment of MDS in a series of trials conducted by the Cancer and Leukemia Research Group B (CALGB, see below). These trials established the clinical activity of azacitidine and were instrumental in FDA approval for its use in myelodysplastic syndromes.

S.D. Gore (✉)
Department of Oncology/Hematologic Malignancies, Johns Hopkins University, Baltimore, MD, USA
e-mail: gorest@jhmi.edu

15.2 Is MDS an Epigenetically Driven Disease?

Though it has long been accepted that tumorigenesis is driven by genomic mutations and chromosomal abnormalities, epigenetic modifications are another mechanism to silence tumor suppressor genes. According to Knudson's two hit hypothesis, the first hit rendering one gene ineffective is either a germline (inherited) or somatic (sporadic) mutation.[58] The second hit can be caused by mutation or chromosomal deletions, but also by methylation of cytosine in gene promoter regions that lead to transcriptional silencing of the tumor suppressor gene. The CpG-rich regions (CpG islands) in promoters of transcriptionally active genes are usually hypomethylated. Methylation at these sites recruits proteins and leads to transcriptionally repressive chromatin conformation. Histone modifications determine the transcriptional state of DNA. Hyperacetylation of histone lysine tails is associated with transcriptionally active DNA, whereas deacetylation of histones by histone deacetylases closes the chromatin and leads to transcriptional silencing. Histone deacetylases also interact with proteins that are involved in cell cycle control and apoptosis and thereby affect cell proliferation.[43]

Similar to other malignancies, myelodysplastic syndromes exhibit common chromosomal abnormalities and mutations. Common cytogenetic abnormalities in myelodysplastic syndromes include deletions of the long arms of chromosome 5, 7, and 20.[79] Mutations in

H.I. Roach et al. (eds.), *Epigenetic Aspects of Chronic Diseases*,
DOI: 10.1007/978-1-84882-644-1_15, © Springer-Verlag London Limited 2011

specific genes can also be linked with myelodysplastic syndromes. Genes commonly mutated include FLT3, AML/RUNX1, p53, RAS, with N-RAS being the most frequent.[72,79] In de novo MDS, cytogenetic abnormalities are not detected in 50% of patients; however, the frequency of genetic additions and losses may go much higher with the use of screening techniques such as array comparative genomic hybridization (CGH) and single nucleotide polymorphism (SNP) arrays.[69]

In addition to important genetic modifications, DNA methylation is frequent in MDS and may play a crucial role in MDS pathogenesis. Aberrant hypermethylation at the CpG islands of promoter regions of genes involved in apoptosis, cell cycle control, DNA repair, proliferation, signal transduction, differentiation, and adhesion is common in myelodysplastic syndromes.[49,79] When aberrant methylation was compared in patients with advanced MDS (Refractory anemia with excess blasts [RAEB] and AML) with that in patients with low-risk MDS, aberrant CpG DNA methylation was significantly higher in RAEB/AML patients.[49] Genes frequently silenced in association with promoter hypermethylation in myelodysplastic syndromes include p15^{INK4b}, E-cadherin (CDH1), human cancer 1(HIC1), calcitonin, and estrogen receptor (ER).[1,46,77,79,83,103] Promoter hypermethylation in myelodysplastic syndrome is distinctly different in CD34+ bone marrow cells derived from normal donors as compared to malignant cells. Furthermore, aberrant methylation in MDS/secondary AML is more extensive than in de novo AML. This suggests that epigenetic profiles may differ among phenotypically similar hematologic malignancies.[26] The increase of aberrant methylation in gene promoters in myelodysplastic syndromes as the disease progresses clearly indicates an epigenetic component in disease manifestation and progression.

15.3 Clinical Impact of DNA Methyltransferase Inhibitors in MDS

15.3.1 *Azacitidine*

High doses of azacitidine were administered as a conventional cytotoxic agent in clinical trials in 1967 in Europe and in 1970 in the USA.[107] On the basis of reports of in vitro differentiating activity by azacitidine, treatment of hematologic malignancies began to focus on the use of lower doses of the drug as a putative differentiating agent.

Several pilot studies, where low doses of azacitidine were either infused or injected subcutaneously, have demonstrated complete or partial hematologic responses and improvement in 49–53% of MDS patients.[85,91] Two randomized trials subsequently confirmed the efficacy of azacitidine in this patient population.

CALGB 9221was a randomized, controlled trial of 191 patients that were either treated with azacitidine (75 mg/m^2/day of azacitidine subcutaneously) for 7 days every 28 days or received supportive care.[92] All patients were evaluated after 4 months on study; patients randomized to best supportive care were allowed to cross over to the treated group if their disease had progressed. Complete responders received three additional cycles of azacitidine following remission; patients with lesser responses continued therapy until disease progression. The most common toxicity in this and previous trials with azacitidine was myelosuppression. Analysis of these findings with the use of standardized response criteria[11] led to an overall response rate of 47% in patients initially treated with azacitidine and of 35% in patients who crossed over to azacitidine treatment. Of the total of 150 patients that received azacitidine initially or after crossover, 13 responses were complete, 3 were partial, and 49 patients improved hematologically.[93]

A median of three cycles was required to observe a response. Of the 40 azacitidine patients (initial or crossover) that depended on red blood cell transfusions and who responded to treatment, 35 became transfusion-independent. In contrast, only 17% of patients treated with the best supportive care responded. The median duration of clinical response to azacitidine was 15 months.[93] Patients receiving azacitidine had improved quality of life whether originally assigned to azacitidine or crossed over.[59] The crossover design precluded analysis of the impact of azacitidine on patient survival, inasmuch as 55% had crossed over. However, the group assigned to initial azacitidine treatment took longer before their disease progressed to AML (defined by 30% bone marrow blast cells) or death (median 19 vs 8 months). These findings suggest that administration of azacitidine may impact the natural history and progression of MDS.

A randomized study in patients with high-grade MDS (Pharmion/Celgene study AZA-001) demonstrated the impact of azacitidine on survival. Patients were assigned to one of three standard regimens: best supportive care, low-dose cytarabine, or intensive cytarabine-based chemotherapy. Once assigned, patients were randomized to the FDA-approved dose schedule of azacitidine or the pre-assigned conventional care regimen.[25] The median survival time of azacitidine-treated patients was 24.5 months versus 15 months in the combined conventional care group. Furthermore, time to AML progression was 18 months versus 11 months in the conventional care group, with 2 year survival of the azacitidine group double that of the conventional care group. Although more convenient dose schedules of azacitidine have been explored in lower risk patients,[68] the FDA-approved schedule of azacitidine represents the only approach that currently improves survival in high-risk MDS.

15.3.2 Decitabine

The azacitidine congener decitabine (2′-deoxy-5-azacytidine) has undergone parallel studies in hematologic malignancies. Early Phase I and II studies established responses in patients with MDS ranging from 32% to 54%.[47,82,111,112,116] The largest studies administered 15 mg/m^2 intravenously over 3 h every 8 h for nine doses. Cycles were repeated every 6 weeks. FDA approval is based on the outcome of a Phase III trial; that compared the scheduled treatment with decitabine to the best supportive care.

The registration trial randomized 170 patients with MDS to receive best supportive care or decitabine. The response rate noted in this trial was consistent with response rates in previous phase II studies. Of 89 patients receiving decitabine, eight had complete responses and seven had partial responses. Median time to response was 3.3 months, and median duration was 10 months. Critically, no patients on best supportive care attained partial or complete responses. Twelve additional decitabine patients showed hematologic improvement compared to six patients receiving best supportive care. More decitabine-treated patients became red blood cell transfusion–dependent, but this was not the case in patients on supportive care.[53]

As with azacitidine, the most common toxic reaction to treatment was myelosuppression. In contrast to the azacitidine trial, the time to AML or death was not improved in the decitabine group. A major difference between this trial and the two Phase III trials of azacitidine was that in the decitabine trial, the median number of cycles administered was two to three, whereas the median number was nine in the CALGB9221 and AZA-001 groups.[25,93] Additionally, in the decitabine trial, the maximum number of cycles administered was eight (by design), whereas in the azacitidine trial, most patients received therapy until disease progression.

As with azacitidine, alternative dosing schedules of decitabine have been explored. A phase I trial in patients with myelodysplastic syndrome, acute myeloid leukemia, chronic myelogenous leukemia, or acute lymphocytic leukemia studied the effect of lower daily doses (5–20 mg/m^2) being administered longer (10–20 days). Conceptually, such dosing schedules would allow for more effective methylation reversal due to reduced cell cycle inhibition and exposure to drug for a greater number of cell divisions. The dose of 15 mg/m^2/day for 10 days was selected on the basis of clinical responses and maximum effect on DNA methylation.[47] A 5-day schedule of decitabine studied in two Phase II studies has been widely adopted in the USA.[54,96] As with the alternative dosing study of azacitidine, response rates appear similar to more intensive schedules; however, the impact of this schedule of decitabine on survival of MDS patients has not been studied.

15.4 Do Azacitidine and Decitabine Work Through Epigenetic Mechanisms in MDS?

Their clinical effectiveness notwithstanding, the exact mechanism by which azacitidine and decitabine act is unclear. As cytidine analogs, both drugs are incorporated into DNA. Unlike decitabine, a major portion of azacitidine is incorporated into RNA, the biological significance of which is uncertain. Once these compounds have been incorporated into DNA, DNA methyltransferases bind during DNA replication and attempt to reproduce the methylation pattern of the parent strand. The methyltransferase forms an irreversible inactive complex with the incorporated cytidine analog in the DNA, resulting in depletion of active DNA methyltransferase; thus methylation patterns can no longer be transmitted to daughter cells. This can lead to the re-expression of genes that had been silenced in the parent cells. If this mechanism is responsible for the therapeutic effects of azanucleosides, then expression of biologically important tumor suppressor genes following treatment is a possible basis for the clinical response.[52,74,106]

Findings examining this hypothesis have since proved inconclusive. A major challenge is the long duration of therapy to achieve clinical improvement. Sampling of bone marrow or blood at the time of the clinical response may provide a cell population that does not represent the original clone. In fact, in responding patients, the sample may contain mostly normal cells. Ideally, molecular events should be studied soon after administration of the drugs, but it may be difficult to correlate early molecular events with clinical responses that occur weeks or months later.

MDS patients treated with decitabine have hematologic responses, whether or not promoter methylation of the tumor suppressor gene $p15^{INK4B}$ had been reversed.[16] In all three patients that responded fully, the methylation of the gene had been reversed.[16] In leukemic patients who had responded to decitabine, p15 expression was greater than in nonresponders.[54] A possible correlation between methylation reversal of $p15^{INK4B}$ and/or CDH-1 can be inferred from a study in which azacitidine treatment of patients with MDS or AML was followed by a treatment with sodium phenylbutyrate, a histone deacetylase (HDAC) inhibitor. Of 12 patients that exhibited hypermethylation of either promoter prior to treatment, six developed methylation reversal in response to treatment and later a hematologic response to treatment. The other six patients had no methylation reversal, and did not respond to treatment.[38]

With further study, more genes silenced by methylation may be identified. Phosphoinositide-phospholipase C beta 1 (PI-PLCbeta1) is involved in lipid signaling and appears to be important in cell growth and differentiation.[28] From a total of 18 patients treated with azacitidine, 10 responded. Nine of the responding patients showed a decrease in promoter methylation and an increase in PI-PLCbeta1 gene expression. Patients receiving best supportive care did not show these differences. Surprisingly, patients who did not respond to treatment exhibited an increase in promoter methylation and a decrease in gene expression. This change may not be significant, however, inasmuch as it took 3–4 months of treatment before the more pronounced changes were observed. Post-treatment samples may be derived from normal cells (without methylated PI-PLCbeta1) that have replaced the malignant cells with methylated PI-PLCbeta 1.[28] A similar design problem limits another study which claims that methylation reversal of p21 may be influential in clinical response.[76] Administration of decitabine to induce maximum increase in gene re-expression (p15 and/or estrogen receptor [ER]) linked clinical response with increased ER expression during the first therapy cycle. While reduction in promoter methylation might have been responsible for ER re-expression in some patients, a subset with no pretreatment promoter methylation also demonstrated re-expression of this gene.[2]

These studies suggest that changes in epigenetic profiles may be linked to response to treatment and constitute indicators of response. On the other hand, in a trial with patients receiving a low dose of decitabine, it was not possible to correlate p15 methylation at baseline or after treatment with a clinical response. Of 48 patients, 29 were analyzed for p15 methylation. Six in this group responded to treatment, but only two had methylation at baseline and by day 12, after therapy, no changes in methylation were noted.[47] Similarly, a study combining azacitidine with the histone deacetylase inhibitor entinostat intending to reverse promoter methylation and re-express silenced tumor suppressor genes (TSG) observed baseline promoter methylation and gene expression in $p15^{INK4B}$, CDH-1, SOCS-1 and DAP-K, four TSGs commonly methylated in myeloid malignancies. While methylation of all four genes reversed on days 15 and 29 of treatment, the extent of methylation reversal was similar in both clinical responders and nonresponders.[22] Using a genome-wide array-based methylation assay, methylation reversal was detected during the first cycle of therapy in more than 600 genes; however, neither baseline methylation nor methylation reversal differentiated clinical responders from nonresponders.[26] Moreover, changes in expression of several frequently methylated genes during cycle one of therapy were not related to clinical response.[22] Another approach utilized a methylation profile constructed from ten genes that grouped patients by high (more than 15% methylation) and low methylation. Those with high methylation tended to have less progression-free or overall survival. In decitabine-treated patients, methylation had decreased by 11.2% after 4 months on therapy, whereas those receiving best supportive care had an increase of 20.1%. In terms of clinical response, greater decreases in methylation were associated with complete and partial responses, whereas methylation increases were linked with no changes or progression in disease. These findings indicate that reduction in methylation in the course of treatment is important to clinical outcome.[90]

An alternative explanation is that administration of azanucleosides induces DNA damage that ultimately leads to cell death. Decitabine treatment of cell lines induces expression of the variant phosphorylated histone gamma H2AX, as well as of p53- and ATM-dependent induction of p21$^{WAF1/CIP1}$.[50] Gamma H2AX expression in blood and bone marrow cells increases in patients treated with azacitidine and decitabine.[2,38] Combination treatment with azacitidine and entinostat caused gamma H2AX to increase in 13 of 23 patients, but was the same in responders and nonresponders.[22] Administration of azanucleosides thus can be seen to induce reversal of promoter methylation and of DNA damage. Further studies are needed to show whether either or both of these molecular changes lead to clinical responses in patients with hematologic malignancies.

15.5 Histone Deacetylase Inhibitors

The transcriptionally repressive state associated with histone deacetylation may affect genes involved in the expression of the malignant phenotype. Fusion proteins in acute promyelocytic leukemia and AML have been found to associate with HDAC–protein complexes that lead to transcriptional repression.[66,110] HDACs may play an important role in the differentiation of hematopoietic cells. Furthermore, overexpression of HDACs has been noted in primary AML cell cultures, as well as in AML cell lines.[108] Histone deacetylases also aid in monitoring the apoptotic pathways.[64]

The clinical development of HDAC inhibitors in hematologic malignancies began in the latter half of the 1990s and has been limited largely to Phase I and Phase II trials. HDAC inhibitors fall into several chemical classes and are further categorized as to whether they are "pan-HDAC inhibitors," i.e. inhibit both Class I and Class II histone deacetylases. Another group of HDAC inhibitors are putatively Class I-specific. Class I HDACs include HDAC 1,2,3, and 8: these enzymes are intranuclear and partly regulate chromatin conformation.[6,15,21,41,45,71,99,104,113,114] Class II HDACs include HDACs 4,5,6,7,9, and 10: these enzymes are cytoplasmic and have multiple protein targets including tubulin, HSP90, and MEF2.[27,42,55,56,60,73,105,109,117,118] As information regarding the large number of non-histone targets of HDACs has accumulated, it has been proposed that these enzymes be called "protein deacetylases." The discussion to follow will focus on clinical development of HDAC inhibitors according to targets and chemical class (Table 15.1).

15.6 Class I-Specific HDAC Inhibitors

15.6.1 Short-Chain Fatty Acids

Sodium phenylbutyrate and valproic acid are both short chain fatty acids that have HDAC inhibitory activity at millimolar concentrations. These drugs are Class I HDAC-specific.[65,78,115] Sodium phenylbutyrate has been used to induce expression of gamma globin in sickle cell anemia patients.[19,20] Addition of sodium phenylbutyrate to AML cell cultures has induced cellular differentiation and inhibited proliferation.[18,37] In Phase I studies of MDS and AML patients, sodium phenylbutyrate was continuously infused for 7 days of a 28-day cycle, 7 days of a 14-day cycle, and 21 days of a 28-day cycle. Responses were limited to clinical improvement.[39] Dose limiting toxicity was encephalopathy; the maximum tolerated dose was 375 mg/kg/day.

The neuroleptic drug, valproic acid (VPA), has similar in vitro effects on cellular differentiation and proliferation.[40,80] Unlike sodium phenylbutyrate, oral VPA is bioavailable, with the doses used to treat neurologic indications approaching millimolar. In a study comparing valproic acid alone or in conjunction with all-trans retinoic acid in MDS and AML, eight patients (44%) receiving valproic acid alone responded to therapy, with one showing a partial response and the others exhibiting hematologic improvement. Sufficient valproic acid was administered for serum concentrations to attain 346 and 693 μM (50 and 100 μg/mL). Those patients also receiving all-trans retinoic acid (ATRA) were given 80 mg/m^2/day in two doses, days 1–7, on alternate weeks.[61,63] In a subsequent study with a larger patient population of 119 patients, the response rate was 21% (one complete and one partial response, with 23 patients showing hematologic improvement).[62]

Myelodysplastic syndrome typically affects older individuals, a group that tends to have fewer treatment

Table 15.1 Clinical studies of histone deacetylase inhibitors

HDAC specificity	Histone deacetylase inhibitor	Class of compound	Study	Trial phase	Disease	Cohort size	Response
Class I	Sodium phenylbutyrate	Short chain fatty acid	Gore 2002[39]	Phase I	MDS, AML	23	2 HI
Class I	Valproic acid	Short chain fatty acid	Kuendgen 2004[63]	Phase I	MDS, AML	23	1 PR, 7 HI
			Kuendgen 2005[62]	Phase II	MDS, AML	75	18 HI
			Pilatrino 2005[81]	Phase II	MDS, AML	20	6 HI
			Raffoux 2005[84]	Phase I	AML	11	1CR, 2CRi, 2HI
			Cimino 2006[12]	Phase I	AML, CML	8	2 HI
			Bug 2005[4]	Phase I	MDS, AML	26	1PR, 1HI
Class I	MGCD0103	Benzamide	Garcia-Manero 2008[30]	Phase I	MDS, AML	29	3 CR
Class I	Entinostat (MS-275)	Benzamide	Gojo 2007[36]	Phase I	MDS, AML	39	NR, 12 SD, 15 RPWBC
Class I	Romidepsin	Benzamide	Klimek 2008[57]	Phase I	MDS, AML	12	1 CR
Pan	Vorinostat	Hydroxamic acid	Garcia-Manero 2008[32]	Phase I	MDS, AML, ALL, CML, CLL	41	2 CR, 2 Cri, 3 HI
			Schaefer 2009[87]	Phase II	AML	37	1 CR
Pan	Panobinostat	Hydroxamic acid	Giles 2006[33]	Phase I	MDS, AML, ALL	15	1 HI
Pan	Belinostat	Hydroxamic acid	Gim sing 2008[35]	Phase I	MM, NHL, CLL	16	NR, 5 SD
			Schlenk 2008[88]	Phase I/II	MDS, AML	22	1 CR. 3 CRi

MDS myelodysplastic syndrome, *AML* acute myeloid leukemia, *CML* chronic myelomonocytic leukemia, *ALL* acute lymphocytic leukemia, *CLL* chronic lymphocytic leukemia, *MM* multiple myeloma, *NHL* non-Hodgkins lymphoma, *CR* complete response, *CRi* complete response with incomplete platelet recovery, *PR* partial response, *HI* hematologic improvement, *SD* stable disease, *RPWBC* 50% reduction in peripheral white blood cells, *NR* no responses seen

options. Therefore, assessing the efficacy of treatment in this cohort is particularly important. Six patients in a group (20 total, 11 evaluable) ranging in age from 60 to 85 years old with MDS/AML showed hematologic improvement on a regimen of valproic acid and ATRA.[81] In another study of elderly patients, treated similarly, three patients had a full response, but two had incomplete platelet recovery and two others showed hematologic improvement.[84] In both studies, participants received valproic acid to reach a serum concentration between 50 and 100 μg/mL. Once desirable serum concentrations were obtained, ATRA was administered at 45 mg/kg/day and valproic acid treatment was started a week later.

Response to treatment with valproic acid in AML patients has been linked to hyperacetylation of histones H3 and H4. A shift toward a more mature cellular population and greater involvement of differentiation genes were associated with an increase in histone acetylation in some cases.[12] However, these findings did not necessarily represent a clinical response. The white blood cell count increased in two patients showing hematologic response.[12] The same was noted in another study of AML patients, in the majority of whom, however, a rapid increase in white blood cells required intervention with AraC or hydroxyurea.[4] This may be due to valproic acid being able to maintain leukemic progenitor cells in culture and to support leukemic colony growth.[5]

15.6.2 Benzamides

The benzamide HDAC inhibitors also inhibit Class I HDACs selectively. MGCD0103 is an amino phenylbenzamide small molecule histone deacetylase inhibitor that was synthesized specifically to inhibit histone deacetylase enzymes 1, 2, 3, and 11.[29,75] The drug showed activity in a Phase I study, but its development was aborted because it caused clinically significant serositis and pericarditis.[30]

Entinostat (formerly known as MS-275) is a synthetic benzamide derivative[98] that functions as a HDAC class 1 inhibitor. Cancer cell line studies have shown that entinostat inhibits proliferation and is a cytotoxic agent.[48,86] Entinostat is orally bioavailable and has a long half-life; this makes it possible to dose weekly. In a phase I study in patients with high-risk myelodysplastic syndrome and acute myeloid leukemia, the maximum weekly tolerated dose of entinostat was 8 mg/m^2, administered for 4 weeks with a 2-week break. Treatment stabilized the disease, decreased dependence on transfusions, reduced differentiation of myeloid lineage and reduction of bone marrow blasts. The most frequent adverse symptoms were fatigue, anorexia, vomiting, and nausea.[36] Entinostat caused histone acetylation to increase, as determined by Western Blot and flow cytometry in peripheral blood and bone marrow mononuclear cells. Histone acetylation appeared to persist for several weeks after the drug had been discontinued. Entinostat administration induced expression of p21$^{WAF1/CIP1}$ and activity of caspase 3. This may link increased histone acetylation with apoptosis.[36]

15.6.3 Romidepsin

Romidepsin is a cyclic depsipeptide that has recently been approved by the FDA for treatment of cutaneous T cell lymphoma (CTCL). However, clinical responses in chronic lymphocytic leukemia, MDS, and AML have not been encouraging. The approved dose schedule in CTCL is intravenous infusion of 14 mg/m^2 on days 1, 8, and 15. Neither complete nor partial responses were seen in CLL or AML patients treated similarly.[7] An increased dose of 18 mg/m^2 on days 1 and 5 of a 21-day treatment cycle led to a complete response in one and stable disease in six patients.[57] Nausea, thrombocytopenia, and febrile neutropenia were fairly common adverse symptoms.

15.6.4 Pan-HDAC Inhibitors

15.6.4.1 Hydroxamic Acids

A variety of hydroxamic acids that inhibit HDAC are currently being evaluated for treatment of hematologic malignancies.

Vorinostat (suberoylanilide hydroxamic acid [SAHA]) has been the first HDAC inhibitor to receive FDA approval for treatment of cutaneous T cell lymphoma.[32] In a phase I study, patients with advanced leukemia or myelodysplastic syndrome were given oral vorinostat two or three times daily for 14 days, followed by 7 days of rest. The maximum tolerated doses were either 200-mg oral vorinostat twice daily for 14 days or 250 mg three times daily for 14 days.[32] Nausea, anorexia, vomiting, fatigue, and thrombocytopenia were the most commonly noted adverse symptoms. Of 41 patients, a total of 7 (17%) responded to vorinostat treatment. Four patients with acute myeloid leukemia had complete responses to vorinostat treatment, two others responded, but with an incomplete blood count. An additional three patients showed hematologic improvement. The clinical response did not appear dose-dependent. Histone acetylation, analyzed in peripheral blood and bone marrow cells, had increased in the course of treatment, but returned to baseline levels during the week of rest. The level of histone acetylation did not correspond to the clinical response; however, in patients with a response to treatment, expression levels of proliferation-associated genes had fallen.[32]

The clinical activity of vorinostat in AML was not confirmed in a randomized Phase II trial of vorinostat in which a dose of 200 mg, administered three times daily for 14/21 days (arm A), was compared with a dose of 400 mg, administered daily (arm B). One patient on arm B developed a complete response; however, 15 patients on arm A and 21 on arm B did not respond.[87] Resistance to vorinostat in culture has been linked to the expression of genes involved with antagonizing oxidative damage.[32,70]

Panobinostat (LBH589) affects histone acetylation and cell cycle and induces apoptosis in leukemic cell lines.[33] A phase I study of intravenous panobinostat in patients with AML, acute lymphocytic leukemia, and MDS, who received the drug daily for 7 days every 21 days, showed that doses above 9 gm/m^2 caused unacceptable QTc prolongation. One patient, however, achieved hematologic improvement with a dose 11.5 mg/m^2 of Panobinostat. Analysis of leukemic blast cells showed that acetylation of histones H2B and H3 increased.[33]

Belinostat (PXD101), whether administered parenterally or enterally, induces expression of several genes, including p21 and thymidylate synthase.[34] In a phase I study conducted in patients with multiple myeloma, non-Hodgkin's lymphoma, and chronic lymphocytic leukemia,[35] belinostat was administered intravenously for 5 consecutive days during a 21-day cycle. Patients tolerated the drug well, with adverse symptoms similar to those seen in other histone deacetylase inhibitors (nausea, vomiting, fatigue, and flushing). In patients with solid tumors, the highest dose administered was 1,000 mg/m^2.[95] The disease was stabilized in five patients, but no patients achieved complete or partial remission. Ongoing phase I/II trials of belinostat alone or in combination with idarubicin have shown better responses than therapy with belinostat alone.[34] In this trial,[34] MDS and AML patients received belinostat by intravenous infusion in 30-min sessions for 5 days or continuously over 1–2 days. In the patients who received belinostat for 5 days, idarubicin was supplied on the fourth and fifth days; in the second group who received it continuously for 1 or 2 days, idarubicin was administered at the same time as the belinostat, i.e., 24 or 48 h. Both groups responded to belinostat alone or in combination with idarubicin.[34,38]

15.7 Why Combine Epigenetic Drugs?

In vitro, expression of genes with dense promoter methylation cannot be induced with HDAC inhibitors alone. This suggests that DNA methylation plays a dominant role in histone acetylation and transcriptional repression.[8,43,97] DNA methyltransferases and histone deacetylases are important to the transcriptional state of genes, interacting with and recruiting other proteins important in chromatin conformation.[43] The combination of DNA methyltransferase inhibitors and histone deacetylase inhibitors may lead to synergistic re-expression of methylated genes. Experiments with colorectal carcinoma (RKO) and leukemia (KG1a) cell line cultures have shown that treatment with a histone deacetylase inhibitor did not lead to re-expression of genes with hypermethylated promoter regions, though gene expression increased in the absence of hypermethylation. Administration of a DNA methyltransferase inhibitor followed by a histone deacetylase inhibitor led to greater gene re-expression than by DNA methyltransferase inhibitor alone.[8] This observation has led to clinical trials with combinations of DNMT inhibitors and HDAC inhibitors.

15.8 Combinations of Epigenetic Drugs

Four Phase I studies have combined DNMT inhibitors with small chain fatty acids. In the first such trial, various dose schedules of azacitidine treatment of patients with MDS and AML were followed by a 7-day continuous infusion of sodium phenylbutyrate. Responses appeared to correlate with reversal of p15 and or CDH1 methylation. Histones 3 and 4 were acetylated, whether after treatment with azacitidine alone or after combination with sodium phenylbutyrate.[38]

A phase I/II study of azacitidine in combination with valproic acid (VPA) and all-trans retinoic acid (ATRA) had an overall response rate of 42% in patients with myelodysplastic syndrome and acute myeloid leukemia. Bisulfite pyrosequencing of normally heavily methylated non-coding LINE sequences showed a transient decrease in global methylation with treatment which had rebounded by the next treatment cycle. p15 and p21 mRNA levels increased a little, but this seemed to have no relationship with the clinical response. Acetylation of histones H3 and H4 occurred in 54% of patients, but this did not appear to correlate with either the dose of valproic acid or the clinical response. As the levels of free and bound valproic acid were higher in patients that responded to treatment, valproic acid blood levels may be important to the clinical response.[94]

Decitabine at 15 mg/m^2/day was administered for 10 days in combination with oral valproic acid at doses ranging from 20 to 50 mg/kg/day.[3] Ten patients of 41 treated with 50 mg/kg of VPA responded. Similar to the trial with azacitidine and valproic acid, there was evidence of H3 and H4 histone acetylation, but there was no association between histone acetylation and response.[31]

One study aimed to determine the optimum biologic dose of decitabine and to use that dose in combination studies with valproic acid. Decitabine at 20 mg/m²/day over 1 h for 10 days produced the largest increase in re-expression of p15 and the estrogen receptor. This dose was combined with increasing doses of valproic acid, with a maximum tolerated dose of 20 mg/kg/day on days 5–21. The overall response rate was 52% in 21 assessable patients (4CR, 4Cri, 3PR). Responses were noted in both treatment phases. Treatment appeared to be more effective in patients whose estrogen receptor gene expression had increased. Histone acetylation increases were not linked with response, but valproic acid addition led to a further increase in H3 acetylation and may have led to a more robust response.[2]

Azacitidine and entinostat were combined in a Phase I study of MDS, AML, and chronic myelomonocytic leukemia. Fourteen of the 30 treated patients responded to treatment; this included three complete and four partial responses. Azacitidine was administered for 10 days at 30, 40, or 50 mg/m²/day. Entinostat was administered orally on days 3 and 10 of each cycle.[22]

These studies have demonstrated the feasibility of combining DNMT inhibitors and HDAC inhibitors in patients with myeloid malignancies. Randomized trials are needed to determine whether addition of the HDAC inhibitor contributes to the clinical outcome. A randomized Phase II trial of azacitidine with and without entinostat in patients with MDS, AML, and chronic myelomonocytic leukemia has been completed, but results are not yet available.

15.9 Where Are We going?

The future of epigenetic therapies in myelodysplastic syndrome is full of possibilities due to ongoing research of DNA methyltransferases and histone deacetylases. Zebularine is a nucleoside analog that was created to be more stable and less toxic than its counterparts azacitidine and decitabine. Although it traps DNMTs, toxicity was minimal when given orally. However, Zebularine may not be clinically useful as it has poor bioavailability.[9,10,23,44] Non-nucleoside analogs may be a solution to the lack of specificity of nucleoside analogs. Based on the proposed mechanism of azacitidine and decitabine, attempts to make direct enzyme inhibitors of DNA methyltransferases are under way. DNA methyltransferase 1(DNMT1), appearing to be the most active, has been targeted first. A small molecule inhibitor 2-(1,3-dioxo-1,3-dihydro-2H-isoindol-2-yl)-3-(1H-indol-3-yl) propanoic acid or RG108 was designed based on a model of the catalytic domain of DNMT1.[3] As the DNA methyltransferases share a conserved catalytic domain, RG108 may have activity against all DNA methyltransferases. The manner by which azacitidine and decitabine bind and trap DNA methyltransferases is thought to be associated with its toxicity, but in vitro studies indicate RG108 does not trap and reduce the level of DNA methyltransferases. The cytotoxic effect of trapping DNA methyltransferases may be integral to their effectiveness.

Hydralazine and procainamide are two FDA-approved drugs that may block DNA methyltransferase activity. Procainamide, by binding to CG-rich sequences of DNA, acts as a competitive inhibitor for DNA methyltransferases.[14,23,89] EGCG or (–)-Epigallocatechin-3-gallate, a green tea alkaloid, may also have the capacity to inhibit DNA methyltransferases.[23,24] While continued development of DNA methyltransferase inhibitors and histone deacetylase inhibitors is necessary to improve treatment options, deeper understanding of the mechanism that brings about epigenetic silencing in this and other syndromes is also needed. Further characterization of hypermethylated genes in these diseases may provide more informative targets for prognosis. It may also be useful to consider targeting proteins that contribute to chromatin formation. For example, methyl cytosine-binding proteins are not only recruited to sites of DNA methylation, but also induce transcriptional silencing on their own.[43] A more complete understanding of the molecular events underlying this disease may present new epigenetic targets. There may also exist markers of treatment response not linked with epigenetic silencing. On the basis of completed trials, combinations of inhibitors of DNA methyltransferase and histone deacetylase remain a promising treatment for MDS and AML. As the mechanisms behind actions of these drugs become better understood, there may arise new opportunities for more effective epigenetic treatments in both hematologic and other malignancies.

References

1. Aggerholm A, Holm MS, Guldberg P, Olesen LH, Hokland P. Promoter hypermethylation of p15INK4B, HIC1, CDH1, and ER is frequent in myelodysplastic syndrome and predicts poor prognosis in early-stage patients. *Eur J Haematol.* 2006;76:23-32.
2. Blum W, Klisovic RB, Hackanson B, et al. Phase I study of decitabine alone or in combination with valproic acid in acute myeloid leukemia. *J Clin Oncol.* 2007;25:3884-3891.
3. Brueckner B, Garcia Boy R, Siedlecki P, et al. Epigenetic reactivation of tumor suppressor genes by a novel small-molecule inhibitor of human DNA methyltransferases. *Cancer Res.* 2005;65:6305-6311.
4. Bug G, Ritter M, Wassmann B, et al. Clinical trial of valproic acid and all-trans retinoic acid in patients with poor-risk acute myeloid leukemia. *Cancer.* 2005;104:2717-2725.
5. Bug G, Schwarz K, Schoch C, et al. Effect of histone deacetylase inhibitor valproic acid on progenitor cells of acute myeloid leukemia. *Haematologica.* 2007;92:542-545.
6. Buggy JJ, Sideris ML, Mak P, Lorimer DD, McIntosh B, Clark JM. Cloning and characterization of a novel human histone deacetylase, HDAC8. *Biochem J.* 2000;350:199-205.
7. Byrd JC, Marcucci G, Parthun MR, et al. A phase 1 and pharmacodynamic study of depsipeptide (FK228) in chronic lymphocytic leukemia and acute myeloid leukemia. *Blood.* 2005;105:959-967.
8. Cameron EE, Bachman KE, Myohanen S, Herman JG, Baylin SB. Synergy of demethylation and histone deacetylase inhibition in the re-expression of genes silenced in cancer. *Nat Genet.* 1999;21:103-107.
9. Cheng JC, Matsen CB, Gonzales FA, et al. Inhibition of DNA methylation and reactivation of silenced genes by zebularine. *J Natl Cancer Inst.* 2003;95:399-409.
10. Cheng JC, Weisenberger DJ, Gonzales FA, et al. Continuous zebularine treatment effectively sustains demethylation in human bladder cancer cells. *Mol Cell Biol.* 2004;24:1270-1278.
11. Cheson BD, Bennett JM, Kantarjian H, et al. Report of an international working group to standardize response criteria for myelodysplastic syndromes. *Blood.* 2000;96:3671-3674.
12. Cimino G, Lo-Coco F, Fenu S, et al. Sequential valproic acid/all-trans retinoic acid treatment reprograms differentiation in refractory and high-risk acute myeloid leukemia. *Cancer Res.* 2006;66:8903-8911.
13. Constantinides PG, Jones PA, Gevers W. Functional striated muscle cells from non-myoblast precursors following 5-azacytidine treatment. *Nature.* 1977;267:364-366.
14. Cornacchia E, Golbus J, Maybaum J, Strahler J, Hanash S, Richardson B. Hydralazine and procainamide inhibit T cell DNA methylation and induce autoreactivity. *J Immunol.* 1988;140:2197-2200.
15. Dangond F, Hafler DA, Tong JK, et al. Differential display cloning of a novel human histone deacetylase (HDAC3) cDNA from PHA-activated immune cells. *Biochem Biophys Res Commun.* 1998;242:648-652.
16. Daskalakis M, Nguyen TT, Nguyen C, et al. Demethylation of a hypermethylated P15/INK4B gene in patients with myelodysplastic syndrome by 5-aza-2′-deoxycytidine (decitabine) treatment. *Blood.* 2002;100:2957-2964.
17. Degos L, Dombret H, Chomienne C, et al. All-trans-retinoic acid as a differentiating agent in the treatment of acute promyelocytic leukemia. *Blood.* 1995;85:2643-2653.
18. DiGiuseppe JA, Weng LJ, Yu KH, et al. Phenylbutyrate-induced G1 arrest and apoptosis in myeloid leukemia cells: structure-function analysis. *Leukemia.* 1999;13:1243-1253.
19. Dover GJ, Brusilow S, Charache S. Induction of fetal hemoglobin production in subjects with sickle cell anemia by oral sodium phenylbutyrate. *Blood.* 1994;84:339-343.
20. Dover GJ, Brusilow S, Samid D. Increased fetal hemoglobin in patients receiving sodium 4-phenylbutyrate. *N Engl J Med.* 1992;327:569-570.
21. Emiliani S, Fischle W, Van Lint C, Al-Abed Y, Verdin E. Characterization of a human RPD3 ortholog, HDAC3. *Proc Natl Acad Sci USA.* 1998;95:2795-2800.
22. Fandy TE, Herman JG, Kerns P, et al. Early epigenetic changes and DNA damage do not predict clinical response in an overlapping schedule of 5-azacytidine and entinostat in patients with myeloid malignancies. *Blood.* 2009;114:2764-2773.
23. Fandy TE, Carraway HM, Gore SDM. DNA demethylating agents and histone deacetylase inhibitors in hematologic malignancies. [Article]. *Cancer J.* 2007;13:40-48.
24. Fang MZ, Wang Y, Ai N, et al. Tea polyphenol (-)-epigallocatechin-3-gallate inhibits DNA methyltransferase and reactivates methylation-silenced genes in cancer cell lines. *Cancer Res.* 2003;63:7563-7570.
25. Fenaux P, Mufti GJ, Hellstrom-Lindberg E, et al. Efficacy of azacitidine compared with that of conventional care regimens in the treatment of higher-risk myelodysplastic syndromes: a randomised, open-label, phase III study. *Lancet Oncol.* 2009;10:223-232.
26. Figueroa ME, Skrabanek L, Li Y, et al. MDS and secondary AML display unique patterns and abundance of aberrant DNA methylation. *Blood.* 2009;114:3448-3458.
27. Fischle W, Emiliani S, Hendzel MJ, et al. A new family of human histone deacetylases related to *Saccharomyces cerevisiae* HDA1p. *J Biol Chem.* 1999;274:11713-11720.
28. Follo MY, Finelli C, Mongiorgi S, et al. Reduction of phosphoinositide-phospholipase C beta1 methylation predicts the responsiveness to azacitidine in high-risk MDS. *Proc Natl Acad Sci.* 2009;106:16811-16816.
29. Fournel M, Bonfils C, Hou Y, et al. MGCD0103, a novel isotype-selective histone deacetylase inhibitor, has broad spectrum antitumor activity in vitro and in vivo. *Mol Cancer Ther.* 2008;7:759-768.
30. Garcia-Manero G, Assouline S, Cortes J, et al. Phase 1 study of the oral isotype specific histone deacetylase inhibitor MGCD0103 in leukemia. *Blood.* 2008;112:981-989.
31. Garcia-Manero G, Kantarjian HM, Sanchez-Gonzalez B, et al. Phase 1/2 study of the combination of 5-aza-2′-deoxycytidine with valproic acid in patients with leukemia. *Blood.* 2006;108:3271-3279.
32. Garcia-Manero G, Yang H, Bueso-Ramos C, et al. Phase 1 study of the histone deacetylase inhibitor vorinostat (suberoylanilide hydroxamic acid [SAHA]) in patients with advanced leukemias and myelodysplastic syndromes. *Blood.* 2008;111:1060-1066.
33. Giles F, Fischer T, Cortes J, et al. A phase I study of intravenous LBH589, a novel cinnamic hydroxamic acid analogue histone deacetylase inhibitor, in patients with refractory

hematologic malignancies. *Clin Cancer Res*. 2006;12:4628-4635.

34. Gimsing P. Belinostat: a new broad acting antineoplastic histone deacetylase inhibitor. *Expert Opin Investig Drugs*. 2009;18:501-508.
35. Gimsing P, Hansen M, Knudsen LM, et al. A phase I clinical trial of the histone deacetylase inhibitor belinostat in patients with advanced hematological neoplasia. *Eur J Haematol*. 2008;81:170-176.
36. Gojo I, Jiemjit A, Trepel JB, et al. Phase 1 and pharmacologic study of MS-275, a histone deacetylase inhibitor, in adults with refractory and relapsed acute leukemias. *Blood*. 2007;109:2781-2790.
37. Gore SD, Samid D, Weng LJ. Impact of the putative differentiating agents sodium phenylbutyrate and sodium phenylacetate on proliferation, differentiation, and apoptosis of primary neoplastic myeloid cells. *Clin Cancer Res*. 1997;3:1755-1762.
38. Gore SD, Baylin S, Sugar E, et al. Combined DNA methyltransferase and histone deacetylase inhibition in the treatment of myeloid neoplasms. *Cancer Res*. 2006;66: 6361-6369.
39. Gore SD, Weng LJ, Figg WD, et al. Impact of prolonged infusions of the putative differentiating agent sodium phenylbutyrate on myelodysplastic syndromes and acute myeloid leukemia. *Clin Cancer Res*. 2002;8:963-970.
40. Gottlicher M, Minucci S, Zhu P, et al. Valproic acid defines a novel class of HDAC inhibitors inducing differentiation of transformed cells. *EMBO J*. 2001;20:6969-6978.
41. Gray SG, Ekström TJ. The human histone deacetylase family. *Exp Cell Res*. 2001;262:75-83.
42. Grozinger CM, Hassig CA, Schreiber SL. Three proteins define a class of human histone deacetylases related to yeast Hda1p. *Proc Natl Acad Sci USA*. 1999;96:4868-4873.
43. Herman JG, Baylin SB. Gene silencing in cancer in association with promoter hypermethylation. *N Engl J Med*. 2003;349:2042-2054.
44. Holleran JL, Parise RA, Joseph E, et al. Plasma pharmacokinetics, oral bioavailability, and interspecies scaling of the DNA methyltransferase inhibitor, zebularine. *Clin Cancer Res*. 2005;11:3862-3868.
45. Hu E, Chen Z, Fredrickson T, et al. Cloning and characterization of a novel human class I histone deacetylase that functions as a transcription repressor. *J Biol Chem*. 2000; 275:15254-15264.
46. Ihalainen J, Pakkala S, Savolainen ER, Jansson SE, Palotie A. Hypermethylation of the calcitonin gene in the myelodysplastic syndromes. *Leukemia*. 1993;7:263-267.
47. Issa JP, Garcia-Manero G, Giles FJ, et al. Phase 1 study of low-dose prolonged exposure schedules of the hypomethylating agent 5-aza-2′-deoxycytidine (decitabine) in hematopoietic malignancies. *Blood*. 2004;103:1635-1640.
48. Jaboin J, Wild J, Hamidi H, et al. MS-27-275, an inhibitor of histone deacetylase, has marked in vitro and in vivo antitumor activity against pediatric solid tumors. *Cancer Res*. 2002;62:6108-6115.
49. Jiang Y, Dunbar A, Gondek LP, et al. Aberrant DNA methylation is a dominant mechanism in MDS progression to AML. *Blood*. 2009;113:1315-1325.
50. Jiemjit A, Fandy TE, Carraway H, et al. p21WAF1//CIP1 induction by 5-azacytosine nucleosides requires DNA damage. *Oncogene*. 2008;27:3615-3623.
51. Jones PA, Taylor SM. Cellular differentiation, cytidine analogs and DNA methylation. *Cell*. 1980;20:85-93.
52. Jones PA, Taylor SM. Hemimethylated duplex DNAs prepared from 5-azacytidine-treated cells. *Nucleic Acids Res*. 1981;9:2933-2947.
53. Kantarjian H, Issa JP, Rosenfeld CS, et al. Decitabine improves patient outcomes in myelodysplastic syndromes: results of a phase III randomized study. *Cancer*. 2006;106: 1794-1803.
54. Kantarjian H, Oki Y, Garcia-Manero G, et al. Results of a randomized study of 3 schedules of low-dose decitabine in higher-risk myelodysplastic syndrome and chronic myelomonocytic leukemia. *Blood*. 2007;109:52-57.
55. Kao HY, Downes M, Ordentlich P, Evans RM. Isolation of a novel histone deacetylase reveals that class I and class II deacetylases promote SMRT-mediated repression. *Genes Dev*. 2000;14:55-66.
56. Kao HY, Lee CH, Komarov A, Han CC, Evans RM. Isolation and characterization of mammalian HDAC10, a novel histone deacetylase. *J Biol Chem*. 2002;277:187-193.
57. Klimek VM, Fircanis S, Maslak P, et al. Tolerability, pharmacodynamics, and pharmacokinetics studies of depsipeptide (Romidepsin) in patients with acute myelogenous leukemia or advanced myelodysplastic syndromes. *Clin Cancer Res*. 2008;14:826-832.
58. Knudson AG. Two genetic hits (more or less) to cancer. *Nat Rev Cancer*. 2001;1:157-162.
59. Kornblith AB, Herndon JE II, Silverman LR, et al. Impact of azacytidine on the quality of life of patients with myelodysplastic syndrome treated in a randomized phase III trial: a cancer and leukemia Group B Study. *J Clin Oncol*. 2002;20:2441-2452.
60. Kovacs JJ, Murphy PJM, Gaillard S, et al. HDAC6 regulates Hsp90 acetylation and chaperone-dependent activation of glucocorticoid receptor. *Mol Cell*. 2005;18:601-607.
61. Kuendgen A, Gattermann N. Valproic acid for the treatment of myeloid malignancies. *Cancer*. 2007;110:943-954.
62. Kuendgen A, Schmid M, Knipp S, et al. Valproic acid (VPA) achieves high response rates in patients with low-risk myelodysplastic syndromes. *ASH Annu Meet Abstr*. 2005; 106:789.
63. Kuendgen A, Strupp C, Aivado M, et al. Treatment of myelodysplastic syndromes with valproic acid alone or in combination with all-trans retinoic acid. *Blood*. 2004;104: 1266-1269.
64. Lane AA, Chabner BA. Histone deacetylase inhibitors in cancer therapy. *J Clin Oncol*. 2009;27:5459-5468.
65. Lea MA, Tulsyan N. Discordant effects of butyrate analogues on erythroleukemia cell proliferation, differentiation and histone deacetylase. *Anticancer Res*. 1995;15:879-883.
66. Lin RJ, Nagy L, Inoue S, Shao W, Miller WH, Evans RM. Role of the histone deacetylase complex in acute promyelocytic leukaemia. *Nature*. 1998;391:811-814.
67. Lubbert M. DNA methylation inhibitors in the treatment of leukemias, myelodysplastic syndromes and hemoglobinopathies: clinical results and possible mechanisms of action. *Curr Top Microbiol Immunol*. 2000;249:135-164.
68. Lyons RM, Cosgriff TM, Modi SS, et al. Hematologic response to three alternative dosing schedules of azacitidine in patients with myelodysplastic syndromes. *J Clin Oncol*. 2009;27:1850-1856.

69. Maciejewski JP, Tiu RV, O'Keefe C. Application of array-based whole genome scanning technologies as a cytogenetic tool in haematological malignancies. *Br J Haematol.* 2009;146:479-488.
70. Marks PA. Thioredoxin in cancer – role of histone deacetylase inhibitors. *Semin Cancer Biol.* 2006;16:436-443.
71. Marks PA, Rifkind RA, Richon VM, Breslow R, Miller T, Kelly WK. Histone deacetylases and cancer: causes and therapies. *Nat Rev Cancer.* 2001;1:194-202.
72. Mihara K, Takihara Y, Kimura A. Genetic and epigenetic alterations in myelodysplastic syndrome. *Cytogenet Genome Res.* 2007;118:297-303.
73. Miska EA, Karlsson C, Langley E, Nielsen SJ, Pines J, Kouzarides T. HDAC4 deacetylase associates with and represses the MEF2 transcription factor. *EMBO J.* 1999;18:5099-5107.
74. Momparler RL. Molecular, cellular and animal pharmacology of 5-AZA-2′-deoxycytidine. *Pharmacol Ther.* 1985;30: 287-299.
75. Moradei O, Leit S, Zhou N, et al. Substituted *N*-(2-aminophenyl)-benzamides, (E)-*N*-(2-aminophenyl)-acrylamides and their analogues: novel classes of histone deacetylase inhibitors. *Bioorg Med Chem Lett.* 2006;16:4048-4052.
76. Mund C, Hackanson B, Stresemann C, Lubbert M, Lyko F. Characterization of DNA demethylation effects induced by 5-aza-2′-deoxycytidine in patients with myelodysplastic syndrome. *Cancer Res.* 2005;65:7086-7090.
77. Nakamaki T, Bartram C, Seriu T, et al. Molecular analysis of the cyclin-dependent kinase inhibitor genes, p15, p16, p18 and p19 in the myelodysplastic syndromes. *Leuk Res.* 1997;21:235-240.
78. Newmark HL, Lupton JR, Young CW. Butyrate as a differentiating agent: pharmacokinetics, analogues and current status. *Cancer Lett.* 1994;78:1-5.
79. Nolte F, Hofmann WK. Myelodysplastic syndromes: molecular pathogenesis and genomic changes. *Ann Hematol.* 2008;87:777-795.
80. Phiel CJ, Zhang F, Huang EY, Guenther MG, Lazar MA, Klein PS. Histone deacetylase is a direct target of valproic acid, a potent anticonvulsant, mood stabilizer, and teratogen. *J Biol Chem.* 2001;276:36734-36741.
81. Pilatrino C, Cilloni D, Messa E, et al. Increase in platelet count in older, poor-risk patients with acute myeloid leukemia or myelodysplastic syndrome treated with valproic acid and all-trans retinoic acid. *Cancer.* 2005;104:101-109.
82. Pinto A, Zagonel V, Attadia V, et al. 5-aza-2′-deoxycytidine as a differentiation inducer in acute myeloid leukaemias and myelodysplastic syndromes of the elderly. *Bone Marrow Transplant.* 1989;4(Suppl 3):28-32.
83. Quesnel B, Guillerm G, Vereecque R, et al. Methylation of the p15INK4b gene in myelodysplastic syndromes is frequent and acquired during disease progression. *Blood.* 1998;91:2985-2990.
84. Raffoux E, Chaibi P, Dombret H, Degos L. Valproic acid and all-trans retinoic acid for the treatment of elderly patients with acute myeloid leukemia. *Haematologica.* 2005;90: 986-988.
85. Raj K, Mufti GJ. Azacytidine (Vidaza(R)) in the treatment of myelodysplastic syndromes. *Ther Clin Risk Manag.* 2006;2:377-388.
86. Saito A, Yamashita T, Mariko Y, et al. A synthetic inhibitor of histone deacetylase, MS-27-275, with marked in vivo antitumor activity against human tumors. *Proc Natl Acad Sci USA.* 1999;96:4592-4597.
87. Schaefer EW, Loaiza-Bonilla A, Juckett M, et al. A phase 2 study of vorinostat in acute myeloid leukemia. *Haematologica.* 2009;94:1375-1382.
88. Schlenk R, Sohlbach K, Hutter ML, et al. Interim results of a phase I/II clinical trial of belinostat in combination with idarubicin in patients with AML not suitable for standard intensive therapy. *ASH Annu Meet Abstr.* 2008;112:1953.
89. Segura-Pacheco B, Trejo-Becerril C, Perez-Cardenas E, et al. Reactivation of tumor suppressor genes by the cardiovascular drugs hydralazine and procainamide and their potential use in cancer therapy. *Clin Cancer Res.* 2003;9: 1596-1603.
90. Shen L, Kantarjian H, Guo Y, et al. DNA methylation predicts survival and response to therapy in patients with myelodysplastic syndromes. *J Clin Oncol.* 2010;28:605-613.
91. Silverman LR, Holland JF, Weinberg RS, et al. Effects of treatment with 5-azacytidine on the in vivo and in vitro hematopoiesis in patients with myelodysplastic syndromes. *Leukemia.* 1993;7(Suppl 1):21-29.
92. Silverman LR, Demakos EP, Peterson BL, et al. Randomized controlled trial of azacitidine in patients with the myelodysplastic syndrome: a study of the cancer and leukemia group B. *J Clin Oncol.* 2002;20:2429-2440.
93. Silverman LR, McKenzie DR, Peterson BL, et al. Further analysis of trials with azacitidine in patients with myelodysplastic syndrome: studies 8421, 8921, and 9221 by the cancer and leukemia group B. *J Clin Oncol.* 2006;24: 3895-3903.
94. Soriano AO, Yang H, Faderl S, et al. Safety and clinical activity of the combination of 5-azacytidine, valproic acid, and all-trans retinoic acid in acute myeloid leukemia and myelodysplastic syndrome. *Blood.* 2007;110:2302-2308.
95. Steele NL, Plumb JA, Vidal L, et al. A phase 1 pharmacokinetic and pharmacodynamic study of the histone deacetylase inhibitor belinostat in patients with advanced solid tumors. *Clin Cancer Res.* 2008;14:804-810.
96. Steensma DP, Baer MR, Slack JL, et al. Multicenter study of decitabine administered daily for 5 days every 4 weeks to adults with myelodysplastic syndromes: the alternative dosing for outpatient treatment (ADOPT) trial. *J Clin Oncol.* 2009;27:3842-3848.
97. Suzuki H, Gabrielson E, Chen W, et al. A genomic screen for genes upregulated by demethylation and histone deacetylase inhibition in human colorectal cancer. *Nat Genet.* 2002;31:141-149.
98. Suzuki T, Ando T, Tsuchiya K, et al. Synthesis and histone deacetylase inhibitory activity of new benzamide derivatives. *J Med Chem.* 1999;42:3001-3003.
99. Taunton J, Hassig CA, Schreiber SL. A mammalian histone deacetylase related to the yeast transcriptional regulator Rpd3p. *Science.* 1996;272:408-411.
100. Taylor SM, Jones PA. Changes in phenotypic expression in embryonic and adult cells treated with 5-azacytidine. *J Cell Physiol.* 1982;111:187-194.
101. Taylor SM, Jones PA. Cellular differentiation. *Int J Obes.* 1985;9(Suppl 1):15-21.

102. Taylor SM, Jones PA. Multiple new phenotypes induced in and 3T3 cells treated with 5-azacytidine. *Cell*. 1979;17: 771-779.
103. Uchida T, Kinoshita T, Nagai H, et al. Hypermethylation of the p15INK4B gene in myelodysplastic syndromes. *Blood*. 1997;90:1403-1409.
104. Van den Wyngaert I, de Vries W, Kremer A, et al. Cloning and characterization of human histone deacetylase 8. *FEBS Lett*. 2000;478:77-83.
105. Verdel A, Khochbin S. Identification of a new family of higher eukaryotic histone deacetylases. *J Biol Chem*. 1999;274:2440-2445.
106. Veselý J. Mode of action and effects of 5-azacytidine and of its derivatives in eukaryotic cells. *Pharmacol Ther*. 1985;28:227-235.
107. Von Hoff DD, Slavik M, Muggia FM. 5-Azacytidine. A new anticancer drug with effectiveness in acute myelogenous leukemia. *Ann Intern Med*. 1976;85:237-245.
108. Wada T, Kikuchi J, Nishimura N, Shimizu R, Kitamura T, Furukawa Y. Expression levels of histone deacetylases determine the cell fate of hematopoietic progenitors. *J Biol Chem*. 2009;284:30673-30683.
109. Wang AH, Bertos NR, Vezmar M, et al. HDAC4, a human histone deacetylase related to yeast HDA1, is a transcriptional corepressor. *Mol Cell Biol*. 1999;19:7816-7827.
110. Wang J, Hoshino T, Redner RL, Kajigaya S, Liu JM. ETO, fusion partner in t(8;21) acute myeloid leukemia, represses transcription by interaction with the human N-CoR/mSin3/HDAC1 complex. *Proc Natl Acad Sci USA*. 1998;95:10860-10865.
111. Wijermans P, Lubbert M, Verhoef G, et al. Low-dose 5-aza-2′-deoxycytidine, a DNA hypomethylating agent, for the treatment of high-risk myelodysplastic syndrome: a multicenter phase II study in elderly patients. *J Clin Oncol*. 2000;18:956-962.
112. Wijermans PW, Krulder JW, Huijgens PC, Neve P. Continuous infusion of low-dose 5-aza-2′-deoxycytidine in elderly patients with high-risk myelodysplastic syndrome. *Leukemia*. 1997;11:1-5.
113. Yang WM, Inouye C, Zeng Y, Bearss D, Seto E. Transcriptional repression by YY1 is mediated by interaction with a mammalian homolog of the yeast global growth regulator RPD3. *Proc Natl Acad Sci USA*. 1996;93:12845-12850.
114. Yang WM, Yao YL, Sun JM, Davie JR, Seto E. Isolation and characterization of cDNAs corresponding to an additional member of the human histone deacetylase gene family. *J Biol Chem*. 1997;272:28001-28007.
115. Yu KH, Weng LJ, Fu S, Piantadosi S, Gore SD. Augmentation of phenylbutyrate-induced differentiation of myeloid leukemia cells using all-trans retinoic acid. *Leukemia*. 1999;13:1258-1265.
116. Zagonel V, Lo RG, Marotta G, et al. 5-aza-2′-deoxycytidine (decitabine) induces trilineage response in unfavourable myelodysplastic syndromes. *Leukemia*. 1993;7(Suppl 1): 30-35.
117. Zhang Y, Li N, Caron C, et al. HDAC-6 interacts with and deacetylates tubulin and microtubules in vivo. *EMBO J*. 2003;22:1168-1179.
118. Zhou X, Marks PA, Rifkind RA, Richon VM. Cloning and characterization of a histone deacetylase, HDAC9. *Proc Natl Acad Sci USA*. 2001;98:10572-10577.

Index

MIX
Papier aus verantwortungsvollen Quellen
Paper from responsible sources
FSC® C105338

If you have any concerns about our products,
you can contact us on
ProductSafety@springernature.com

In case Publisher is established outside the EU,
the EU authorized representative is:
Springer Nature Customer Service Center GmbH
Europaplatz 3, 69115 Heidelberg, Germany

Printed by Libri Plureos GmbH
in Hamburg, Germany